FLORENCE NIGHTINGALE

'I have been most enthralled by this portrait drawn with extraordinary unobtrusive skill of one of the most remarkable human beings that can ever have lived.'
Lord David Cecil

'This monumental and absorbing biography is a truly remarkable achievement. Its firmness of construction, its immense and unfaltering mastery of detail, its sound sense and the vitality of the writing, the understanding and the good sense, make it memorable.'
Evening News

'Good biographies are rare and this, beyond doubt, is among the best. A great woman, finely portrayed.'
Birmingham Post

CECIL WOODHAM-SMITH

Florence Nightingale
1820-1910

Collins
FONTANA BOOKS

First published 1951
Revised and slightly abridged version published in
Penguin Books 1955
First issued in Fontana Books 1964
Second Impression, October 1968
Third Impression, November 1969
Fourth Impression, November 1970

Cover photograph reproduced by kind
permission of the Radio Times Hulton
Picture Library

Printed in Great Britain
Collins Clear-Type Press
London and Glasgow

To

G. I. W.-S.

Acknowledgements

In writing this biography I have been given the opportunity of presenting what I believe to be a complete picture of Miss Nightingale for the first time.

When Sir Edward Cook wrote his admirable official life immediately after Miss Nightingale's death, there was a large body of material which, for family and personal reasons, was either not available to him or he was asked not to use. He did not, for instance, see the Verney Nightingale papers: he saw only part of the collection I have described as the Herbert papers; and there was a great deal of other correspondence of which he was asked to make only a limited use. I have been fortunate enough to be given access to this material.

My thanks are due, first and foremost, to Sir Harry Verney, Bart., who most generously placed at my disposal the very important Verney Nightingale papers comprising the domestic correspondence and private papers of Miss Nightingale's mother, Frances, her sister Parthenope, Lady Verney, and the Nightingale family circle. I am deeply indebted to Lord Herbert for allowing me to use unpublished material from the Herbert papers, establishing, among other important points, the true nature of the relationship between Miss Nightingale and Lord and Lady Herbert of Lea. I should like to thank the late Mrs Salmon, Sir Harry Verney's sister, for unpublished letters and private information, and I owe very much to the late Lady Stephen, not only for unpublished letters and reminiscences, but for her kindness, too, in procuring me access to family papers.

I am indebted to Sir Ralph Verney, Bart., for the correspondence of his father, Mr Frederick Verney, with Miss Nightingale; to Sir Maurice Bonham-Carter for permission to make use of family papers; and to Mr Leigh Smith for information of importance from the Leigh Smith papers. Sir Shane Leslie has allowed me to use two letters from his biography of Cardinal Manning, and Messrs Hodder and Stoughton have kindly given me permission to quote extracts from letters published in Miss Elizabeth

7

Haldane's *Mrs Gaskell and Her Friends*. I should like also to thank Mr James Pope-Hennessy and Mrs Vaughan Nash. I owe, as every biographer of Miss Nightingale must owe, a debt to Sir Edward Cook, who in his official life performed the task of tracing a path through the mass of the Nightingale papers – the enormous collection of private and official letters and documents left by Miss Nightingale at her death and deposited at the British Museum in 1940. I should like to thank the Museum authorities, through whose co-operation I was enabled to examine the Nightingale papers at the National Library of Wales when, during the war, it was not possible to see manuscripts in London. Finally I should like to record my obligation to Mr Michael Sadleir, without whose support, criticism, and constant assistance this book could not have been written.

CECIL WOODHAM-SMITH

NOTE TO THE FONTANA EDITION

In preparing this edition the author has shortened some of the references to Miss Nightingale's work in connexion with welfare and sanitation in India which she regarded as having a limited interest for the general reader. The pen and ink drawing reproduced on page 15 is included by kind permission of Sir Harry Verney, Bart.

I

I t was something new to call a girl Florence. Within fifty years there would be thousands of girls all over the world christened Florence in honour of this baby, but in the summer of 1820, when Fanny Nightingale fixed on the name for her daughter, it was new.

Novelty was the fashion in 1820. Europe was still rejoicing in the liberty which followed the end of the Napoleonic wars. Years of restriction had bred a longing for change, and now that freedom to travel had returned the roads and cities of Europe were thronged with travellers. Fanny and William Nightingale had been travelling in Europe since their marriage in 1818. They already had one daughter, born in Naples in 1819 and christened by the Greek name for her birthplace, Parthenope. For her second confinement Fanny chose Florence. She loved gaiety, and Florence had the reputation of being the gayest city in Europe. The Nightingales took a large furnished villa, the Villa Colombaia near the Porta Romana, where a second girl was born on May 12, 1820. Fanny decided she too should be named after her birthplace, and on July 4, 1820, she was christened Florence in the drawing-room of the villa.

It would have been better if Florence had been a boy. Though William Edward Nightingale, or as he was always called W. E. N., was rich, there were complications attached to his property. He had inherited from an uncle, and, under his uncle's will, if W. E. N. should have no son, the property passed on his death to his sister and next to her eldest son. However, the Nightingales felt no anxiety. Not all W. E. N.'s fortune was involved. He had inherited when a minor; a lead mine discovered on the property had greatly increased its value, and during his minority a large sum had been invested on his behalf which was absolutely his. The Nightingales had been married for two years, and Fanny had already had two healthy children. The next baby would be a boy.

Though they were both handsome, agreeable, and intelligent, they were not a well-matched couple. Only a few months before their engagement Fanny had been anxious to marry another man; and she was six years older than W. E. N. In 1820 she was thirty-two, extremely beautiful, generous, and extravagant. She had great vitality, was indefatigable in the pursuit of pleasure, never tired unless bored, always good-natured unless thwarted, always kind unless her obstinacy were aroused. In the art of making people comfortable, in the arrangement of a house, the production of good dinners, she possessed genius. She intended, when she returned to England, to make herself a position as a hostess.

She came from a remarkable family. Her grandfather, Samuel Smith, had been a well-known character, celebrated for the riches he had amassed as a London merchant and for his humanitarian principles. He had come to the assistance of Flora Macdonald when she was a penniless prisoner in the Tower in spite of the fact that he was a strong Hanoverian. To show his sympathy with the struggle of the American colonists for freedom in the War of Independence he had relinquished his title to a large part of the city of Savannah. His son William Smith, Fanny's father, devoted his wealth to collecting pictures and fighting lost causes. For forty-six years he sat in the House of Commons, fighting for the weak, the unpopular, and the oppressed. He was a leading Abolitionist; he championed the sweated factory workers; he did battle for the rights of Dissenters and Jews.

His children did not inherit his altruism. At Parndon Hall in Essex, in his London house in Park Street, his political and humanitarian activities were carried on against a background of ceaseless junketings. With tireless energy the young Smiths danced, went on pleasure parties and picnics, played parlour games, got up amateur theatricals. There were ten children, five sons and five daughters, all good-looking, all with immense zest for living and amazing health. William Smith himself at eighty wrote that he had 'no recollection whatever of any bodily pain or illness'. None of his ten children died before the age of sixty-nine, six lived to be over eighty, and Fanny lived to be ninety-two. Looking back on her youth fifty years later, Fanny described family life as a 'hurly burly'. 'We Smiths never thought of anything all day long but our own ease and pleasure', she wrote.

Fanny was the beauty of the family; yet Fanny did not marry. Her sister, Anne, married an immensely rich Mr Nicholson of Waverley Abbey near Farnham, the house which gave Scott the title for the Waverley novels; her sister, Joanna, married Mr

Bonham Carter, eldest son of a well-known Hampshire family, and settled near Winchester. Fanny remained at home with Patty and Julia who held 'advanced' views and suffered from nerves. Association with her father's friends had left its mark on Fanny; she had acquired a passion for good conversation, 'had a preference for clever elderly gentlemen, and was comparatively indifferent to gay young ones', and despised the junketings of Parndon. She had already 'arrived at the age when the world acquits those parents who suffer their daughters to act for themselves', when, in 1816, she fell in love with the Honourable James Sinclair, a younger son of the Earl of Caithness. His character was allowed to be good and his intentions disinterested, but he possessed no income beyond the pay of a captain in the Ross-shire Militia, and no expectations. In immense letters, full of worldly wisdom, kindness, and unanswerable common sense, William Smith pointed out the absurdity of a woman of Fanny's habits contemplating life on an income of scarcely four hundred pounds sterling a year and declined, in justice to his other children, to assume the support of her future family. Fanny pleaded in vain that her affections were entirely given away and that losing James would quite break her down. By 1817 the affair was at an end.

Fanny was now nearly thirty, and William Edward Nightingale was nearly twenty-four. She had known him since he was a boy; he had been at school with her younger brother Octavius and had been coming to the house for years, an awkward lanky schoolboy, immensely tall, immensely thin, with a habit of always standing upright propped against mantelpieces and doors because he disliked folding himself into a chair.

Originally his name had been Shore, but at twenty-one, when he came into the fortune left him by his uncle, he changed his name to Nightingale. He went up to Cambridge with an income of between seven and eight thousand pounds sterling a year, and Cambridge transformed him. He proved, though lazy, to be clever. He gained a reputation for wit. His looks improved; his height and a remote and gentle manner gave him distinction. He developed into a dilettante, rich, appreciative, indolent, charming.

It was an unexpected result. Wild blood ran in W. E. N.'s veins. The uncle, his mother's brother, from whom he inherited, had been an eccentric sporting squire, known throughout Derbyshire as mad Peter Nightingale. Peter Nightingale had been a dare-devil horseman, a rider in midnight steeplechases, a layer of wagers, given to hard drinking and low company.

In 1817 W. E. N. became engaged to Fanny. He was very much in love. Fanny's rich beauty warmed his reserved temperament, and for a short time he thawed. The period was brief. Normally, as Fanny wrote later, 'Mr Nightingale is seldom in the melting mood.'

Fanny's family did not approve. They were fond of W. E. N., but they had no faith in his character. He was clever but he was indolent, hated making up his mind, hated taking action – he was not the husband for Fanny. Within six months they were married and had gone abroad.

Fanny believed she would be able to mould W. E. N. She intended him to become one of the prosperous, cultivated, and liberal-minded country gentlemen who played an important part in English public life. They would have a beautiful house, a fine library, maintain an interest in the arts, and entertain.

After nearly three years in Italy Fanny began to feel it was time they came home. W. E. N., she wrote, would have been content to idle in Italy for the rest of his life. As long as he had books and conversation he was indifferent to other pleasures. However, Fanny prevailed, and in 1821, when Florence was a year old, the Nightingales returned to England.

The first necessity was to house themselves. The Nightingales had no family place. Peter Nightingale had inhabited a tumble-down building, half manor, half farm, totally inadequate for the needs of Fanny, W. E. N., and two babies accompanied by maids, footmen, valet, coachman, and cook.

Before they left Italy, W. E. N. had decided to abandon the old house and had made a flying trip to England to have work started on a new house on higher ground. He was an amateur architect and himself produced the designs from which the plans were drawn. He gave his house mullions, a steep pitched roof, a vaguely Gothic air. The effect is not unpleasing, and the situation of the house is unrivalled. Lea Hurst stands high above a rolling country, terraced gardens fall steeply away on every side, and the view from the windows is immensely wide, so that the house, as Mrs Gaskell wrote, seems to be floating in air.

But no sooner was Lea Hurst finished than Fanny realized she had made a mistake. As a family place Lea Hurst was inadequate; as a house in which to entertain it was impossible. The only attraction was a wonderful view. The situation was inaccessible, the house cold. The Nightingales attempted one winter there, and both children got bronchitis. Above all, Lea Hurst was much too small.

Fanny's standards of accommodation descended to her daughter. Twenty years later at a dinner-party Florence denied that Lea Hurst was anything but a small house. 'Why', she said, 'it has only fifteen bedrooms.'

By 1823 Fanny had convinced W. E. N. that Lea Hurst, except in summer, was impossible. Certainly they would keep up a property where the Nightingales had been rooted for generations, but they must also have another house, a larger house, and in a warmer part of the country than Derbyshire.

In 1825 W. E. N. bought Embley Park, near Romsey, in Hampshire, on the borders of the New Forest. It was a good-sized plain square house of the late Georgian period, London was reasonably near, and Fanny's two married sisters, Mrs Nicholson at Waverley Abbey near Farnham, and Mrs Bonham Carter at Fairoaks near Winchester, were within easy reach. Above all, in contrast to the uncivilized remoteness of Lea Hurst, Embley was in the centre of a 'good neighbourhood'.

By the time Florence was five, the pattern of the Nightingale's life was fixed. The summer was passed at Lea Hurst, the remainder of the year at Embley Park, and twice a year during the spring and autumn seasons a visit was paid to London. Fanny would have liked a house in London, but W. E. N. refused.

He did, however, proceed to turn himself into an English country gentleman. He shot, he fished, he hunted, did a great deal for his tenants, supported a free school at great expense near Lea Hurst, and in Hampshire took an active part in local politics. Fanny looked forward to the day when he would stand for Parliament. W. E. N. was a Whig and in favour of Parliamentary Reform. 'How I hate Tories, all Beer and Money', he wrote to Fanny in 1830.

Fanny's life ran smoothly. If she fretted after the son who failed to appear, she did not record it. The only shadow was cast by Florence. Florence was not an easy child.

The two little girls were not called by their full names. Florence was shortened to Flo, Parthenope to Parthe or Pop. Flo was much the prettier. Neither of the girls inherited their mother's outstanding beauty, but Flo promised to grow up more than

13

ordinarily good-looking. She was lightly built, singularly graceful, with thick bright chestnut hair and a delicate complexion.

Both Fanny and W. E. N. loved children. All the closely related families of the Nightingale circle – Smiths, Shores, Nicholsons, Bonham Carters – delighted in children. A stream of cousins spent their holidays at Embley and Lea Hurst, and almost invariably Fanny had a couple of family babies in the house, enjoying a change of air and being fed upon country butter and eggs and cream. 'Kiss all babies for me' is a frequent ending to the first letters Flo wrote home. Her childhood was filled with gardens to play in, ponies to ride, and a succession of dogs, cats, and birds to be looked after.

And yet Flo was not happy. If she had been an ordinary naughty child, Fanny would have understood her, but she was not naughty. She was strange, passionate, wrong-headed, obstinate, and miserable.

In an autobiographical note Miss Nightingale records that as a very young child she had an obsession that she was not like other people. She was a monster. That was her secret which might at any moment be found out. Strangers must be avoided, especially children. She worked herself into an agony at the prospect of seeing a new face, and to be looked at was torture. She doubted her capacity to behave like other people and refused to dine downstairs, convinced she would betray herself by doing something extraordinary with her knife and fork.

Realization of the gulf which separated her from everyone round her came hand in hand with the dawnings of conscious thought. At first she was overwhelmed with terror and guilt. Surely she ought to be like everyone else? What might not people do to her if they found out the truth? But almost before she had grown out of babyhood, guilt and terror were succeeded by discontent. She wrote that as early as the age of six she was aware that the rich smooth life of Embley and Lea Hurst was utterly distasteful to her. She ceased to be terrified; she resisted, disliked, and despised it.

She began, like many imaginative children, to escape into dreams. She told herself stories in which she played a heroine's part, and for hours at a time transferred herself completely to a dream world.

Though she shrank from meeting people, she was not self-sufficient. She was a child who craved for sympathy and attached

14

herself with embarrassing vehemence to anyone whom she felt to be sympathetic. Her childhood was a series of passions – for her governess Miss Christie, for W. E. N.'s younger sister 'Aunt Mai', for a beautiful older cousin. When Miss Christie left, when Aunt Mai married, when the beautiful cousin got tired of her devotion, the violence of her feelings made her physically ill.

She did not attach herself to her mother. The companion of her childhood was W. E. N. Among the Verney Nightingale papers is preserved a sketch by Julia Smith, Fanny's unmarried younger sister, of W. E. N. and the two little girls. The trio have their backs to the artist. W. E. N., in frock-coat and top-hat, is in the middle, tall and thin as a hop pole; the two children, one on each side, wear pantalettes and broad brimmed hats. Parthe, as Aunt Julia points out in a note scribbled below the sketch, clings to her father's coat-tail while Flo 'independently stumps along by herself'.

W. E. N. was a man to enchant a child. He loved the curious and the odd, and he loved jokes; he had a mind stored with information and the leisure to impart it. He had great patience and he was never patronizing. Partly as a result of marrying Fanny, partly by temperament, he was a lonely man, and it was with intense pleasure he discovered intellectual companionship in his daughters. Both were quick; both were unusually responsive; both learned easily, but the more intelligent, just as she was the prettier, was Flo.

It was a difficult situation for Parthe. She was the elder, the plainer, the less intelligent, the less remarkable. Flo, strange, passionate, uncomfortable little thing, had something about her which struck people as exceptional. Flo dominated. Flo led, and Parthe followed, but Parthe followed resentfully. She was possessive toward Flo, she adored Flo, but she was bitterly envious of Flo. Fanny made a practice of sending the children to stay with their relatives separately. In 1830 Flo wrote to Parthe from Fairoaks: 'Pray dear Pop, let us love each other better than we have done. It is the will of God and Mamma particularly desires it.'

W. E. N.'s plan for their education brought about the final division between the girls. To find a governess proved impossible. The world did not contain a woman who united the intellectual equipment required by W. E. N. with the standard of elegance and breeding demanded by Fanny. In 1832 he determined to teach the girls himself. A governess was engaged for music and drawing, and the girls learned Greek, Latin, German, French, Italian, history, and philosophy from their father. The time-table was formidable and W. E. N. exacting; the girls were required to work long hours, and Parthe rebelled. She left her sister struggling with Greek verbs and joined her mother or escaped into the garden.

Miss Nightingale and her father were deeply in sympathy. Both had the same regard for accuracy, the same cast of mind at once humorous and gloomy, the same passion for abstract speculation.

Parthe did not want to toil at Greek, but she resented the companionship between her father and sister, and in the summer of 1834 she wrote him a protest. He replied in a characteristic letter – involved, vague, and curiously reminiscent of a soliloquy in a poetic drama. 'My dear Pop – not one word . . . among my

16

waking dreams I sometimes fancy that you and I have not made half as much of each other's society as we might have done. . . . I have more subjects than one in hand, or in mind, which are likely enough to lend themselves to our future intercourse. In the meantime I feel that you are satiating yourself (*perhaps* usefully) with many matters which suit your infantine and merry days of 15 – or is it 16? – and, thro' a nervousness of interfering with them, I curtail my letter to a simple expression of my rejoicing at your merriments and your happiness. – W. E. N.'

Change of subject did not produce power to concentrate in Parthe. She continued to be bored, resentful, and cross, and W. E. N. became angry with her. In 1835 Parthe wrote to him when he was in London. 'Flo in bed, coughing, told me all. . . . I am properly punished, if you knew how very bitterly I feel your messages through her and your acknowledgement in your own letter, that you have ceased to care enough for my society, to be sorry I behave so ill.'

To send messages of reproof to an elder sister through a younger did not make the elder less jealous, but it made the younger self-righteous. At fifteen Florence could be sanctimonious. 'I hope our matutinal moments may not have been quite unprofitably spent, though we may not have improved our minds as we ought', she wrote to Fanny from the Isle of Wight in 1835. Her correspondence tended to be a record of her own good deeds. She had learned to eat sandwiches, which was an effort; she had practised curling her hair, as her mother desired; she had been devoting most of her spare time to looking after a baby cousin – 'Dear little Robert I am sure I never love him the less for being ugly'. By the time Florence was sixteen, the family had divided. She was W. E. N.'s companion in the library; Parthe was Fanny's companion in the drawing-room. Fanny was always busy; there were flowers to arrange, an increasing number of friends to be entertained, and innumerable letters to be written to the vast circle of the Nightingale family's connexions.

On Florence's fourteenth birthday W. E. N. calculated she had already twenty-seven first cousins and nearly two dozen aunts and uncles by blood and marriage. In the centre of this circle were the energetic handsome Smiths, who had the strongest possible family feelings. As each married, the circle was enlarged by the addition of a whole new family. Husbands of aunts, wives of uncles brought in their brothers and sisters and their wives and hus-

17

bands; even the brothers and sisters of grandmothers with their train of children and grandchildren were corresponded with, visited, kept informed, consulted. With the single exception of Fanny's eldest brother, who maintained domestic arrangements which were described as decidedly improper, not one of the huge clan was anything but respectable and prosperous. Enormous numbers of letters were written. Not only major events, weddings, births, deaths, but the choice of a place for a holiday, the advisability of taking a holiday at all, the dismissal of a coachman or cook, the selection of a dress or a carpet provoked correspondence and consultations with aunts, uncles, cousins, and grandmothers.

To Miss Nightingale letters and consultations were an intolerable waste of time. 'I craved', she wrote, 'for some regular occupation, for something worth doing instead of frittering time away on useless trifles.' Parthe had a large correspondence and numerous intimate friends among her cousins; but to Florence only three families were of importance – the Nicholsons of Waverley Abbey, the Bonham Carters of Fairoaks, and the family of Aunt Mai, now Mrs Sam Smith of Combe Hurst, Surrey.

Aunt Mai was a person of importance to the Nightingales. She was W. E. N.'s sister; and, should he have no son, the property would pass to her. In 1827 she married Fanny's younger brother, Sam Smith. It was then seven years since the birth of Florence, Fanny was nearly forty, and there was no sign of another child. It was almost certain that, if Aunt Mai had a son, he would eventually inherit Embley and Lea Hurst, and the marriage which linked the two families more closely together was welcomed. In 1831 a son was born, and Fanny, in whom all hope of another child must now have died, behaved admirably. The situation was not easy for her. Not only was Aunt Mai mother of the heir Fanny had failed to produce, she was also the object of the extravagant devotion of Fanny's difficult little daughter Flo. Nevertheless, Fanny's affectionate relations with Aunt Mai were unclouded. Aunt Mai's son was accepted as the heir and given a privileged position in the Nightingale family. Florence was known to have a special gift with babies, and when he was a few days old he was laid in her arms: 'My boy Shore', the eleven-year-old Flo proudly called him. Shore was recognized as being her special property, and devotion to Shore, pride in him, and Shore's devotion to her became one of the most important relationships in her life.

So Florence grew into girlhood in a life that seemed all smoothness and peace. At Embley and Lea Hurst there were comfort, security, and affection; there were intelligence and companionship. And yet beneath the surface there was no peace. Florence was brought up in a hot-house of emotion.

It was the result of a literary fashion. The wave of romanticism which had swept Europe had penetrated English domestic life, and ordinary wives and mothers were reproducing the behaviour of the heroines of Byron and Chateaubriand. Since a rigid respectability governed their behaviour, their emotions had to be expended on the commonplace events of everyday life. The naughtiness of a child, a misunderstanding between friends, the non-arrival of a letter necessitated smelling salts, a darkened room, a soothing draught. Women prided themselves on being martyrs to their excessive sensibility, and 'delicacy' was universal. Fanny, Parthe, and Florence were all considered 'delicate', though Fanny lived to be ninety-two, Florence ninety, and Parthe seventy-five.

Miss Nightingale grew up in this age and was indelibly impressed by it. Though her extraordinary mind owed its quality to uncompromising clarity and realism, her character contained the contradiction that she was also emotional, prone to exaggeration, and abnormally sensitive. The atmosphere in which she was brought up prevented her from achieving balance; throughout her life, when feelings were in question, she entered another world – violent, exaggerated, and unreasoning.

In the summer of 1834 she and Parthe were at Cowes with their governess, and W. E. N. wrote to tell them he had been invited to stand for Parliament as candidate for the Andover division. The girls' emotion approached hysteria. 'What extraordinary news you have sent us', wrote the young Flo. 'It quite convulsed our quiet little world. ... Parthe, after a deep reading of the letter in which she neither saw nor heard anything which passed around, screamed out "Papa is going to be M.P. for Andover!" Miss White and I stood aghast! ... I could not sleep after it. I slept so lightly that I had the feeling on my mind that something very extraordinary, or dreadful, had happened and kept starting up to find out what it was.'

W. E. N.'s candidature for Andover proved a turning-point in the Nightingales' lives. He was a fervent supporter of the Reform Bill of 1832, and had refused to enter political life until the Bill became law, when he believed a new age of political integrity would dawn. The election, in 1835, was the first to be held at Andover under the new franchise, and he entered the contest full of enthusiasm and hope. Fanny saw her plans maturing. They were to have a house in London.

He was not only defeated but profoundly disillusioned. The seat was lost because he refused to bribe the voters. A main object of the Reform Bill had been to end the purchase of votes, but the newly enfranchised electors of Andover took the view that the possession of a vote had always meant hard cash and that the extended franchise merely brought what had been the perquisite of a few within reach of the many. W. E. N.'s first contact with practical politics left him disgusted, and he resolved never to be persuaded to attempt an entry into political life again.

Fanny was defeated. W. E. N. ceased to adapt himself to the character she had planned. He gave up hunting and took long ambling rides in the New Forest; avoided political meetings and attended congresses of learned societies; spent more time teaching Florence and took to passing the greater part of each day in his library. An immensely tall desk was made for him, and he read standing up, he wrote standing up, he meditated standing up, contemplating for hours at a stretch such abstract subjects as the nature of moral impulses, the relation of ethics to aesthetics, and the proofs of the existence of an immortal soul in mortal man.

He would have been content to pass his life in tranquillity, allowing day after day to slide gently by. His natural home, he was fond of saying, was in 'the quiet and the shadows'. But life in the quiet and the shadows was unbearable to Fanny. She had been forced to give up her plans for W. E. N., but she did not resign herself. She transferred her plans and her ambitions to her daughters.

They were sixteen and seventeen. Next year, or the year after at latest, the girls must be launched in society. Since there was to be no house in London, everything must be done from Embley. The history of Lea Hurst repeated itself. The house was discovered to be entirely inadequate. Six more bedrooms must be added, new kitchens built, the exterior remodelled, the interior completely redecorated.

W. E. N. was tempted. He had made a certain reputation with

Lea Hurst and contemplated with pleasure the task of getting out designs to convert the Georgian plainness of Embley to a fashionable Gothic outline. The expense was bound to be considerable, and Fanny proposed that, while the alterations were carried out, they should make an extended tour abroad. It would do the girls good to see something of the world. They could hear music, practise their languages, go to a few parties, buy clothes in Paris. Europe was so cheap that the tour would be an economy. W. E. N. agreed. He loved travelling, loved Europe, had many friends in Italy and France. He got out his own design for a travelling carriage, the alterations to Embley were started at once, and the Nightingales fixed a date to leave in September 1837.

At this moment, in the midst of bustle, plans, discussions, Miss Nightingale received a call from God.

It is possible to know a very great deal about Miss Nightingale's inner life and feelings because she had the habit of writing what she called 'private notes'. She was unhappy in her environment, she had no one to confide in, and she poured herself out on paper. She hoarded paper (every odd scrap, every half sheet was preserved), and a very large number of her private notes exists. She wrote them on anything that came to her hand – on odd pieces of blotting-paper, on the backs of calendars, the margins of letters; sometimes she dated them, sometimes not. Sometimes they cover several foolscap pages, sometimes consist of one sentence. Occasionally she used a private note as the basis of a letter. Frequently she repeated a note several times at different dates with only a slight variation in wording. From time to time she also kept diaries; but it was in her private notes, written from girlhood to old age, that she recorded her true feelings, her secret experiences, and her uncensored opinions.

Her experience was similar to that which came to Joan of Arc. In a private note she wrote: 'On February 7th, 1837, God spoke to me and called me to His service.' It was not an inward revelation. She heard, as Joan of Arc heard, an objective voice, a voice outside herself, speaking to her in human words.

She was not quite seventeen, and she was already living largely in a dream world, which was often more actual to her than the real world. But the voices which spoke to her were not a phenomenon of adolescence. Nearly forty years later, in a private note of 1874, she wrote that during her life her 'voices' had spoken to her four times. Once on February 7, 1837, the date of her call; once in 1853 before going to her first post at the Hospital for Poor Gentlewomen in Harley Street; once before the Crimea in 1854; and once after Sidney Herbert's death in 1861.

Her path was not made clear. The voices which spoke to Joan told her to take a definite course of action; Miss Nightingale was told nothing definite. God had called her to His service, but what form that service was to take she did not know. The idea of nursing did not enter her mind. She doctored her dolls; she nursed sick pets; she was especially fond of babies. Her protective instincts were strong, but they had not yet led her to the knowledge that God had called her to the service of the sick.

Meanwhile she knew herself to be God's, and she was at peace. Her call had filled her with confidence and faith. God had spoken to her once; presently He would speak to her again.

On September 8, 1837, the Nightingales crossed from Southampton to Le Havre. They took with them Fanny's maid, a courier, and the girls' devoted disapproving old nurse, Mrs Gale. W. E. N. refused to take his valet.

2

THE travelling carriage W. E. N. had designed was enormous. Five years later, when he lent it to Fanny's sister, it held Mrs Bonham Carter and six of her daughters, besides a tutor and a maid. On the roof were seats for servants and for the family to enjoy the air and admire the scenery in fine weather. Six horses drew the carriage, ridden by postillions.

On the morning of September 9 the Nightingales left Le Havre to travel through France to Italy. The weather was brilliant; the girls sat on the roof; the postillions laughed, sang, and cracked their long whips as the carriage lumbered down the straight roads of France.

Florence was in transports of delight. She kept a diary of the tour and recorded that at Chartres she sat all night at her window enchanted by the beauty of moonlight on the cathedral. Her head was full of legends, and she turned romantic landscapes into the background for imaginary dramas. As they approached Narbonne, a lurid sunset flamed in the sky, and the city, half hidden by strangely shaped rocks, seemed sinister. She shuddered and imagined herself a traveller entering a city stricken by plague.

Yet for all her rhapsodies she remained precise. Each day she noted in her diary the exact hours of departure and arrival and the exact distance covered. Her letters to her favourite girl cousin, Hilary Bonham Carter, were not merely transports, for she was capable of mature observation. When they went over the castle at Blaye, she was struck by the fact that the custodian, a Napoleonic veteran, 'seemed to have fought everyone but to feel rancune against none'. At Bosuste, a village on the French-Spanish border which had had the misfortune to become a battleground in the Carlist war, she was impressed not by the horrors of war but by 'the indifference which misery brings'.

On December 15, 1837, the Nightingales drove into the gay and pretty town of Nice. There was a large English colony at Nice, balls, and concerts. With startling suddenness cathedrals, moonlight, and scenery vanished from Florence's diary and letters. She developed a passion for dancing and wrote to Hilary Bonham Carter that at the biggest ball of the season she danced every quadrille. Her letters took on a new tone; she began to attempt witticisms and to coin phrases.

When the time came to leave Nice, on January 8, 1838, she was heartbroken. As they climbed up the Corniche, she would not look at the famous view or admire the gold and silver lights over the sea. She sat inside the carriage shedding tears. 'The worst of travelling is that you leave people as soon as you have become intimate with them, often never to see them again', she wrote in her diary.

Her tears dried themselves with remarkable speed. The Nightingales reached Genoa on January 13, 1838, and on the 17th Florence wrote to Hilary Bonham Carter that of all towns in the world Genoa was the one she liked best. It was 'like an Arabian Nights dream come true'. In 1838 Genoa was still one of the richest and most splendid cities in Europe. Its palaces and gardens, its opera and theatres, its fountains and statues, had earned it the name of 'Genova la Superba'.

Florence went to more balls, and at 'the most splendid ball of the season' she had so many partners that she became confused. An officer came up and challenged her 'in a rage' because, after refusing to dance with him, she sat out with someone else and there was an *embrouillement*.

On February 14, after giving a large evening party, the Nightingales left Genoa for Florence. They halted at Nervi (described by Florence in one of her phrases as 'a town of palaces inhabited by washerwomen'), and again at Pisa, where they went to a ball given by the Grand Duke of Tuscany and a morning entertainment which included luncheon and an inspection of the Grand Duke's camels. On February 27 they reached Florence.

In 1838, thanks to the liberal policy of the Grand Duke, Florence was the intellectual capital of Italy. Fashion and learning united. There were parties for Fanny, learned conversaziones for W. E. N., educational opportunities for the girls. The Nightingales settled in a handsome suite at the Albergo del Arno, near the Ponte Vecchio. They had a *salon* fifty feet long, a dining-room with a terrace overlooking the Arno, and bedrooms which were magnificently (Mrs Gale, the girls' old nurse, said indecently) frescoed. Fanny wrote in March 1838 to her sister Julia Smith that the Grand Duke was 'exceedingly distinguishing and polite', the balls at the Grand Ducal Palace 'exceedingly fine', cards had been sent by the Grand Duke not only for public functions but for private entertainments, and Florence had been 'much noticed'.

Florence became (as she told Hilary Bonham Carter) 'music mad'. The opera in Florence was one of the best in Europe. Grisi and Lablache, the two most famous singers of the day, performed there. Florence lived for opera, persuaded Fanny to take her three times a week, and declared she would like to go every night. But she did more than go into transports. With laborious patience she kept a book in which she made a detailed comparison, in the form of a table, of the score, libretto, and performance of every opera she heard. Instinctively she reached out for facts. Transports, ecstasies, were not enough. Her mind demanded something hard to bite on, and the romantic extravagance of her emotion crystallized surprisingly into figures.

Italy not only gave her music: in Florence she learned an enthusiasm for the cause of Italian freedom. Italy had been handed over to Austria, the military despot of Europe, by the Congress of Vienna and was being crushed into subjection. Like thousands of her contemporaries, Florence Nightingale was seized with a passion for Italy and Italian freedom as violent as falling in love. To her, as to the Brownings and to George Meredith, the cause of Italian freedom was more than a political conviction; it was a religion, a faith, the embodiment of the struggle of good against the powers of darkness. W. E. N. and Fanny had friends in England intimately connected with the inner circle of Italian patriots. Through these friends the Nightingales were brought into the heart of the movement to set Italy free.

Since childhood Fanny had known a remarkable family, the Allens of Cresselly, in Pembrokeshire. Fanny Allen was one of the intellectual women who were the advance guard of the feminist movement. Her niece Emma married Charles Darwin. Her sister Jessie married the Italian historian Sismondi. W. E. N. and Sismondi were introduced by the Allens. A friendship sprang up, and it was agreed that, after leaving Florence and making a tour of the Italian lakes in July and August, the Nightingales would visit Geneva, where, for reasons of prudence, Sismondi was living as an exile.

They reached Geneva in the first week of September 1838, and found the city overflowing with political refugees. The Austrian Government was determined to crush independent thought in

Italy; every man of talent felt himself in danger, and a horde of doctors, scientists, educationists, scholars, writers, and poets poured across the frontier to safety in Geneva.

The change was startling. The world which the Nightingales entered was a world of poverty, learning, and sacrifice. Florence instantly responded, and balls and palaces passed from her mind. She became the disciple of Sismondi. He was very short, almost a dwarf, and extremely ugly – so ugly that his wife had hesitated before marrying him – but his nature was charming and his conversation enchanting. He could not bear to see unhappiness or to cause pain to any living creature, fed the mice in his study while he worked, and in Italy had a crowd of 300 beggars permanently encamped outside his door. Through Sismondi she met well-known figures of the Italian movement: Ugoni and Madame Calandrini, who had been ruined and exiled for opening progressive schools; Ricciardi, a young nobleman whom the Austrians had shut up among lunatics with the object of destroying his mind; Confalioneri, who had been practically buried alive for fifteen years – 'he still walks as if he had chains on his legs', wrote Florence.

W. E. N. would have stayed indefinitely in Geneva. He had found society which suited him among the professors of the University of Geneva and had struck up a friendship with de Candolles, the celebrated botanist. But, without warning, a crisis arose. Louis Napoleon Bonaparte, afterwards Napoleon III, after being expelled from Europe by the French, returned to Switzerland to see his dying mother. The French Government demanded his surrender. The Swiss, maintaining their right to asylum, refused. The French issued an ultimatum; the Swiss refused to give way, and French troops began to march on the Swiss. War seemed only a few days distant, and the Nightingales hastily prepared to leave for Paris. As they packed, every man and woman in Geneva was toiling at erecting barricades in the streets.

To leave Geneva was not easy. There were no horses. The vast carriage stood loaded, but every horse in Geneva had been requisitioned for artillery. W. E. N. scoured the town for horses while Fanny and the girls sat in the *salon* forbidden to go out, listening to the sound of barricades being erected in the street. Next day he managed to obtain inferior horses at an exorbitant price, and the Nightingales left. Sismondi saw them off, bursting into tears when he said farewell.

26

A few days later the crisis ended. The English Government mediated, Louis Napoleon voluntarily left Switzerland, and the French Government accepted the suggestion that he should be allowed to live in England. By the time the Nightingales reached Paris, the Genevese, hysterical with relief, were singing, dancing, and embracing each other in the streets. The experience made a deep impression on Florence. There were realities in Europe to which England, safely entrenched behind the Channel, was indifferent and blind. 'At home in England changes and revolutions are like storms one only hears', she wrote to Hilary Bonham Carter in November 1838.

W. E. N. proposed to spend four months in Paris and had taken an apartment in the Place Vendôme. It was, wrote Fanny, 'extremely splendid'. The rooms were vast and richly decorated; the dining-room had gilt mirrors, velvet draperies and carved chairs, the *salon* crimson satin and ebony cabinets. The windows framed a view of Napoleon's statue on the Vendôme column.

Fanny intended if possible to enter intellectual society and had an introduction, given her by her sister Patty, of which she had great hopes; it was to one of the most celebrated women in Paris – Miss Mary Clarke.

Without money, influence, or beauty, Mary Clarke had made herself a major figure in the political and literary world of Paris. In her hands the *salon* was revived, and every Friday night Cabinet Ministers, Dukes of France, English peers, bishops, scholars, and writers of international reputation crowded the drawing-room of her apartment in the former hotel of the Clermont-Tonnerre family, 120 rue du Bac.

Mary Clarke was not a Bohemian. She loved society, the great world, great houses, and great people. Her family connexions were excellent. The Clarkes were an old Scottish Jacobite family, Lord Dalrymple was her cousin, and her sister married Mr Frewen Turner, a Member of Parliament and owner of the famous Elizabethan mansion of Cold Overton in Leicestershire. Her personal appearance was odd. She was very small, with the figure and height of a child; her eyes were startlingly large and bright, and at a period when women brushed their hair smoothly she wore hers over her forehead in a tangle of curls. Guizot, who was devoted to her, said that she and his Yorkshire terrier patronized the same coiffeur. Yet though she had no ordinary feminine attractions, men were devoted to her, and many men wished to

27

marry her. Ampère, son of the celebrated electrician and her intimate friend, wrote: 'Her great charm lay in the absence of it. I never knew a woman so devoid of charm in the ordinary sense of the word and yet so fascinating. She was hardly a woman at all.' Her effect on her friends was very great. No one, wrote Miss Nightingale, ever had so much influence in forming character, but her candour sometimes took her friends aback. 'She had never a breath of posing or of "edifying" in her presentation of herself', added Miss Nightingale, 'even when it would have been almost desirable. ... She was always undressed – naked in full view. A little clothing would have been decent.'

Mary was launched on her career by Madame Récamier. About 1830, her mother, who suffered from bronchitis, was forbidden ever to go into fresh air again; the whole of Mrs Clarke's life must be spent indoors, and Madame Récamier who admired her character and attainments invited her to come and live in her house, the Abbaye-aux-Bois. (Mrs Clarke obeyed these instructions and lived to be ninety-two.) Madame Récamier was now fifty-four and her life had narrowed to a single object – the diversion of Chateaubriand, upon whom ennui had seized with the frightful effect of an incurable disease. Mary was invited to meet him. Her freshness amused him; she was invited again and conquered him completely. He declared he 'delighted in her'. 'La jeune Anglaise is like no one else in the world. Boredom is impossible where she is.'

Mary became Madame Récamier's close friend, visited her daily, was always present at her evenings; and when in 1838 the Clarkes moved to 120 rue du Bac, the intimacy continued.

She entered another distinguished circle through her close friendship with Claude Fauriel, the medieval scholar to whom the preservation of old French and Provençal literature is largely due. In 1837 Mary and Fauriel had been on terms of closest intimacy for more than fifteen years. They met daily; they travelled together; he dined with her almost every night, was invariably present at her parties, and behaved as master of the house; he had great respect for her mind, always read her his poems to criticize, and had left her his manuscripts in his will. Their intimacy was unconcealed and accepted; yet Mary's reputation was unblemished; she was the friend of bishops and deans, her *salon* was 'serious', and she refused to receive George Sand on account of her irregular life.

There was in fact no cause for scandal. They were friends, not lovers. But friendship was Fauriel's choice, not Mary's. He was devoted to her, but he had never fallen in love with her. He could have married her, but he had never asked her to marry him. She was in love with him, and jealousy tormented her.

While Mary Clarke was in love with Fauriel, who had only devoted friendship to give her, Fauriel's great friend Julius Mohl was hopelessly in love with Mary, who felt only friendship for him. M. Mohl was as much at home in the rue du Bac as Fauriel and, like him, dined with Mary almost every night. The younger son of a well-known German aristocratic family, he had become a naturalized Frenchman because he loved Mary Clarke and wished to live near her in Paris. When Queen Victoria asked him why he had given up his native country for France, 'Ma foi, madame, j'étais amoureux', he told her. He was an Oriental scholar of very great distinction. 'M. Mohl', wrote Miss Nightingale, 'was ... consulted, though a staunch Protestant, by the Jesuit Missions Étrangères in Paris as an authority superior to their own.' In addition to collecting material for a history of religion, he was engaged on a translation of the Persian epic *Shah Nameh* by Firdausi, for which he received a yearly grant from the French Academy. His character was charming, and he was a celebrated conversationalist. Nassau Senior, the American scholar, selected his conversation to record in one of *Senior's Conversations*. He spoke and wrote English perfectly and passionately admired English political principles.

The Nightingales were not the kind of connexion which appealed to Mary Clarke. She did not care for young ladies; in fact, she did not care for women at all. 'I don't like young ladies', she wrote to a friend who had asked permission to make an introduction to her, 'I can't abide women. Why don't they talk about interesting things? Why don't they use their brains? My dear, they have no manners. I can't abide them in my drawing-room. What with their shyness and their inability to hold their tongues, they ain't fit for decent company. If your friend is a man, bring him without thinking twice about it, but if she is a woman – think well.' She was, however, 'absurdly fond' of children and regularly gave children's parties, and she acknowledged Fanny's letter of introduction by inviting the Nightingales to a 'children's *soirée*'.

One afternoon near Christmas they drove up to 120 rue du Bac. No servants were visible, but a clamour came from above. They walked up into a front drawing-room crowded with dancing, singing children; no one took any notice of them, and they went through to the back drawing-room, where two impressive and eminent-looking gentlemen were boiling a large black kettle over a log fire. One was Claude Fauriel, the other Julius Mohl.

In the midst of the children, dancing, singing, and clapping her hands, was a strange little figure no bigger than a child herself, whom Florence realized must be the celebrated Miss Mary Clarke. The children began to play blind man's bluff, and without further ado Florence picked up her skirts and joined in. It was the happiest possible introduction. She was never so unself-consciously gay as with children – indeed, all the Nightingales were past-masters in the art of amusing the young.

Immediately after the children's *soirée* Mary Clarke invited the Nightingales to one of her celebrated Friday evenings. She had fallen in love with the whole family, most of all with Florence, but also with W. E. N.'s remote charm, Fanny's rich beauty and overflowing kindness, and Parthe's elegance. They christened her 'Clarkey', and she turned their stay in Paris into a carnival. They met almost every day and, as Florence wrote to Hilary Bonham Carter, 'tore about' together. Clarkey went everywhere and knew everyone. She took the girls to private parties, to studios, to galleries and concerts, to the opera, to the theatre, to receptions and balls. She introduced them to Madame Récamier. They were asked several times to the Abbaye-aux-Bois, met Chateaubriand, and were paid the very great compliment of being invited to hear Chateaubriand read his memoirs. And so when the famous readings from the *Mémoires d'outre-tombe* were given by Chateaubriand at the Abbaye-aux-Bois in January 1839, Florence Nightingale was in the audience.

She was wildly happy. She had a 'passion' for Clarkey, she was beginning an important friendship with Julius Mohl, and for the first time in her life she was breathing the air of freedom.

Fanny smiled on her infatuation. She wanted Clarkey to be the family's intimate friend. Clarkey was to be the source from which she intended to collect 'notabilities' to add lustre to her parties at Embley. Clarkey was unconventional, but Clarkey was accepted by the best society, and Fanny was satisfied. But the young Florence was receiving impressions of which Fanny never dreamed. One of the deepest was the impression made by Clarkey's friendship with Claude Fauriel. She observed that

Clarkey and Fauriel met daily, that they were devoted and made no secret of their devotion, that Fauriel had the greatest possible respect for Clarkey's mental powers and treated her as an equal, above all, that this close intimacy was accepted without disapproval by everyone, even by so conventional a woman as her mother.

She acquired a belief in the possibility of a daily intimacy, a close friendship between a man and a woman on terms which did not include passion, and which did not provoke scandal. It was a belief she never lost and one which regulated her conduct throughout her life.

The Nightingales had now been abroad for eighteen months, and the alterations to Embley were due to be finished by June. Before they went down to the country, Fanny wished to spend part of the season in London and have the girls presented at Court.

In April 1839 the family left for London. Fanny was well satisfied. The tour had shown her that in Florence she possessed a daughter who promised to be exceptional. She began to concentrate on Florence. Her pride in her was immense, her hopes for her brilliant.

They were doomed. Florence's conscience was awake, and the brief halcyon period was over. It was two years since God had spoken to her. Why had He not spoken again? The answer was evident – she was not worthy. She had forgotten God in the pleasure of balls and operas, in the vanity of being admired. She loved pleasure too much; she loved society too much; she must school herself to turn her back on it. In March, 1839, before she left Paris, she wrote in a private note that, to make herself worthy to be God's servant, the first temptation to be overcome was 'the desire to shine in society'.

3

WITH the return of the Nightingales to London the first great struggle of Miss Nightingale's life began. It was divided into two stages and lasted fourteen years. First she groped within herself for five years before she reached the certainty that her 'call' was to nurse the sick; next a bitter conflict with her family followed, and nine more years passed before she was able to nurse.

In April 1839 Fanny and William Nightingale had no inkling of Florence's secret life of agony, aspiration, and despair. They congratulated themselves on the possession of a charming and gifted young daughter destined for a brilliant social success. She was graceful, witty, vividly good-looking. Among the Verney Nightingale papers are two small oblong packets, carefully wrapped in several thicknesses of black paper, which contain two tresses of hair tied with silk and labelled in Fanny's writing 'Flo 1839' and 'Parthe 1839'. The hair has been so well protected that it might have been cut yesterday. Florence's hair is of unusual beauty, bright chestnut in colour, thick, glossy, and wavy. In middle age her hair became dark, but at nineteen it was golden-red. Parthe's hair has less life and colour; it is light brown, almost blonde, fine, soft, and straight.

The Nightingales reached London on April 6. W. E. N. went down to Embley and found the alterations would not be finished by June; Fanny then decided to spend the whole season in London. Her sister, Mrs Nicholson, was bringing out her two elder girls, and the families united to take a floor of the Carlton Hotel. On May 24 Florence and Parthe were presented at the Queen's birthday Drawing Room. Florence, wearing a white dress bought in Paris, looked, wrote Fanny, 'very nice', and 'was not nearly as nervous as she expected'.

The girls were caught up in a whirl of gaiety. Wherever the Nicholsons went, they met hosts of friends; carpets were rolled up for dancing; the air rang with screams of laughter; servants ran about; impromptu meals appeared. W. E. N. called it the

'Waverley Saturnalia'. 'The *piano*,' he remarked, 'is not their *forte*.'

Once more her 'call' vanished from Florence's mind; once more she became absorbed in parties and dresses and partners and balls. The summer was hot, the hotel noisy; she was exhausted, deliriously happy, perpetually excited.

She had been seized by a 'passion' for her cousin, Marianne Nicholson. 'I never loved but one person with passion in my life and that was her,' she wrote in 1846. Marianne was dazzlingly beautiful – 'that brilliant face is almost as the face of an angel.' She had exceptional musical gifts, an exquisite soprano voice, and possessed a confidence in her own charm which enabled her to dare anything. She would even take hold of W. E. N. and shake him. 'I was internally screaming with laughter,' wrote Florence to Hilary Bonham Carter. 'I should think no one ever shook my Papa's sacred person before.'

But to love Marianne was dangerous. Her moods were unpredictable; she was angel and devil, pointlessly cruel, pointlessly kind, generous or mean, malicious or good-natured, truthful or a liar without reason or motive.

Marianne's capacity to love was reserved for her family. She adored her brothers and sisters. All the Nicholsons adored each other and stood by each other through thick and thin, and of all her family the one Marianne loved best was her brother Henry.

By an unhappy chance Henry fell in love with Florence. She did not love him, but she encouraged him because he brought her closer to Marianne.

In September, W. E. N. decided that, finished or not, Embley must be occupied. The move was a series of disasters. They arrived to find men still working in the house and retreated to one of the lodges. Several days passed in discomfort. At last the move was fixed. That day a hurricane broke, the worst storm of a stormy autumn. The Nightingales waited all the morning, but not a soul appeared. W. E. N. went up to the house to 'do housemaid', and late in the afternoon Fanny and Florence determined to follow. The drive was under water, and they waded to the house. It was deserted; the servants had failed to arrive. W. E. N. lighted a fire in the servants' hall, and hungry and shivering they peered into the larder. It contained nothing but joints of raw meat. Darkness fell, and still no one came. At last the sound of wheels was heard. Florence rushed out. It was a cart containing 'a man in a cloak carrying a looking glass'. W. E. N. took a candle from a carriage lamp and stuck it on a spike, and they sat by its glimmering light. W. E. N. was in excellent spirits – he found their situation diverting. When the servants at last appeared, their explanations were unsatisfactory. The steward spoke 'with tears in his eyes', but the truth was that the servants refused to start until they had had their tea, and there was a further difficulty because Mrs Gale thought it beneath her dignity to be conveyed in a cart.

Embley was now a handsome house, 'able to receive five able bodied females with their husbands and belongings', wrote Florence to Clarkey in October 1839. Fanny's descriptions of the drawing-room mention fawn-coloured walls 'pale and cool with gold mouldings', a blue ceiling 'as skiey as possible', 'purple silk cushions ornamented with gold fleur de lys', and a set of chairs in tapestry, 'with a blue and white ground but worked in a variety of colours with red predominating in the groups of flowers or figures'. The sofas were covered in red silk damask. The carpet was 'green of a yellowish tint' and considered 'a great prize'. Fanny had fallen in love with the 'veloutés' she had seen in Paris and had had a carpet specially woven at Axminster in the same design; this carpet is now in the Victoria and Albert Museum, London. W. E. N. had the library enlarged and put in new shelves of oak elaborately carved in a Gothic style with each section divided by a caryatid. On top of the shelves stood antique busts, and the windows were hung with heavy crimson curtains intricately draped and looped.

Straightening the house and settling the servants filled the next two months. It was not until the New Year that Florence could draw breath. She was deeply, furiously discontented, with life and with herself. Her infatuation for Marianne was perpetual torture. She had let Henry fall more in love with her than ever. Toward making herself worthy, toward justifying her 'call' she had done nothing. Her life at home was hateful; impossible that God should have bestowed the gift of time on His female creatures to be used as Fanny wished her to use it. 'Faddling twaddling and the endless tweedling of nosegays in jugs,' Clarkey called it.

Miserable, irritable, bored, she became unwell. She was rescued by Aunt Mai. A long visit at Christmas had transformed their relationship. They were now devoted friends, no longer fond aunt and adoring little niece but equals revelling in closest intimacy. Aunt Mai was W. E. N.'s sister and possessed many of his

34

qualities – intellectual curiosity, humour, interest in abstract speculation. She, too, had a leaning toward the metaphysical and the transcendental, and her love for Florence had a mystical quality. In spite of their difference in age she worshipped Florence with the worship of a disciple for a master. She placed Florence above ordinary humanity, above the claims even of her husband and her children, and became her protector, interpreter, and consoler. Aunt Mai's tact, her energy, her flow of words were inexhaustible, and in innumerable letters, almost always undated, written on flimsy paper with a thin pen, criss-crossed on every page, she endeavoured in a flood of apologies, explanations, excuses, to make life easier for Florence. In January 1840 she persuaded Fanny that 'Flo would be all the better for a little change', and, at the end of the month, Florence was allowed to pay a visit to Combe Hurst.

At once her spirits soared. London was buzzing with gossip of Queen Victoria's wedding, and she wrote a lively account of the ceremony on February 10. 'There were but 3 Tories there. Ld Melbourne pressed the Queen to ask more, told her how obnoxious it was. Queen said "It is MY marriage and I will only have those who can sympathize with me." Mr Harcourt told Lord Colchester that there was a great levee to receive the Prince and they were all standing with the Queen ready to receive him. When his carriage was announced, she walked out of the room. Nobody could conceive what she was going to do, and before anyone could stop her, she had run downstairs and was in his arms.'

Florence went to several dinners and to the opera, received a flattering number of Valentines, and spent a great deal of her time with Aunt Mai's children. On the surface she was happy enough, charming enough, gay enough, but beneath the surface was agony and despair. It was three years since she had been 'called', and she still did not know to what. That was the frightful dilemma – what had she been called to do? The first necessity was to improve herself, to become worthy. God was waiting for her to become worthy before He could give her instructions. But how was she to make herself worthy?

She had written to Clarkey that mathematics gave her certainty; mathematics required hard work, and perhaps she would find life more satisfactory, be more satisfactory herself, if she studied mathematics. She confided in Aunt Mai, and they began to work together, getting up before it was light to avoid disturbing the routine of the house. She became wildly happy. If only her parents would let her have lessons, if only they could be persuaded to let her study mathematics instead of doing worsted work and practising quadrilles.

In March 1840 Aunt Mai wrote Fanny a cautious letter. 'Flo and I have a good deal of talk about the employment of time and so forth. I am much impressed with the idea that hard work is necessary to give zest to life in a character like hers, where there is great power of mind and a more than common inclination to apply. *So* I write to ask you if you in any way object to a mathematical master, if one can find a clean middle aged respectable person ... of course shall not do anything without your permit. ...'

Fanny did not approve; home duties were not to be neglected for mathematics. Aunt Mai hastened to assure her there was no neglect: 'Flo and I have begun getting up at 6, lighting our fire and sitting very comforably at our work, and I think if she had a subject which required *all* her powers and which she pursued regularly and vigorously for a couple of hours she would be happier all day for it ... she is disappointed at her want of success in music. ... I allow her quadrille playing is bad.'

Her daughter's destiny, Fanny answered, was, she sincerely hoped, to marry, and what use were mathematics to a married woman? 'If she throws up her mathematics,' Aunt Mai replied, 'in the more active and interesting pursuits of future life in which I hope I may live to see her powers engaged, whatever she may have done in that line will have benefited her character.' In a burst of sincerity she added a postscript: 'I don't think you have any idea of half that is in her.'

Fanny would not give way. Letters flew backward and forward. Where were Florence's lessons to be given? If at her grandmother's in Bedford Square there would be a problem about Aunt Patty. Patty was getting peculiar. There were 'outbreaks', and it would not be suitable for Florence to meet her. Fanny's brother, Mr Octavius Smith, offered the use of his library. Fanny raised difficulties about the master. Would it not be more suitable to have a clergyman; was the master proposed a married man; was there not a class; who would be present during the lessons? Aunt Mai persevered, soothing, reassuring, producing a married man, a clergyman, one accustomed to teaching young ladies, a chaperon. W. E. N. then entered the correspondence with an entirely new consideration. 'Why Mathematics? I cannot see that Mathematics would do great service. History or Philosophy, natural or moral, I should like best.' In reply Aunt Mai reported a dialogue. 'I told Flo this preference.' 'I don't think I shall succeed so well in anything that requires quickness as in what requires only work.' 'Then you prefer mathematics?' 'Yes.'

All Aunt Mai could obtain was a compromise. She enlisted the aid of the Octavius Smiths. Mrs Octavius Smith had been ill and asked that Florence, whose gift for managing children was well known, should be allowed to come and stay with her and help with the children. Fanny agreed that she might go for a month, and during part of April and May, 1840, she stayed with the Octavius Smiths and had a lesson in mathematics twice a week in their library. In the middle of May she went back to Embley and the mathematics lessons came to an end.

In the summer of 1840 Fanny had a series of house-parties at Embley. Clarkey came for a long visit, and Fauriel and Julius Mohl. Fanny's circle had lately been extended by a friendship with the Palmerstons. In 1839 Lord Palmerston married Lady Cowper, widow of Earl Cowper and sister of Lord Melbourne, and settled at Broadlands, near Romsey, a few miles from Embley. The Palmerstons took a fancy to the Nightingales, and a friendship sprang up. 'The Palmerstons ask us to dine *en famille*, I mention this merely to show how friendly they are,' W. E. N. had written to Florence in April. During the summer the Palmerstons were constantly at Embley, bringing with them their son-in-law Lord Ashley, better known by his subsequent title of Lord Shaftesbury, the reformer and philanthropist, founder of the Ragged School Union.

And still Florence was not satisfied. The previous autumn she had complained that she had no one to talk to; now she complained she had too many people to talk to. She had said she must have intelligent conversation and exchange of ideas; now she said she must have time for study.

Among her cousins her most intimate friend was Hilary Bonham Carter, eldest of the six daughters of Fanny's sister Joanna. Hilary, who was a year younger than Florence, had been devoted to her from childhood. She was unusually pretty and had a talent for painting; a self-portrait shows a charming little pointed face framed in heavy bands of soft hair, a wide forehead, large eyes under delicate brows, a sensitive mouth, and an expression of intelligence and sweetness. When in 1838 her father had died, she had become the support of her mother, a nervous unpractical woman overwhelmed by the responsibility of bringing up a large family unaided. Florence made Hilary the confidant of her difficulties, pouring out her soul in enormous letters, telling almost all – but not all. The story of her 'call' on February 7, 1837, she confided to no one.

She had a feeling of oppression, she felt herself pursued by servants, guests, relations in a clutching, demanding horde. 'There are hundreds of human beings always crying after ladies,' she wrote to Hilary Bonham Carter in 1841. 'Ladies' work has always to be fitted in, where a man is, his business is the law.' She wrote the phrase again in a private note: 'Hundreds of human beings always crying after ladies,' and added, 'I must have some leisure to find out a few things.'

She had no leisure. Christmas, 1841, was spent at Waverley. Fanny described the festivities in a letter to W. E. N., who had refused to leave home, as 'awesome'. Eighty people slept in the house. There was a huge masked ball which went on until five o'clock in the morning, succeeded the following night by an amateur performance of *The Merchant of Venice*. Henry played Shylock, rushed up to London where Macready was performing the part, interviewed him in his dressing-room, and secured directions for the interpretation of the part. He had come down from Cambridge, was preparing to be called to the Bar, and was still desperately in love with Florence. In March the Nightingales went to London for the season and took rooms at the Burlington Hotel, Old Burlington Street.

Florence was very gay. Though she was only twenty-two, she was becoming a figure in intellectual society. Her demure exterior concealed wit. She danced beautifully, yet possessed a surprising degree of learning, had great vitality and was an excellent mimic. She was 'very much noticed', Fanny wrote, by the new Prussian Ambassador, the Chevalier Bunsen, and his wife. The Bunsens united intellect, good breeding, and wealth. The Chevalier (he was created Baron in 1857) was a Biblical scholar of European reputation who shared with his friend Lepsius the credit of being the world's leading Egyptologist; he had married an Englishwoman of good family, was extremely rich, and had a house in Carlton House Terrace besides a place in Sussex. The Bunsens were close friends of the Queen and Prince Albert and were liberal Evangelicals. Florence, who went constantly to their house, was addressed by the Chevalier as 'My favourite and admired Miss Nightingale'; he lent her books and discussed archaeology and religion with her.

She had achieved a success and could not help feeling satisfaction; yet she reproached herself bitterly. 'All I do is done to win admiration,' she wrote in a private note. She cared too much for lights, pretty clothes, glitter, the allurements she called in her private notes 'the pride of life'. Over and over again she told herself that before she could hope to be worthy enough for God to reveal the path of service the temptation to shine in society must be conquered.

In the summer of 1842 the temptation to shine was greatly increased. In May at a dinner-party given by the Palmerstons at Broadlands, Florence was introduced to Richard Monckton Milnes. In 1842, Richard Monckton Milnes was thirty-three. He was the only son of Mr Richard Pemberton Milnes and heir to the estate of Fryston in Yorkshire. He had achieved a brilliant success in London society and was predicted an important political career.

He wrote talented poetry himself but had an even greater talent for discerning poetic genius. In 1848 he was responsible for the collection and publication of the first collected edition of the poems of Keats. The breakfasts he gave in his rooms in Pall Mall were famous. He invited everyone in the public eye whether famous or notorious, whether known to him or not. Carlyle, when asked what he thought would be the first thing to happen if Christ came to earth again, said: 'Monckton Milnes would ask him to breakfast.'

He diffused amiability. 'He always put you in a good humour with yourself,' said Thackeray; and his wit was never malicious – life was his target, not humanity. Life, he was fond of saying, is a jest not witty but humorous. His kindness and generosity were based on love for his fellow men. 'No one who knew Richard ever hesitated to ask him a favour,' wrote one of his friends. He was 'a good man to go to in distress,' wrote another. 'He treated all his fellow mortals as if they were his brothers and sisters,' said Florence.

His humanity expressed itself in philanthropic work. He loved children and worked for many years, against ceaseless opposition, to improve the treatment of young criminals. It was largely owing to his efforts that juvenile offenders ceased to be sent to jail with adult criminals and were sent to reformatories instead.

But he had another side to his character. The humane lover of children, the connoisseur of literature was also the man who introduced Swinburne to the works of the Marquis de Sade.

During the summer Richard Monckton Milnes came several times to Embley. He was falling in love with Florence, and he made himself the friend of Fanny, Parthe, and W. E. N. By the end of July, when as usual the Nightingales went north to Lea Hurst, he was treated as one of the family.

Fanny had connexions in northern society; earlier in the year she had met the Duke of Devonshire and been asked to dine. Now she, W. E. N., and the two girls were invited to stay at Chatsworth to meet H.R.H. the Duke of Sussex. All the society of the north assembled, and in honour of the Royal guest the huge house was crammed with 'Howards, Cavendishes, Percys, Greys, all in gala dress with stars, garters, diamonds and velvets,' wrote Fanny to Clarkey in August 1842. The entertainment was planned on an enormous scale. Mr Joseph Paxton, later designer of the Crystal Palace, was head gardener at Chatsworth, and he had erected a vast glasshouse in the Park. 'An omnibus,' wrote Fanny, 'plied at the gates of Chatsworth every evening to take those who could not walk so far to the monster conservatory, which covers an acre of ground, and where groves of palms and bananas are making all haste to grow to their natural size.' One evening the huge glasshouse was brilliantly lit for a 'promenade'. Another evening there was a magnificent ball to open a new banqueting hall.

Florence was indifferent to the splendours of Chatsworth; the devotion of Richard Monckton Milnes left her unmoved. Some time in the summer of 1842 she had taken the first step toward the fulfilment of her destiny. She had become conscious of the world of misery, suffering, and despair which lay outside her little world of ease and comfort.

Eighteen forty-two was a terrible year for the people of England. The country was in the grip of what has passed into history as 'the hungry forties'. In villages, as in towns, there were starvation, sweated labour, ignorance, and dirt. Diseased scarecrows swarmed not only in the airless undrained courts of London, but in the 'black filth' of rural cottages; workhouses, hospitals, and prisons were overflowing. In the summer of 1842, Florence wrote in a private note: 'My mind is absorbed with the idea of the sufferings of man, it besets me behind and before ... all that poets sing of the glories of this world seems to me untrue. All the people I see are eaten up with care or poverty or disease.'

She had progressed. She knew now that her destiny lay among the miserable of the world, but what form that destiny was to take she still had no idea.

In the autumn of 1842 she called on the Bunsens, and Baroness Bunsen, in her Memoir of her husband, records that she asked him a question couched more or less in these words: 'What can an individual do towards lifting the load of suffering from the helpless and miserable?' In reply, he mentioned the work of Pastor Fliedner and his wife at Kaiserswerth, on the Rhine, where Protestant Deaconesses were trained in the hospital of the Institution to nurse the sick poor. Florence's attention was not arrested; she had not yet begun to think of nursing.

Meanwhile her mother steadily progressed to social success, and at Embley party followed party. 'Pray send him a sly line that he will find *notabilities* here on the 24th,' wrote W. E. N. to Clarkey in October 1843, arranging a visit from Ranke the historian, '– to wit the Speaker [Shaw Lefèvre], the Foreign Secretary [Palmerston], the Catholic Weld [future owner of Lulworth and nephew of the Cardinal of that Ilk], and mayhap a Queen's Equerry or two, a Baron of the Exchequer ... and a couple of Baronets. He should think well on this. Yours quizzically but faithfully, W. E. N.' Florence scribbled a postscript: 'Papa is quizzing the baronets who are not wise ones. Provided *you* come I care for nobody, no not I, and shall be quite satisfied. As M. de Something said to the Staël "Nous aurons à nous deux de l'esprit pour quarante; vous pour quatre et moi pour zéro".'

She had formed her own circle. She saw Richard Monckton Milnes constantly and had become interested in his philanthropic work. She had a new friend, Miss Louisa Stewart Mackenzie, later the second wife of Lord Ashburton. The Palmerstons were devoted to her, and she was very friendly with Lord Ashley.

But when at the end of July 1843 the Nightingales once more went north to Lea Hurst, her whole being was concentrated on the poor and sick. She began to spend the greater part of her day in the cottages, began to badger her mother in and out of season for medicines, food, bedding, clothes. Fanny, who was generous in distributing charity, felt Florence was unreasonable. 'Perhaps if we got a Sœur de Charité Flo would let us rest in some peace,' she wrote to W. E. N. during the summer of 1843.

When the time came to go to Embley, Florence wanted to stay behind at Lea Hurst. Fanny would not hear of her remaining, and she had to come south. 'It breaks my heart to leave Lea Hurst,' she wrote to Aunt Mai in September 1843.

When her mother had succeeded in getting Florence away, fresh difficulties arose. One of her friends died in childbirth, leaving a daughter. Florence demanded permission to cancel her engagements, give up London for the autumn season, and look after the baby. When Fanny refused, she fretted herself into an illness. Forced to give way, Fanny compromised by allowing her to go and look after the baby for a few weeks – at the height of the season when she should have been in London going to parties. Fanny was bitterly disappointed, and misery resulted on both sides.

Florence's misery was a thousand times increased by a terrifying discovery. She records in a private note that in the autumn of 1843 she suddenly realized the extent to which the habit she called 'dreaming' had enslaved her. She fell into 'trance-like' states in the midst of ordinary life, while, for instance, she was making conversation with the Ashburtons at Sir William Heathcote's dinner. She could not control herself, and she gave way with the shameful ecstasy of the drugtaker.

The whole world went wrong for her in the winter of 1843. Henry Nicholson was pressing her to become engaged to him, Marianne was beginning to be angry with her for not accepting him. When Christmas came and she found herself at Waverley with Henry and Marianne – the strain became too great, and she broke down. As she lay in bed listening to the sounds of revelry floating up from downstairs, she despaired. Was there nothing for her but dreaming? Had she better close her eyes and find what satisfaction she could in a false paradise of consoling visions? And then, she wrote in a private note, 'an acquaintance with a woman to whom all unseen things seemed real and eternal things near, awakened me.'

Miss Hannah Nicholson, sister of Mr Nicholson of Waverley, called Aunt Hannah by the Nightingales, was a deeply religious woman with the gentleness, purity, and limited vision of a nun. She did not understand Florence; there were depths, violences, capacities in her she was unable to grasp, but she knew Florence was ill, not on good terms with her family, and unhappy. She believed she could provide the solution. Union with God would bring reconciliation with earthly life. A close intimacy sprang up. Days were spent discussing the life of the soul and the way of the soul to God. Florence had an enormous amount of unexpended affection. In a private note, written at this time, she speaks of 'those I love – and no one knows *how* I love'. She adored Aunt Hannah, and became her disciple.

There was, however, an essential difference between them which Miss Nicholson did not appreciate. Though Florence sought union with God, she did not seek that state as an end in itself. Union with God was a necessary qualification for the performance of God's work, a preparation for action, not submission. Aunt Hannah believed that once Florence's soul was one with God she would be reconciled to the state of life to which it had pleased Him to call her.

In January 1844 Florence went back to Embley and began to write Aunt Hannah letters of enormous length. 'When I write the floodgates of my egotism are opened by your sympathy.' She called her letters 'my outpourings'. But she did not write either of her 'call' or of her dreams, and at the very moment when she seemed to be most thoroughly under Aunt Hannah's influence she took a secret decision of the greatest importance entirely opposed to everything Miss Nicholson hoped.

Sometime in the spring of 1844 the knowledge came to her that her vocation lay in hospitals among the sick. At last, seven years after her 'call', her destiny was clear. 'Since I was twenty-four,' she wrote in a private note thirteen years later, '. . . there never was any vagueness in my plans or ideas as to what God's work was for me.'

In June Dr Samuel Gridley Howe, the American philanthropist, and his wife Julia Ward Howe, later to become celebrated as the author of the 'Battle Hymn of the Republic', came to stay at Embley. Dr Howe's daughter describes in a book of reminiscences how after dinner on the night of his arrival Florence came up to him in the drawing-room. Would he meet her privately in the library for a few moments before breakfast? Dr Howe consented. When husband and wife were alone, Mrs Howe reminded him that they had heard the younger Miss Nightingale described as an exceptional girl likely to make an exceptional career for herself, though her mother would prefer her to lead a

43

more conventional life. In the library next morning Florence went straight to the point: 'Dr Howe, do you think it would be unsuitable and unbecoming for a young Englishwoman to devote herself to works of charity in hospitals and elsewhere as Catholic sisters do? Do you think it would be a dreadful thing?' He gave a sincere answer: 'My dear Miss Florence, it would be unusual, and in England whatever is unusual is thought to be unsuitable; but I say to you "go forward", if you have a vocation for that way of life, act up to your inspiration and you will find there is never anything unbecoming or unladylike in doing your duty for the good of others. Choose, go on with it, wherever it may lead you and God be with you.'

She had reached the turning-point of her life, but she confided in no one. The word 'hospital' had not yet been uttered to her family; she was well advised to hesitate before introducing it; it was a dread word. She must think out some method by which her parents might be brought to consent to their daughter entering a hospital. Throughout the summer she meditated in secret. 'I dug after my little plan in silence,' she wrote.

It was an unsatisfactory summer. Marianne and Henry Nicholson came to stay; Marianne was cold, and the visit was not a success. Claude Fauriel died in July, and Clarkey did not feel equal to paying her usual visit to Embley, to Florence's deep disappointment. The illness of the previous winter, misery over Marianne, the weight of the shameful secret of her 'dreams', the perpetual frustration of her life at home, brought her low, and she wrote Clarkey an unhappy letter. 'Oh do not say that "you will not cloud young people's spirits". Do you think young people are so afraid of sorrow, or that if they have lively spirits, which I often doubt, they think these are worth anything, except in so far as they can be put at the service of sorrow? ... When one thinks there are hundreds and thousands of people suffering ... when one sees in every cottage some trouble which defies sympathy – and there is all the world putting on its shoes and stockings every morning all the same – and the wandering earth going its inexorable treadmill through those cold hearted stars, in the eternal silence, as if nothing were the matter; death seems less dreary than life at that rate.'

That summer when she went north to Lea Hurst there was scarlet fever in the cottages, and she was forbidden to go near them. All through the autumn she was ailing, and when, at

44

Christmas, Fanny and Parthe went to Waverley, she was too ill to go. She stayed in bed at Embley pouring out letters, notes, analyses, plans; striving to find a way to get away from home to a hospital; striving to find a solution to her relations with Marianne; striving to achieve the state of union with God in which Aunt Hannah assured her all difficulties would vanish.

On New Year's Eve 1844 she was unable to leave her room and sat writing late at night with 'a little black tea pot on the hob'. Outside it was freezing hard with a brilliant moon. She watched three hares playing on the whitened grass of the lawn; in the stillness the world seemed to be dead except for those three hares. At Waverley at this moment there was a ball. She sighed after the ball and the dress she had been going to wear, a pink dress with black lace flounces, ruefully aware that the 'pride of life' was by no means dead. 'I am convinced of it when I think of my black lace flounces,' she wrote to Aunt Hannah.

In a fortnight or so she was convalescent, and Hilary Bonham Carter came to stay. She was another victim of family life. In the previous year Hilary had met Clarkey. With Clarkey she had attended an atelier and was pronounced to have genuine talent. Clarkey had implored Mrs Bonham Carter to let Hilary work seriously. But Hilary could not be 'spared'. Now she was spending her life housekeeping, teaching her younger sisters, doing the flowers, and, as a concession, attending a 'ladies' atelier' in London where so little was expected that lessons were taken 'when social engagements permitted'.

The girls were alone for two days; the weather was fine and still, and they made long expeditions in the New Forest, walking from breakfast until sunset, and talking all day. Florence poured out her heart on the subject of Marianne, but she spoke neither of her determination to work in hospitals nor of the shameful secret of her 'dreams' – that ever-growing terror. During the forced inaction of her illness and the idleness of her convalescence she had found herself more enslaved than ever before, and she was beginning to fear for her mental balance.

That February Aunt Mai's son Shore, 'my boy Shore', the heir to Embley, now aged fourteen, came to convalesce at Embley after measles. She looked after Shore entirely and had a month of freedom from shameful visions. 'Whilst he is with me all that is mine is his,' she wrote to Aunt Hannah in February 1845, 'my head and hands and time.' With Shore she took little walks on the gravel paths, hunted for snowdrops, read aloud and had 'a great deal of conversation about dogs'. At night, when she had put him to bed and given him his medicine, they had serious talks. She 'warned him against lying long in bed, and the temptations of the world, liking to be praised and admired and a general favourite more than anything else and we were both very much affected'.

In March, Shore went home, and the Nightingales started once more on their round, up to London, back to Embley for parties in June, up to the North for July and August, back to Embley again for the shooting, up to London in November, back again to Embley at Christmas time. And as week followed week, Florence became more wretchedly unhappy. Nearly a year had passed since her interview with Dr Howe, and she was no farther forward; eight years had passed since her 'call', and not merely had she accomplished nothing, she had slipped backwards – she had lost the sense of walking with God. In private notes, in enormous letters to Aunt Hannah, she reproached herself with frantic bitterness. Again the strain was too great. Though she went with her family to London in February, as soon as she arrived she was ill. On March 1, 1845, she was in bed in the Burlington Hotel, suffering from bronchitis, unable to go anywhere and writing to Clarkey in deep depression. Outside was thick yellow fog; candles were lighted though it was only two in the afternoon, but in spite of them and a large fire the fog hung in the room. Clarkey had suggested that she should express herself through writing, but she had no desire to write. She wanted to act, to work, to perform deeds. 'You ask me why I do not write something. ... I had so much rather live than write – writing is only a substitute for living. ... I think one's feelings waste themselves in words, they ought all to be distilled into actions and into actions which bring results.'

Before she left London in the spring of this year, she received a shattering blow: Henry Nicholson proposed and insisted on a definite answer. She refused him. Henry was heartbroken, and the Nicholsons were furious. Florence had, they said with justice, encouraged Henry; Marianne ended her friendship with Florence, and the Nightingales and the Nicholsons ceased to be intimate. To Florence the loss of Marianne was a catastrophe; through the summer she suffered tortures. 'I have walked up and down all these long summer evenings in the garden,' she wrote to Hilary in July, 'and could find no words but "My God, my God, why hast Thou forsaken me".' She did not blame Marianne; she wrote no recriminations; she blamed only herself. 'I was not a worthy friend for her. I was not true either to her or to myself in our friendship. I was afraid of her: that is the truth.'

Embley was especially full of visitors during the summer. At Whitsuntide Fanny had had a large party, 'the picked and chosen of society assembled,' she wrote triumphantly to Clarkey. Florence, heartbroken and miserable, moved in a dream. Richard Monckton Milnes stayed both at Embley and at Broadlands with the Palmerstons during the summer, but she was hardly conscious of him. Nothing was real to her but her suffering over Marianne.

She was approaching a mental collapse when two serious illnesses in the family saved her.

In August she went with her father to visit her grandmother, Mrs Shore, found her seriously ill, and was allowed to stay and nurse her. Hardly was Mrs Shore convalescent when Mrs Gale, the girls' old nurse, was taken ill at Lea Hurst. Again Florence was allowed to nurse Gale, and when she seemed too ill to be moved to Embley at the end of the summer Fanny prepared to give up her winter gaieties and stay at Lea Hurst in order that the old nurse might not be separated from 'her children'. Gale insisted, however, on being taken to Embley, and there, a week or so later, she died, sitting upright in her chair with Florence beside her holding her hand.

For a short time Florence and her mother drew closer; her heart was melted by Fanny's kindness, and one of the few intimate letters she ever wrote to her mother described Gale's death: 'Did I tell you one night she was very suffering and I was doubting whether I should speak to her, something good about the weary and heavy laden, when she said quite distinctly "Oh I was so well, quite well till now, but I've been sadly off my teas and breakfasts of late". Oh my dear Mum, life is nothing so much as profoundly ridiculous after all. Is that what the eternal spirit is talking about, when it is ... with the other invisible spirits on the eve of becoming like them?' The old nurse's last words, Florence wrote to Hilary, were to say sharply, 'Hannah, get to your work.' The details of her funeral had been discussed fully by Gale during her illness. She was to be carried across the common, '*not* over the stiles'; everyone on the estate came in a clean smock, and Mr

Hogg, the steward, said he was sure Mrs Gale would have enjoyed it very much.

These two episodes brought a certain amount of emancipation. Since Florence had proved herself entirely capable in nursing her grandmother and Gale, it was difficult to forbid her to continue to nurse. In the autumn there was an unusual amount of sickness in the village of Wellow, and she took an active part. She mentions being present at two death-beds and a difficult birth.

And now she moved forward another step – she realized the necessity of training in nursing. The discovery came as a shock. Neither she herself nor anyone else she had ever met had been taught how to nurse. It was universally assumed that the only qualification needed for taking care of the sick was to be a woman. Ignorance was complete, and its consequences disastrous. 'I saw a poor woman die before my eyes this summer because there was nothing but fools to sit up with her, who poisoned her as much as if they had given her arsenic,' she wrote to Hilary Bonham Carter in December 1845.

In 1844, when she first knew with certainty that her vocation lay among the sick in hospitals, she had not had the actual practice of nursing in her mind. She had spoken to Dr Howe of 'devoting herself to works of charity in hospitals'. She too had thought that the qualities needed to relieve the misery of the sick were tenderness, sympathy, goodness, and patience. Now her short experience had already shown her that only knowledge and expert skill brought relief; and her destiny, which was to lighten the load of suffering, could be fulfilled only if she were armed with knowledge. She must learn how to nurse. How could she learn? There was perhaps one avenue by which she might succeed.

The idea was bold, but since she had achieved a little independence, she had been becoming bolder. Her plan was to persuade her parents to allow her to go for three months to Salisbury Infirmary to learn nursing: Salisbury was only a few miles from Embley, the Infirmary was a well-known hospital, and the head physician, Dr Fowler, was an old friend. He held advanced views, and she thought he might support her.

In December 1845 the Fowlers came to stay at Embley, and Florence proposed her plan. A storm burst. 'Mama was terrified,' she wrote to Hilary. The reason was 'not the physically revolting parts of a hospital but things about the surgeons and nurses which you may guess'. Parthe had hysterics. Florence persisted, and her mother's terror passed into furious anger. Writing twenty years later, Miss Nightingale described a series of scenes. Fanny accused her of having 'an attachment of which she was ashamed', a secret love affair with some 'low vulgar surgeon'. In floods of tears Fanny wept that Florence wanted to 'disgrace herself'.

The Fowlers, embarrassed, 'threw cold water'. W. E. N., coldly disgusted, went away to London. Was it for this he had educated a charming daughter? Was this to be the end of the Latin and the Greek, the poetry and the philosophy, the Italian tour and the Paris frocks? Hilary described meeting him at a dinner-party a few days later. It had been hoped he would give them inside political news – there was a Cabinet crisis and he was known to have seen Palmerston. But he was morose and would talk of nothing but spoiled and ungrateful daughters and forecast the very worst future for a race at the mercy of the modern girl. Florence was left defeated, helpless, hopelessly depressed. 'No advantage that I can see comes of my living on, excepting that one becomes less and less of a young lady every year,' she wrote to Hilary Bonham Carter. 'You will laugh, dear, at the whole plan I daresay; but no one but the mother of it knows how precious an infant idea becomes; nor how the soul dies, between the destruction of one and the taking up of another. I shall never do anything and am worse than dust and nothing. ... Oh for some strong thing to sweep this loathsome life into the past.'

4

I⊤ was not surprising that the Nightingales were horror struck.
In 1845 hospitals were places of wretchedness, degradation, and
squalor. 'Hospital smell', the result of dirt and lack of sanitation,
was accepted as unavoidable and was commonly so overpower-
ing that persons entering the wards for the first time were seized
with nausea. Wards were usually large, bare, and gloomy. Beds
were crammed in, fifty or sixty, less than two feet apart. Even
decency was impossible. Fifteen years later, when some improvement had been made, Miss Nightingale wrote in *Notes on Hospitals:* 'The floors were made of ordinary wood which, owing to
lack of cleaning and lack of sanitary conveniences for the
patients' use, had become saturated with organic matter, which
when washed gave off the smell of something quite other than
soap and water.' Walls and ceilings were 'of common plaster' also
'saturated with impurity'. Heating was supplied by a single fire
at the end of each ward, and in winter, windows were kept closed
for warmth, sometimes for months at a time. In some hospitals
half the windows were boarded up in winter. After a time the
smell became 'sickening', walls streamed with moisture, and 'a
minute vegetation appeared'. The remedy for this was 'frequent
lime washing with scraping', but the workmen engaged in the
task 'frequently become seriously ill'.

The patients came from the slum tenements called 'rookeries',
from hovels, from cellars where cholera lurked. Gin and brandy
were smuggled into the wards, and fearful scenes took place, ending by half-dying creatures attacking each other in frenzy or
writhing in fits of the 'screaming horrors'. In certain hospitals it
was not unknown for the police to be called in to restore order.

The sick came into hospital filthy and remained filthy. In 1854
Miss Nightingale wrote: 'The nurses did not as a general rule
wash patients, they could *never* wash their feet – and it was with
difficulty and only in great haste that they could have a drop of
water, just to *dab* their hands and face. The beds on which the
patients lay were dirty. It was common practice to put a new

patient into the same sheets used by the last occupant of the bed, and mattresses were generally of flock sodden and seldom if ever cleaned.'

Yet physically disgusting conditions were not the real obstacle to her scheme; the insuperable objection was the notorious immorality of hospital nurses. 'It was *preferred*,' wrote Miss Nightingale, 'that the nurses should be women who had lost their characters, i.e. should have had one child.' It was common for nurses to sleep in the wards they nursed, and not unknown for nurses of male wards to sleep in the wards with the men. In a letter written on May 29, 1854, she described the sleeping accommodation provided for nurses in one of London's most famous hospitals. 'The nurses ... slept in wooden cages on the landing places outside the doors of the wards, where it was impossible for any woman of character to sleep, where it was impossible for the Night Nurse taking her rest in the day to sleep at all owing to the noise, where there was not light or air.' The nurse had no other home than the ward; there she lived, slept, and frequently cooked her meals. Discipline and supervision were almost non-existent. A very large number of patients were under the charge of one nurse – in one case a single night nurse had charge of four wards. The level of decency among the patients was almost unbelievably low.

Drink was the curse of the hospital nurse, as of the patients. 'The nurses are all drunkards, sisters and all,' said the physician of a large London hospital in 1851, 'and there are but two nurses whom the surgeons can trust to give the patients their medicine.' In 1854 the head nurse of a London hospital told Miss Nightingale that 'in the course of her large experience she had never known a nurse who was not drunken, and there was immoral conduct practised in the very wards, of which she gave me some awful examples.'

Miss Nightingale herself nursed a nurse who alternated nursing with prostitution. Mrs Gaskell, writing to Catherine Winkworth on October 20, 1854, repeated the story. 'F. N. undressed the woman, who was half tipsy and kept saying, "You would not think it Ma'am but a week ago I was in silk and satins dancing at Woolwich. Yes Ma'am for all I am so dirty I am draped in silk and satins sometimes. Real French silk and satins." This woman was a nurse earning her five guineas a week nursing ladies.'

One of the extraordinary features of Miss Nightingale's life is the passage of time. She starts with a 'call' in 1837. But what has she been called to do? What is her vocation to be? Eight years pass before, in 1845, she finds out. Even then she is only half-way. Eight more years pass before she gains freedom in 1853 to pursue her vocation. Sixteen years in all, sixteen years during which the eager susceptible girl was slowly hammered into the steely powerful woman of genius. The last eight years, the years after her failure in 1845, were years in which suffering piled on suffering, frustration followed frustration, until she was brought to the verge of madness.

Yet she endured year after year. She had the capacity to assert herself, but she did not. The bonds which bound her were only of straw, but she did not break them. Her temperament held her a prisoner. She could act only when she felt moral justification, and she felt no moral justification. Her sense of guilt trapped her. She was convinced that the difficulties which confronted her were God's punishment for her sinfulness; she was unworthy, and by being unworthy she had brought her sufferings on her own head. 'Bless me, too, as poor Esau said,' she wrote to Aunt Hannah on Christmas Eve, 1845, 'I have *so* felt with him and cried with an exceeding bitter cry "Bless me, even me also, Oh my father", but he never has yet and I have not deserved that He should.'

At the end of 1845, she 'went down into the depths. My misery and vacuity were indescribable'. Humiliated, snubbed, lonely, she found relief in writing private notes. On page after page, in tens of thousands of words, she poured out her wretchedness, her fear, and her frustration. 'This morning I felt as if my soul would pass away in tears, in utter loneliness in a bitter passion of tears and agony of solitude.' 'I cannot live – forgive me, oh Lord, and let me die, this day let me die.' 'The day of personal hopes and fears is over for me, now I dread and desire no more.' 'The sorrows of Hell compass me about, pray God He will not leave my soul in Hell.' 'The plough goes over the soul.'

She spent her nights sleepless, wrestling with her soul, seeking with tears and prayers to make herself worthy to receive the kindness of God; she spent her days performing the duties of the daughter at home.

'My life was not painful, but tiresome,' she wrote in a reminiscence thirty years later. W. E. N. liked his two daughters to sit with him in the library after breakfast while he went through *The Times*, reading aloud anything that struck him as good. 'To hear

52

little disjointed bits read out to us out of book or newspaper! Now for Parthe the morning's reading did not matter; she went on with her drawing; but for me, who had no such cover, the thing was boring to desperation.' 'What is my business in this world and what have I done this fortnight?' she wrote on July 7, 1846. 'I have read the "Daughter at Home" to Father and two chapters of Mackintosh; a volume of Sybil to Mamma. Learnt seven tunes by heart. Written various letters. Ridden with Papa. Paid eight visits. Done Company. And that is all.'

'Dreaming' enslaved her more and more. While W. E. N. was reading *The Times*, while she was making conversation with visitors or taking 'a little drive' with Fanny, she escaped into a dream world. Her dreams centred upon Richard Monckton Milnes. She imagined herself married to him, performing heroic deeds with him.

Yet, in spite of her wretchedness, she was making progress. The philosophy which told her to submit did not tell her to relinquish her determination; indeed, it gave her strength to persist, since she believed that as soon as she had attained a worthy state she would be released from submission. She began to equip herself with knowledge against that day.

At Lord Ashley's suggestion she started to study Blue Books and hospital reports – during the past few years the first Blue Books dealing with public health had been published. In 1838, Dr Southwood Smith, Dr Arnott, and Dr Kay presented their report on the condition of the poor in East London to the Poor Law Commissioners. Two years later the Select Committee presented its first report on the Health of Towns. In 1842, the first report on the Sanitary Condition of the Labouring Classes was published, and in 1844 the first report of the Health of Towns Commission.

She worked in secret. She got up before dawn and wrote by candlelight, wrapped in a shawl. Notebook after notebook was filled with a mass of facts, compared, indexed, and tabulated. She wrote privately for reports to M. Mohl in Paris; she procured information on hospitals in Berlin from the Bunsens. In the cold dark mornings she laid the foundation of the vast and detailed knowledge of sanitary conditions which was to make her the first expert in Europe. Then the breakfast-bell rang, and she came down to be the Daughter at Home.

Fanny had put her in charge of still-room, pantry, and linen-room. 'I am up to my chin in linen, glass and china,' she wrote to Clarkey in December 1846, 'and I am very fond of housekeeping. In this too highly educated, too-little-active age, it is at least a practical application of our theories to something – and yet in the middle of my lists, my green lists, brown lists, red lists, of all my instruments of the ornamental in culinary accomplishments, which I cannot even divine the use of, I cannot help asking in my head, "Can reasonable people want all this?" ... "And a proper stupid answer you'll get," says the best Versailles service, "so go and do your accounts; there is one of us cracked." '

Twice a year she went through linen-room, plate-chest, china and glass cupboard, and storeroom, checking, listing damage, replacements, and repairs. In the still-room she supervised preserving, and she wrote to Hilary Bonham Carter in September 1846 that after a hard day's work she was 'surveying fifty-six jam pots with the eye of an artist'.

So month followed month – it seemed without progress or event, but in her character a profound change was taking place. 'I feel,' she wrote in a private note of 1846, 'as if all my being were gradually drawing together at one point.' She decided that her longing for affection, her susceptibility were too powerful for safety and she began deliberately to detach herself from human relationships. Love, marriage, even friendship, must be renounced. So in September 1846 she wrote to Hilary Bonham Carter: 'Are not one's earthly friends too often Atalanta's apple, thrown in each other's way to hinder that course, at the end of which is laid up the crown of righteousness? So, dearest, it is well that *we* should not see too much of each other. ... Farewell my beloved one.' In a private note she wrote: 'Oh God, no more love. No more marriage O God.'

In July the Nightingales had gone north to Lea Hurst, and in the cottages she found peace. In a private note written on July 16 she wrote: 'Rubbed Mrs Spence for the 2nd time. I am such a creeping worm that if I have anything of the kind to do I can do without marriage or intellect or social intercourse or any of the things people sigh after. ... I want nothing else, my heart is filled. I am at home.'

In October the Chevalier Bunsen sent her the Year Book of the Institution of Deaconesses at Kaiserswerth. Four years earlier he had mentioned Kaiserswerth, but she had not then reached the knowledge that nursing was her vocation. Now with overwhelming joy she realized that Kaiserswerth was what she had been seeking. There she could have training in nursing, and the objections raised against English hospitals did not apply. The religious atmosphere, the ascetic discipline placed the nurses above suspicion. On October 7 she wrote in a private note: 'There is my home, there are my brothers and sisters all at work. There my heart is and there, I trust, will one day be my body.'

The Year Book became her treasure, but she did not dare mention Kaiserswerth to her mother. Fanny was busier, more successful than ever, and Embley was filled for autumn parties; and so, she wrote in a private note, whenever she wanted 'refreshment in the midst of this table d'hôte of people at Embley' she went upstairs and secretly read the Year Book of the Deaconesses of Kaiserswerth.

In June 1847 the Nightingales went to Oxford for the meeting of the British Association, and Richard Monckton Milnes went with them. The weather was perfect, the flowering acacias out everywhere, and Florence told Clarkey that she had 'never imagined so much loveliness and learning'. She strolled through college gardens and cloisters with Richard, and in New College cloister she picked a white rose to press 'for a remembrance'. They went to lunch at Christ Church with Professor Buckland, the famous naturalist, who kept animals at liberty in his rooms. Florence 'invited a Bear of 3 months old in to lunch, who climbed like a squirrel for the butter on the table ... which went to his head and he became obstreperous. Mr Buckland put on his cap and gown and rebuked it, at which it became violent and was carried out in disgrace. ... When we came out it was still walking and storming and howling on its hind legs – gesticulating and remonstrating. I spoke to it but Papa pulled me away, fearing it would bite. I said "Let alone, I'm going to mesmerize it." Mr Milnes followed the suggestion and in $\frac{1}{2}$ minute the little bear began to yawn, in less than 3 min. was stretched fast asleep on the gravel.'

From Oxford W. E. N. and Florence went on to pay visits to Lord and Lady Sherborne and Lord and Lady Lovelace. Lady Lovelace, who was Byron's daughter, had a 'passion' for Florence and handed round a set of verses she had written in praise of her 'soft and silver voice', her 'grave and lucid eye'.

And Florence persisted in being steadily more miserable. The old story repeated itself: she feared success because she enjoyed it too much; 'vanity, love of display, love of glory' were still her besetting sins. 'Everything I do is poisoned by the fear that I am not doing it in simplicity and godly sincerity,' she wrote.

In September Clarkey married Julius Mohl, and Florence wrote her a long confused letter on the subject of marriage: '... We must all take Sappho's leap, one way or other, before we attain to her repose – though some take it to death and some to marriage and some again to a new life even in this world. Which of them is the better part, God only knows. Popular prejudice gives it in favour of marriage. ... In single life, the stage of the Present and the Outward World is so filled with phantoms, the phantoms, not unreal tho' intangible, of Vague Remorse, Tears, dwelling on the threshold of every thing we undertake alone, Dissatisfaction with what is, and Restless Yearnings for what is not ... love laying to sleep those phantoms (by assuring us of a love so great that we may lay aside all care for our own happiness ... because it is of so much consequence to another) gives that leisure frame to our mind, which opens it at once to joy.'

Her destiny may have demanded that marriage should be put behind her, but the desire to be loved died hard. She could not rid her heart of longing for 'a love so great that we may lay aside all care for our own happiness ... because it is of so much consequence to another'. Nor could she bring herself to face losing Richard Monckton Milnes. Month after month she temporized, evading the moment when she must give him a definite answer. Fanny passed from impatience to anger, accusing her of godless ingratitude, perversity, and conceit.

At this point Florence found consolation in a new friend. The previous autumn, through Clarkey, she had met Selina Bracebridge, wife of Charles Holte Bracebridge, of Atherstone Hall, near Coventry. Selina understood her. In a retrospect Miss Nightingale wrote: 'She never told me life was fair and my share of its blessings great and that I *ought* to be happy. She did not know that I was miserable but she felt it; and to me, young, strong and blooming as I then was, to me, the idol of the man I adored, the spoilt child of fortune, she had the heart and the instinct to say – "Earth, my child, has a grave and in heaven there is rest".' Selina and her husband became family friends, and she was given a pet name by the Nightingales, the Greek character 'sigma' – Σ – partly in compliment to the Hellenic traits of her character, partly in reference to her love for Greece.

Through the spring of 1847 Florence had a new dream. She imagined Σ was always with her, always waiting for her, beautiful, kind, and loving. She wrote imaginary dialogues in which she put down Σ's part in the conversation as well as her own. The dialogues were never sent; they were too private to be read even by Σ. She feared again for her mental balance. All round her she could see the effects of enforced idleness and frustration – 'I see so many of my kind who have gone mad for want of something to do. People who might have been so happy. Aunt Evans, Aunt Patty,' she wrote. Would she wake up one day to find she was elderly and mad and subject, like Aunt Patty, to 'outbursts'?

In the autumn of 1847 she broke down completely. She wrote to Clarkey that she could not face 'the prospect of three winter months of perpetual row'. She collapsed, took to her bed, coughed.

She was rescued by the Bracebridges, who were going to spend the winter in Rome; Σ persuaded Fanny to let them take Florence. The fuss was enormous: the clothes she was to take, the books she was to read, the sights she was to see, were separately the subject of consideration, reconsideration, letters, interviews, advice from uncles, aunts, grandmothers, and cousins. Solemn farewells were said. Parthe was overcome at the idea of separation, and for the last few days Fanny and W. E. N. withdrew from Embley and left the two sisters alone. Florence was apathetic. 'Dreaming' had enslaved her even further during her illness, and she was terrified. In a private note she wrote: 'I see nothing desirable but death.'

On October 27 the party left England, going overland to Marseilles and thence by sea to Civita Vecchia, the port for Rome.

*

'Oh how happy I was! I never enjoyed any time in my life as much as my time in Rome,' Miss Nightingale wrote. Fifty years later she could still describe every street, every turning, every building in minute detail. One of the great moments of her life was her first sight of the Michelangelo ceiling in the Sistine Chapel. 'I did not think I was looking at pictures but straight into Heaven itself,' she wrote to Parthe on December 17, 1847. She remained alone in the Sistine for the whole of one day and for the rest of her life had prints of the Sistine frescoes hanging in her room.

She danced out the old year of 1847 into 1848, and it was 'the happiest New Year I have ever spent'. In January she wrote: 'This is the most entire and unbroken freedom from dreaming that I ever had.' Her health recovered, and she was well all the six months she was in Rome.

And in Rome, during the winter of 1847, she met Sidney Herbert. Their strange and fatal intimacy began in picture galleries and churches, during strolls in the Borghese gardens and sightseeing expeditions to Tivoli. Each was destined to exercise an extraordinary influence on the other; each in meeting the other had met his and her fate; but no portent indicated that this was the most important moment of their lives. The acquaintance opened with Florence's introduction by Σ to Sidney Herbert's wife, a remarkably beautiful girl who was a close friend, 'almost like a daughter', to Σ. She had recently married Sidney Herbert, half-brother and heir presumptive to the Earl of Pembroke, and they were now wintering in Rome in the course of a postponed wedding tour. She was immediately attracted by Florence, and they became intimate friends.

Liz was a woman of great charm. She was beautiful, with brilliant dark eyes and a glowing olive skin. She had a childlike eagerness, a simple power of enjoyment which made her a delightful companion. After knowing her for a few weeks, Miss Nightingale wrote in a private note of 'the great kindness, the desire of love, the magnanimous generosity' which distinguished her character.

Fate had heaped blessing upon blessing on Sidney Herbert's head. He was astonishingly good-looking – 'a tall and graceful figure surmounted by a face of such singular sweetness as to be unforgettable,' wrote Gladstone. His hair was thick, waving and dark chestnut in colour, his eyes dark and shaded with long lashes. Tall, broad-shouldered yet graceful, he was a superb shot, a remarkably good horseman, and an ardent rider to hounds. He had great wealth; he lived at Wilton, one of the most beautiful houses in England which he would eventually inherit; he had a house in Belgrave Square and vast estates in Ireland in addition. He was brilliantly clever; his wit and social talents were famous; yet he secretly belonged to an association the members of which were pledged to give away a large part of their incomes in private charity.

And yet – with so much goodness, brilliance, and beauty he

was without zest for life. Parliamentary success and philanthropic achievements were dust and ashes in his mouth, and in spite of his gifts, his capacity for bestowing happiness, and his good fortune, he would have preferred never to have lived at all. He longed only for quiet – the peace of Wilton. 'There is not a spot about Wilton which I do not love as if it were a person,' he wrote. 'If one had nothing to do but consult one's own taste and one's own ease I should be too glad to live down here a domestic life.'

It was impossible. Fate heaped on him glittering prize after glittering prize. Riches, high office, power, responsibility, descended on him. He found the burden almost intolerable and turned for consolation to religion – both Sidney Herbert and his wife were devout Christians who consecrated their lives to philanthropic works. He built a new church at Wilton, he worked to improve the condition of the poor, he was in process of building and endowing a convalescent home, he was interested in a plan for emigrating sweated workers, and into these and other plans Liz threw herself heart and soul, worshipping her husband and desiring to share his every activity and thought.

In Rome the Herberts had a small circle of friends who met almost daily. One was Doctor Manning, Archdeacon of Chichester, who was wintering in Rome to improve his health, which had broken down under the stress of religious doubts; another was Mary Stanley, sister of Doctor Stanley, then Canon of Canterbury, later famous as Dean Stanley of Westminster. It was the time when the Oxford Movement was shaking the Church of England to its foundations, and the Herberts and their friends belonged to the reforming High Church party, popularly called Puseyites. Would the path which they were following lead them to the Roman Catholic Church? To help them decide this point, they had come to Rome. Miss Nightingale, however, was indifferent. She formed a friendship with Mary Stanley because Mary Stanley was interested in nursing and had visited hospitals in England and Europe, and Mary Stanley developed a 'passion' for her. With Manning and with Sidney Herbert she discussed social work and schemes of philanthropy. Religious doctrines or the claims of one church against another meant nothing to her.

In April 1848 she left Rome. The city was in arms again, and Garibaldi was riding into Rome to defend the city against the Austrians. She heard with indignation the suggestion that Rome should be surrendered without a blow in order that its monuments might be preserved. 'They must carry out their defence to the last,' she wrote to Clarkey. 'I should like to see them fight the streets, inch by inch, till the last man dies at his barricade, till St Peter's is level with the ground, till the Vatican is blown into the air. ... If I were in Rome I should be the first to fire the Sistine ... and Michael Angelo would cry "Well done" as he saw his work destroyed.'

She reached home to find Fanny and Parthe occupied by the excitement and fuss of a family wedding: Laura Nicholson, Marianne's youngest sister, married Jack Bonham Carter, Hilary's eldest brother. The celebrations, in the Waverley manner, were colossal. Goodwill was in the air, differences were forgotten, and Florence and Parthe were bridesmaids.

Her friendship with the Herberts, a source of profound satisfaction to Fanny, grew closer. As soon as they returned to London in May, she dined with them. The next month she went with them to the opening of their convalescent home at Charmouth, staying at Wilton on her way back. She met a circle of intelligent, socially impeccable, extremely influential people intensely interested in hospital reform. Public opinion was awakening; the Herberts and their friends were eager for information, and Miss Nightingale, who had now been working for more than five years collecting facts on public health and hospitals, had an enormous mass of detailed information at her finger-tips. She gradually became known as an expert on hospitals.

The Herberts knew of her plan to go to Kaiserswerth and approved, and the Bunsens were thinking of sending their daughter there. Once more the fulfilment of her desires seemed within the range of possibility. Who could disapprove of what the Herberts and the Bunsens approved? Surely her mother must allow herself to be convinced. But it was necessary to proceed cautiously – the very word 'hospital' might be fatal.

In September 1848 a heaven-sent opportunity offered. Parthe was ordered to take a cure at Carlsbad, and the Nightingales planned to go on to Frankfurt, where Clarkey and her husband M. Mohl were staying. Kaiserswerth being near Frankfurt, Florence's plan was to leave her family for a week or two to 'visit the deaconesses and perhaps fit in a little training'.

But 1848 was the year of revolution in Europe. When disorders broke out in Frankfurt, W. E. N. thought it wiser to stay in England; and the Nightingales went to Malvern instead of

Carlsbad. 'All that I most wanted to do at Kaiserswerth lay for the first time within reach of my mouth, and the ripe plum has dropped,' Miss Nightingale wrote to Clarkey in October.

Her reaction was violent. Her mother was not concerned, for Fanny knew nothing of the scheme. It was God Himself who had prevented her. God who had cut her off from Kaiserswerth, perhaps for years. The old reasoning tortured her. This misfortune had come upon her because she was sinful. God wanted her to go to Kaiserswerth, but He could not let her go until she had reached a greater state of worthiness. She went down into the depths of depression; the short period of comparative happiness was over. She 'hated God to hear her laugh as if she had not repented of her sin'. The winter season at Embley lay before her. 'My God what am I to do,' she wrote in a private note of October 1848. 'Teach me, tell me. I cannot go on any longer waiting till my situation should change, dreaming what the change should be.'

She dreamed of fame, of Richard Monckton Milnes. To escape from 'dreaming', she sought relief in nursing the poor of Wellow, the village near Embley, and Fanny and Parthe became irritated. W. E. N., who hated dirt, disease, and ugliness, was disgusted. He told Florence she was being theatrical; if she wanted something to do, let her work in the school at Wellow. She did for a time, but she failed. 'I was disgusted with my utter impotence,' she wrote in a private note. 'I made no improvement. I obtained no influence. ... Why should I? ... Education I know is not my genius.'

Aunt Hannah wrote soothingly that anything, even a house-party or a dinner, can be done to the glory of God. 'How can it be to the glory of God,' answered Florence, 'when there is so much misery in the world which we might be curing instead of living in luxury.' Aunt Hannah, who did not pay her usual visit to Embley in 1848, did not answer: Florence was becoming known as a rebel daughter, and Aunt Hannah could not countenance that. The following year she wrote that she found it necessary with 'advancing years and delicate health' to 'confine herself to visits to near relatives', and her correspondence with Florence ceased.

In March 1849 the Nightingales went to London for the season. Miss Nightingale was in a mounting delirium of misery and frustration. 'Dreaming' became uncontrollable. She fell into

61

trances in which hours were blotted out; she lost sense of time and place against her will. In daily life she moved like an automaton, could not remember what had been said or even where she had been. Agonies of guilt and self-reproach were intensified by the conviction that her worst fears were being realized and that she was going insane.

Again and again she made resolutions to end 'dreaming', to 'tear the sin out', to 'stamp it out' – but they were always broken. She turned on herself with savagery, hating herself, despising her weakness. She would have killed herself if she had not thought it mortal sin. On June 7, 1849, she determined at whatever cost to herself to 'crucify' her sin. The 7th of each month was devoted to self-examination because she had received her 'call' on February 7. In this wretched state another blow fell on her. Richard Monckton Milnes would be put off no longer. He insisted on a definite answer – would she marry him or not? She refused him.

It was an act which required extraordinary courage. She was deeply stirred by him; she called him 'the man I adored'; and she renounced him for the sake of a destiny which it seemed impossible she would ever fulfil.

In a private note she analysed her reasons. She wrote several versions; she began it, broke off, returned to it again before she could clarify her emotions. 'I have an intellectual nature which requires satisfaction and that would find it in him. I have a passionate nature which requires satisfaction and that would find it in him. I have a moral, an active, nature which requires satisfaction and that would not find it in his life. Sometimes I think I will satisfy my passional nature at all events, because that will at least secure me from the evil of dreaming. But would it? I could be satisfied to spend a life with him in combining our different powers in some great object. I could not satisfy this nature by spending a life with him in making society and arranging domestic things.'

But she could not always be rational. She wrote again with a pencil that trembled, hesitated, and dug itself into the paper. 'I do not understand it. ... I am ashamed to understand it. ... I know that if I were to see him again ... the very thought of doing so quite overcomes me. I know that since I refused him not one day has passed without my thinking of him, that life is desolate without his sympathy.' At night she dreamed of him. He came and told her that he had arranged for her to go to Kaiserswerth.

And yet desperately as she longed for him, she would not give way. 'I know I could not bear his life,' she wrote, 'that to be nailed to a continuation, an exaggeration of my present life without hope of another would be intolerable to me – that voluntarily to put it out of my power ever to be able to seize the chance of forming for myself a true and rich life would seem to me like suicide.'

Fanny was severely disappointed and furiously resentful. Her obstinacy hardened; she determined that Florence should not have her own ungrateful way, and what had begun as genuine maternal solicitude for her daughter's welfare turned into a contest of wills in which love and kindness were forgotten.

By the autumn Miss Nightingale's mental and physical state was pitiable. She was far from well and fainted on several occasions; sometimes her mind became a blank and she looked at people wildly and vaguely, not hearing what was said to her. Σ once more intervened. The Bracebridges were going to Egypt and then to Greece, and they persuaded Fanny to let them take Florence. But the Nightingale circle disapproved. 'I think you are all martyrs for having consented,' wrote Ellen Tollet, Parthe's favourite cousin. 'I suppose really large minded people think less of space and distance than we do,' commented another. Aunt Patty Smith wrote acidly that 'it was to be hoped that change of air and the satisfaction of doing her duty would do Flo good'.

Miss Nightingale herself commented to Hilary Bonham Carter that as Rome had done her some good the family were going to send her farther afield in the hopes that that would be even better.

A journey to Egypt was an adventure in 1849. But she was in a state when Egypt, the desert, even the brilliant landscapes of the Nile itself meant as little as scenes painted on a backcloth. She was on the verge of mental collapse.

In a small black notebook she recorded her secret agonies; the entries are scribbled in pencil, in phrases which repeat themselves, in writing which wavers and becomes all but indecipherable. The weight of guilt laid on her conscience by 'dreaming' was driving her insane.

Jan. 26.   Went with party to Jenab ... but I spoiled it all with dreaming. Disappointed with myself and the effect of Egypt on me. Rome was better.

Feb. 16.   Where was I all the while – dreaming. Karnak itself cannot save me now.

Feb. 22.   God spoke to me again, sitting on the steps of the portico at Karnak.

March 3.   Ill. Did not get up in the morning.

March 7.   God called me in the morning and asked me would I do good for Him, for Him alone without the reputation.

March 9.   During half an hour I had by myself in my cabin, settled the question with God.

March 15. God has delivered me from the great offence and the constant murderer of all my thoughts.

March 21. Undisturbed by my great enemy.

April 1.   Not able to go out but wished God to have it all His own way. I like Him to do exactly as He likes without even telling me the reason.

May 7.   (In Athens) I have felt here the suspension of all my faculties, I could not write, could not read. ...

May 9.   I cannot even draw a pattern for a few minutes without turning faint.

May 12.   To-day I am 30 – the age Christ began his mission. Now no more childish things. No more love. No more marriage. Now Lord let me think only of Thy Will, what Thou willest me to do. Oh Lord Thy Will, Thy Will.

May 18.   To-morrow is Sacrament Sunday; I have read over all my history, a history of miserable woe, mistake, and blinding vanity, of seeking great things for myself.

May 19.   Whit Sunday. Oh how happy I am to be away from the scene of temptation on this day. I thank Thee Father, three Whitsuntides have I spent torn with temptation and overcome. Here I am not safe. God I place myself in Thy Hands ... if it be Thy Will that I should go on suffering let it be so.

May 21.   (Ill) Now I am 30, the year when I thought I should have accomplished my Kaiserswerth mission ... let me only accomplish the Will of God, let me not desire great things for myself.

June 7.   I thought I would go up the Eumenides cave and ask God there to explain to me what were these Eumenides which pursued me. I would not ask to be released from them – Welcome Eumenides – but to be delivered from doing further wrong. ... This day twelve months ago June 7th 1849 I made that desperate effort, that crucifixion of the sin, in faith that it would cure me. Oh what is crucifixion – would I not joyfully submit to crucifixion, Father, to be rid of this. But this long moral death, this failure of all attempts

64

to cure. What does it signify to me now whether I see this or do that? I never can be sure of seeing it. I may see nothing but my own self practising an attitude.

June 10.	The Lord spoke to me; he said 'Give five minutes every hour to the thought of me. Couldst thou but love Me as Lizzie loves her husband, how happy wouldst thou be.' But Lizzie does not give five minutes every hour to the thought of her husband, she thinks of him every minute, spontaneously.
June 12.	To Megara. Alas it little matters where I go – sold as I am to the enemy – whether in Athens or in London, it is all one to me.
June 17.	After a sleepless night physically and morally ill and broken down, a slave – glad to leave Athens. I had no wish on earth but to sleep. ...
June 18.	I had no wish to be on deck. I let all the glorious sunrises, the gorgeous sunsets, the lovely moonlights pass by. I had no wish, no energy, I longed but for *sleep*. .... My enemy is too strong for me, everything has been tried. .... All, all is in vain.
June 22.	Began to sleep.
June 24.	Here too (Trieste) I was free.
June 29.	Four long days of absolute slavery.
June 30.	[Written faintly and shakily] I cannot write a letter, can do nothing.
July 1.	I lay in bed and called on God to save me.

Again Σ saved her. If Florence continued to be thwarted, she would go out of her mind. Σ acted on her own responsibility. They were to travel home from Greece by land; she chose a route through Prague and Berlin and suggested that she and her husband should spend a fortnight at Düsseldorf while Florence visited Kaiserswerth.

Miss Nightingale was too exhausted, too wretched to be grateful. 'On the brink of my accomplishing my greatest wish,' she wrote, 'with Σ positively planning it for me, I seemed to be unfit, unmanned for it, it seemed to be not the calling for me. ... I did not feel the spirit, the energy, to do anything at Kaiserswerth.' As they travelled from Trieste to Prague and on to Berlin, she wrote that she was 'lost and past redemption, a slave that could not be set free'.

She found relief in the companionship of animals. On the Nile she had had two little chameleons which slept on her bed and had been 'so sorry to part with them, they were such company'. She was travelling now with two tortoises, called Mr and Mrs Hill in honour of two missionaries at Athens, a cicada named Plato, and Athena, a baby owl, which she had rescued from some Greek boys at the Parthenon. Athena was fierce, and Miss Nightingale had had to mesmerize her according to Richard Monckton Milne's method before she could be persuaded to enter a cage, but she became devoted to her mistress and travelled everywhere in her pocket. At Prague Athena ate Plato.

When Miss Nightingale reached Berlin, she was still miserably depressed. 'I had 3 paths among which to choose,' she wrote on July 10, 1850. 'I might have been a married woman, or a literary woman, or a hospital sister. And now it seemed to me as if quiet with somebody to look for my coming back were all I wanted.'

But in Berlin she began visiting hospitals and charitable institutions, and her spirits instantly revived. 'All at once I felt how rich life was,' she wrote on July 15. On July 31 she reached Kaiserswerth. 'With the feeling with which a pilgrim first looks on the Kedron I saw the Rhine dearer to me than the Nile.'

She stayed a fortnight. It was a visit of inspection, and she did not nurse but was shown the work of the Institution and helped with the children.

On August 13 she left Kaiserswerth 'feeling so brave as if nothing could ever vex me again'. She was well, brimming with vitality, her powers of concentration had returned, and she performed the feat of dashing off a pamphlet of thirty-two pages in less than a week; telling the unwanted women kept in 'busy idleness' in England, the women she saw on all sides 'going mad for the want of something to do,' of work, happiness, and comradeship waiting for them at Kaiserswerth. It was printed in 1851 'by the inmates of the Ragged Colonial School at Westminster' and issued anonymously under the title *The Institution of Kaiserswerth on the Rhine for the Practical Training of Deaconesses, under the direction of the Rev. Pastor Fliedner, embracing the support and care of a Hospital, Infant and Industrial Schools, and a Female Penitentiary.*

On August 21 she reached Lea Hurst and 'surprised my dear people, sitting in the drawing-room, with the owl in my pocket. Sat with Mama and Parthe in the nursery. Rode with Papa.' Happiness lasted only a few hours. Fanny was furiously angry; the visit to Kaiserswerth was not to be spoken of, was shameful, a disgrace. The old resentments broke out; the old accusations were repeated. Parthe had hysterics; Fanny raged and wept. Florence must be forced to do her duty, made to stay at home, and engage in the pursuits proper to her upbringing and station.

Five years had passed since her attempt to enter Salisbury Infirmary; she was no longer a girl but a woman of thirty, and she had accomplished nothing. Only her determination persisted. 'Resignation!' she had written in 1847, 'I never understood that word!' A new struggle began, more bitter, more unhappy than ever before.

5

THE conflict changed its character for the worse. Fanny had begun by sincerely wishing for Florence's happiness, sincerely believing what she wished to do would ruin her life. That point was passed. There was justification for Miss Nightingale's sense of guilt: she did evoke the worst from each one of her family; normally kind, normally generous, they behaved to her as to no one else. The most furious opposition came from Parthe. Parthe was thirty-one, and the truth was that she had achieved only moderate success; the successes, the lovers, the popularity were Florence's, not hers.

She cast herself for the part of the adoring indispensable sister who could not be left out. 'Her sense of existence is lost in Florence', wrote Mrs Gaskell. 'I never saw such adoring love.' 'Your own Flo,' wrote W. E. N. to Parthe in 1849. 'Your idolatrized wondrous Flo. . . .' Florence's growing celebrity and success were to be shared, but what if, instead of creating a brilliant, interesting life for Parthe, she went off to lead a sordid existence of her own? The possibility drove Parthe frantic.

Fanny and W. E. N. accused Florence of heartlessness: for nearly a year she had been away, at great expense, doing exactly what she pleased; Parthe had been left behind to mope, and her health had suffered. They demanded that she should devote herself entirely to Parthe for the next six months. 'To this,' wrote Miss Nightingale in a retrospect, 'I acceded. And when I committed this act of insanity had there been any sane person in the house he should have sent for Connolly to me.' Dr Connolly specialized in diseases of the brain.

Until the spring of 1851 she was to be Parthe's slave. Parthe triumphed. She made little scenes and was coaxed out of them; she sketched with Florence, sang with Florence, wandered with her in the garden, chattered of poetry and art. The effect on Miss Nightingale was devastating. She had left Kaiserswerth feeling 'so brave as if nothing could ever vex me again'; within a few weeks she was sunk in the old miseries. 'Dreaming' returned, and never had she been so hopelessly enslaved. Once more nights were spent in agonized self-reproach; once more stupid with misery and frustration she was carried on Fanny's merry-go-round from Lea Hurst to Embley, Embley to London and back to Embley again.

In October the Nightingales received tragic news. Henry Nicholson had been drowned in Spain. It was with 'deepest relief' that Miss Nightingale received a summons to Waverley; Henry's mother wished for the presence of the girl Henry had loved. The house was in a hubbub, and she took charge, accompanying Mrs Nicholson to Henry's chambers in London and packing his possessions. The Nicholsons besought her to stay on, but Parthe insisted she should return. She wrote asking for 'an extension of leave', but was allowed only a week-end. 'Thank you, thank you dear Mum for letting me stay until Tuesday,' she wrote.

After three months she was in despair. 'My present life is suicide,' she wrote on December 31, 1850. 'Slowly, I have opened my eyes to the fact that I cannot now deliver myself from the habit of dreaming which, like gin drinking, is eating out my vital strength. Now I have let myself go entirely. Temporary respite only I have. Henry's death and Waverley was one. ... My God what will become of me? ... I have no desire but to die. There is not a night that I do not lie down on my bed, wishing that I may leave it no more. Unconsciousness is all I desire. I remain in bed as long as I can, for what have I to wake for?' Three months of subjection to Parthe had still to run. 'Oh, how am I to get through this day, to talk all through this day, is the thought of every morning,' she wrote in January, 1851. '... This is the sting of death. In my thirty-first year I see nothing desirable but death.' She did not reproach Parthe – 'she is a child playing in God's garden', she wrote on January 7, 1851, 'and delighting in the happiness of all his works, knowing nothing of life but the English drawing-room, nothing of struggle in her own unselfish nature.' The reproaches were heaped on her own head. 'What is to become of me,' she wrote. 'I can hardly open my mouth without giving dear Parthe vexation – everything I say or do is a subject of annoyance to her.' 'Oh dear good woman,' she wrote of Fanny, 'when I feel her disappointment in me it is as if I were going insane ... what a murderer am I to disturb their happiness. ... What am I that their life is not good enough for me? Oh God what am I? The thoughts and feelings that I have now I can remember since I was six years old. It was not I that made them. Oh God how did they come? ... But why, oh my God cannot I be satisfied with the life that satisfies so many people? I am told that the conversation of all these good clever men ought to be enough for me. Why am I starving, desperate, diseased on it? ... What is the cause of it. ... Oh what do books know of the real troubles of life. Death, why it's a happiness. ... My God what am I to do?'

In the spring of 1851 she unexpectedly met Richard Monckton Milnes at a party given in London by Lady Palmerston. She had not seen him since the day she refused him, and she was shaken. He came across to her and said lightly: 'The noise of this room is like a cotton mill.' She was deeply wounded – how could he speak as if she were an ordinary acquaintance? On March 16, 1851, she met him again. 'Last night I saw him again for the second time,' she wrote in a private note, 'he would hardly speak. ... I was miserable. ... I wanted to find him longing to talk to me, willing to give me another opportunity, to keep open another decision.'

She did not want to recognize the fact that her refusal of Richard was final. But he had waited for her decision for nine years, and he would not reopen the subject. Some weeks later he became engaged to the Honourable Annabel Crewe.

In April the six months of slavery to Parthe ended, and Florence went immediately to Wilton to stay with Liz Herbert. When she returned to Embley she brought with her Dr Elizabeth Blackwell, one of the first women to become a doctor. Dr Elizabeth was the daughter of a Bristol merchant who had emigrated and become a citizen of the United States. She had contrived to get herself accepted as a medical student at Geneva Medical College in the State of New York, had taken a medical degree, and was now studying medicine in Europe. Her experiences confirmed Fanny's worst fears. She had been training at 'La Maternité', the State School of Midwifery in Paris, where life was 'infernal'. It seemed that the female pupils were 'pretty generally the mistresses of the students', and Dr Blackwell's younger sister was going to Paris in 'male attire' to 'avoid improper advances'. Finally, Dr Blackwell had contracted purulent ophthalmia at 'La Maternité' and lost an eye.

One afternoon Dr Elizabeth admired the façade of Embley. 'Do you know what I always think when I look at that row of windows?' said Miss Nightingale. 'I think how I should turn it into a hospital and just where I should place the beds.'

That summer her attitude to life began to change. The absurdity of her six months' slavery to Parthe, Richard Monckton Milne's decisive action, the encouragement of the Herberts forced her eyes open. In a private note she wrote: 'There are knots which are Gordian and can only be cut.' Her sense of guilt lessened, and at long last she saw herself as the victim not the criminal. On June 8, 1851, she wrote a private note on her family in a new vein. 'I must expect no sympathy or help from them. I must *take* some things, as few as I can, to enable me to live. I must *take* them, they will not be given to me. ...'

A fortnight later she had arranged to go to Kaiserswerth.

Opinion had changed since her attempt to enter Salisbury Infirmary in 1845. Interest in hospitals was in the air; Fanny could no longer assert that a plan approved by the Herberts, the Bunsens, and the Bracebridges was shameful. Forced to yield, she gave way with the worst grace. Everything was to be done in secret. Parthe was ailing – Fanny declared that worry over Florence was making her ill – and had been ordered a three months' cure at Carlsbad. Florence was to leave England with Fanny and Parthe, go on to Kaiserswerth, and join them again to come home. Fanny forbade her to tell anyone where she was going or to write any letters from Kaiserswerth – she was not to tell Shore on any account, because young men were so carelessly indiscreet.

W. E. N. stayed at home. He was beginning to find the perpetual conflict in his family unbearable, and he 'retreated into the shadows'. Parthe was transported with fury. Scene followed scene, reaching a climax in the hotel at Carlsbad the night before Miss Nightingale left. 'My sister,' wrote Miss Nightingale in a retrospect, 'threw my bracelets which I offered her to wear, in my face and the scene which followed was so violent that I fainted.' The following evening she reached Kaiserswerth.

In 1833 a young pastor named Theodore Fliedner and his wife placed a bed and a chair in a summerhouse in their back garden and converted it into a refuge for a single destitute discharged prisoner. From this beginning grew the Kaiserswerth Institution. In 1851 it included a hospital with a hundred beds, an infant school, a penitentiary, an orphan asylum, and a normal school for training school mistresses.

Life was Spartan, work rigorously hard, the food such as was eaten by peasants. 'Until yesterday,' wrote Miss Nightingale to Fanny in July 1851, 'I never had time even to send my things to the wash. We have ten minutes for each of our meals, of which we have four. We get up at 5; breakfast ¼ before 6. The patients dine at 11; the Sisters at 12. We drink tea (i.e. a drink made of ground rye) between 2 and 3, broths at 12 and 7; bread at the two former, vegetables at 12. Several evenings in the week we collect in the Great Hall for a bible lesson. ... I find the deepest interest in everything here and am so well in body and mind. This is life. Now I know what it is to live and to love life, and really I should be sorry now to leave life. ... I wish for no other earth, no other world than this.'

She slept in the orphan asylum and worked with the children and in the hospital. She was present at operations, which was considered almost indecent. 'The operation to which Mrs Bracebridge alludes,' she told Fanny, 'was an amputation at which I was present, but which I did not mention to Parthe, knowing that she would see no more in my interest in it than the pleasure dirty boys have in playing in the puddles round a butcher's shop.'

Prayer accompanied every incident at Kaiserswerth. Twenty years later Miss Nightingale told Sir Harry Verney: 'We were all taught to pray aloud extempore before the whole community whenever it was called for. And, at all the little fêtes or whenever he appeared Fliedner and his wife did this themselves about *everything*. It was all prayed out loud to God before everybody. If a child did wrong it was recommended to God before all the others. ... We should all have thought it wrong not only to allege shyness but to feel shyness. ... For the children there were perpetual birthdays ... *every* birthday was fêted, there was dressing up, with flowers, telling stories, singing, every birthday child asked its own guests and I was always asked. My bad German and foreign stories amused them. ... There was of course popping down on knees and praying in the fête.'

Miss Nightingale always denied she had been 'trained' at Kaiserswerth. 'The nursing there was nil,' she wrote in 1897, 'the hygiene horrible. The hospital was certainly the worst part of Kaiserswerth. But never have I met with a higher tone, a purer devotion than there. There was no neglect. It was the more remarkable because many of the Deaconesses had been only peasants – none were gentlewomen (when I was there).'

Toward the end of her stay the Herberts visited her, and Herr

Fliedner told them that 'no person had ever passed so distinguished an examination, or shown herself so thoroughly mistress of all she had to learn as Miss Nightingale.' She was completely satisfied, completely happy; her heart overflowed, and she made one last effort to be reconciled with her mother and Parthe. On August 31, 1851, she wrote a long humble beseeching letter setting out her point of view once again. She repeated what she had tried to explain a hundred times before but never so gently, so affectionately. 'Give me time, give me faith. Trust me, help me. Say to me "Follow the dictates of that spirit within thee." ... My beloved people I cannot bear to grieve you. Give me your blessing,' she wrote.

Neither Fanny nor Parthe responded – she never appealed to them again.

On October 8, 1851, Miss Nightingale joined her mother and sister at Cologne. They were furiously resentful – 'They would hardly speak to me,' wrote Miss Nightingale in a retrospect. 'I was treated as if I had come from committing a crime.' A miserable party travelled toward England. Parthe's health had not improved: the cure had failed, because, she said, she had been so anxious while her sister was at Kaiserswerth. Fanny made light of Parthe's condition but fussed herself into a state of nervous exhaustion over bandboxes. For her part, Miss Nightingale was seething with plans. Kaiserswerth had whetted her appetite. She was wild to train in earnest. She wanted a larger hospital, one of the great London hospitals.

Once more her plans were doomed. She arrived at Embley to find W. E. N. in pain from inflamed eyes. An oculist ordered him at once to Umberslade, in Warwickshire, for a course of cold-water treatment, but he would not go unless Florence went with him.

She could not escape. Parthe was better because her sister was at home. W. E. N. would only undergo his treatment if she were with him; her sense of obligation was enormous, and if the claims of family affection were strong the claims of suffering were stronger. She entered the cage, gave up all she had gained, and submitted once more.

'O weary days – oh evenings that seem never to end – for how many years have I watched that drawing-room clock and thought it never would reach the ten! and for twenty, thirty years more to do this!' she wrote in a private note of 1851. In a long private note headed 'Butchered to make a Roman Holiday', she wrote a furious indictment of family life. 'Women don't consider themselves as human beings at all. There is absolutely no God, no country, no duty to them at all, except family. ... I have known a good deal of convents. And of course everyone has talked of the petty grinding tyrannies supposed to be exercised there. But I know nothing like the petty grinding tyranny of a good English family. And the only alleviation is that the tyrannized submits with a heart full of affection.'

She was not only furious for herself: Hilary Bonham Carter was being sacrificed, betrayed by her unselfishness, gentleness, and sweetness. In 1850, Hilary had spent almost a year in Paris living with Clarkey and working in a studio; though her talents were pronounced remarkable, she had never returned. Her mother 'could not be left alone', she was 'needed at home'.

In 1852, under the title *Cassandra*, Miss Nightingale wrote a description of the life of a girl in a prosperous comfortable home. Cassandra is herself, Cassandra's family the Nightingales. *Cassandra* was not published, but a number of copies were printed privately. She takes Cassandra through a day. The morning is spent 'sitting round a table in the drawing-room, looking at prints, doing worsted work and reading little books.' The afternoon is passed 'taking a little drive'. When night comes, Cassandra declares, women 'suffer – even physically ... the accumulation of nervous energy, which has had nothing to do during the day, makes them feel every night, when they go to bed, as if they were going mad. The vacuity and boredom of this existence are sugared over by false sentiment. Women go about maudling to each other and preaching to their daughters that "women have no passions" ... if the young girls of the "higher classes" who never commit a false step, ... were to speak and say what are their thoughts employed on, their *thoughts* which alone are free, what would they say? ...

'Is not one fancying herself the nurse of some new friend in sickness, another engaging in romantic dangers with him ... another undergoing unheard of trials under the observation of one whom she has chosen as the companion of her dream?' Finally, Cassandra, 'who can neither find happiness in life nor alter it, dies,' slain by her family. ' "Free – free – oh! divine freedom, art thou come at last? Welcome beautiful death!" '

Miss Nightingale never suffered more acutely than at the period when she wrote *Cassandra*, but her suffering was no longer despair; it was rebellion. In herself she was free. 'Dreaming' tortured her no longer. 'I have come into possession of myself,' she wrote in a private note of 1852. On her thirty-second birthday she wrote to W. E. N.: 'I am glad to think that my youth is past and it never never can return – that time of disappointed inexperience when a man possesses nothing, not even himself.' She possessed herself now, and she was at peace. Fanny and Parthe frantically prolonged the struggle for another year, but victory had in fact been won when she went to Kaiserswerth.

W. E. N. became uneasy, for it was borne in on him that his wife and daughter were treating Florence badly. In the early spring of the following year Florence went with him to Umberslade for his eye treatment. When they returned he was secretly her ally. During the summer of 1852, he indicated that as Fanny required all letters to be handed round it might 'avoid enquiry' if Florence wrote to him at the Athenaeum Club instead of at home.

March of this year brought the Nightingales to London for the season once more, and the restrictions imposed on Miss Nightingale reached absurdity. She was treated as a schoolgirl, her movements controlled, her letters read, her invitations supervised; yet she was a woman of over thirty with a distinguished circle of her own. Among her friends in 1852 were Elizabeth Barrett Browning, George Eliot, Lord Shaftesbury, Lord Palmerston, Arthur Stanley, later Dean Stanley, and his sister Mary Stanley, and the poet Arthur Hugh Clough. She was held prisoner by her belief in Parthe, by her affection for W. E. N., by her sense of duty. Where was there an outlet for her? Where in all England was there an opportunity for a woman of her class with her vocation?

In the summer of 1852 it seemed as if she might find what she sought in the Roman Catholic Church. At this time Manning was a priest of the Roman Church remarkable for his devoted work in the poorest districts of the East End. In the course of a charitable investigation Miss Nightingale came on a child of fourteen who was being forced into prostitution, tried to rescue the girl, but found no organization in the Church of England would receive her. The child was Irish and, as far as she had any religion, Catholic. Miss Nightingale applied to Manning, who acted instantly, taking the child under his protection and placing her in the Convent of the Good Shepherd.

She was deeply impressed, and a friendship sprang up which

produced a vast correspondence. She confided her family diffi-culties to Manning; he consoled and advised her, 'writing end-lessly.' She avowed freely that she had 'Catholic yearnings', she signed her letters 'your weary penitent'.

She was fascinated by the organization of the Church of Rome. 'If you knew what a home the Catholic Church would be to me!' she wrote. 'All that I want I should find in her. All my difficulties would be removed. I have laboriously to pick up, here and there, crumbs by which to live. She would give me daily bread. The daughters of St Vincent would open their arms to me. They have already done so, and what should I find there. My work already laid out for me instead of seeking it to and fro and finding none; my home, sympathy, human and divine.' . . . 'Why cannot I enter the Catholic Church at once' she asked 'as the best form of truth I have ever known, and as cutting the Gordian knot I cannot untie?'

But though she seemed on the brink of conversion to Catho-licism, she was engaged on a line of thought totally at variance with Catholic teaching. During the summer of this year she began an attempt to formulate 'a new religion for the Artizans of Eng-land', aiming at demonstrating that free thought is not incompa-tible with belief in God. She described God as the Absolute, the Perfect, the Spirit of Truth. The moral world, she contended, was ruled by laws as fixed in their operation as those which science had recently discovered ruled the physical world. She did not touch on the Christian doctrines of salvation, redemption, and the incarnation of Christ. She was not drawn to the figure of Jesus. Her God was God the Father, not God the Son.

When Manning read her manuscript, he decided that she was not in the requisite state of mind for admission into the Church of Rome. She craved an opportunity to exercise her powers, but she was very far from submission; indeed, she had no conception of submission in the Catholic sense – it was an idea utterly foreign to her. Submission to her meant endurance, not yielding. In the essence of her character she was a chooser, a heretic, and he refused to accept her as a convert.

Nevertheless, he continued to be her friend, and his friendship proved of enormous importance. In spite of the fact that she was a Protestant, he arranged for her to enter a Catholic hospital where the nurses were nuns, and therefore 'moral danger' did not exist. In the summer of 1852, he told her she could be received either by the Sisters of Mercy in Dublin or by the Sisters of Charity in the rue Oudinot, Paris. She wished to do both, to go first for a short time to Dublin, later for a longer time to Paris.

Once again there was a storm. Fanny and Parthe had hysterics. All the old arguments and reproaches were revived: Parthe's health, Florence's heartlessness, Parthe's devotion, Florence's ingratitude. Once more she was forced to abandon her plans.

Her friends became alarmed. Fanny's treatment of her younger daughter was beginning to look like mania. Was Florence's life to be ruined, her remarkable talents wasted because Fanny had an obsession? Aunt Mai and Mrs Bracebridge thought it their duty to interfere, interviewing Fanny separately and together. Fanny, with her world against her, bombarded with what the Herberts thought, the Bunsens thought, the Shaftesburys thought, took refuge in a new policy. She vacillated; she 'could not make up her mind to a definite step'. She would discuss the question of Florence's future, agree to a plan, and next day behave as if she had never heard of any such suggestion in her life before. During the summer of that year Miss Nightingale wrote an imaginary speech to her mother. She wrote gaily, she was not suffering as she had suffered in the past. 'Well, my dear, you don't imagine that with my "talents", and my "European reputation", and my "beautiful letters", and all that, I'm going to stay dangling about my mother's drawing-room all my life! ... You must look at me as your vagabond son. .... I shan't cost you nearly as much as a son would have done. I haven't cost you much yet, except for my visits to Egypt and Rome. Remember, I should have cost you a great deal more if I had been married or a son. ... Well, you must now consider me married or a son. You were willing enough to part with me to be married.'

Parthe was growing steadily worse: in almost daily scenes she attacked Florence with violent reproaches. She declared she was dying – Florence's behaviour was killing her. She complained of suffering agonies from mysterious pains. She was taken to see the Queen's physician, Sir James Clark, who was a personal friend of the Nightingales, and he diagnosed 'rheumatic headaches', adding that she was 'nervous, fanciful and unstable,' but that he could find 'no physical disease'. She did not improve, and in August 1852 he arranged to have her for some weeks under observation at his house in Scotland, Birk Hall, near Ballater.

Miss Nightingale managed to get 'permission for absence' from

Fanny and went with Dr and Mrs Fowler, who in 1845 had been concerned in her attempt to enter Salisbury Infirmary, to Dublin, where she intended to use Manning's introduction and enter the hospital of the Sisters of Mercy.

But she was disappointed. 'The hospital has got a whole holiday and is being repaired so there's nothing to be seen,' wrote Hilary Bonham Carter to Clarkey. Nevertheless, while in Dublin she had, in some hospital, an experience of great importance. She speaks in a 'Memorandum of 1852' of a 'terrible lesson learned in Dublin', and her inclination toward Roman Catholicism vanished.

She was called from Dublin by Sir James Clark. Parthe had had a mental breakdown. There were 'delusions', 'some degree of chronic delirium', and 'extreme irritability'. Sir James, a friend of the Bunsens and the Herberts, admired Miss Nightingale, and he told her that she must separate herself from her sister. Parthe's only chance of regaining normal health and balance was to learn to live without her; for Parthe's sake she must, at any rate for a time, leave home.

Ten years later Miss Nightingale wrote: 'A very successful and justly successful physician once seriously told a sister who was being Devoured that she must leave home in order that the Devouree might recover health and balance which had been lost in the process of devouring. This person was myself.' In a private note she described Sir James Clark's talk to her as 'a terrible lesson which tore open my eyes as nothing less could have done. My life has been decided thereby.' On September 20 Sir James wrote W. E. N. a letter of grave warning. Parthe showed 'alarming indications' of 'extreme irritability, total absorption in self, some degree of chronic delirium ... nervous feelings have been, I venture to say, fostered by overindulgence.' He recommended that Parthe should be separated from her family and placed in the care of 'some judicious kind relative in the family connexion with whom she could reside for some time.'

Fanny's reply was to declare that Sir James was making a fuss about nothing. On October 12 Hilary Bonham Carter wrote to Clarkey from Embley: 'Aunt Fanny does not like Parthe's illness made much of. She says "One need not talk much about a little bilious attack".' None of Sir James's recommendations were followed, and Hilary wrote that Embley was filled with 'an immense amount of company!'

78

After Miss Nightingale had brought Parthe back from Scotland, she stayed only a few days at Embley. The burden of her responsibility for Parthe had been removed, the last chain which held her had been broken, and she began quietly to separate herself from home. By the end of October she had obtained an authorization from the Council of the Sisters of Charity in Paris allowing her to work in their hospitals and institutions.

Manning wrote to his friend the Abbé de Genettes in Paris announcing her arrival: Miss Nightingale, who was in London staying with the Herberts, began to pack her trunks.

At this stage Fanny suddenly spoke as if the visit to Paris was an entirely new idea. 'Flo is thinking of some new expedition perhaps to Paris,' she wrote to Aunt Mai. 'I cannot make up my mind to it.' One evening W. E. N. unexpectedly appeared in the Herberts' drawing-room. He was distraught, saying that life at Embley was unendurable. Parthe was ill and in hysterics and Fanny at her wits' end. A large party of visitors was expected – how could Fanny manage all alone? Florence must leave London and come home.

Before Miss Nightingale could make a decision, Great Aunt Evans was taken ill. The journey to Paris was cancelled and she went to Cromford Bridge House to nurse Great Aunt Evans through her last illness. On New Year's Eve she was back at Embley writing her 'Memorandum for 1852.' 'I am so glad this year is over; nevertheless it has not been wasted I trust. ... I have re-modelled my whole religious belief. ... I have re-cast my social belief. ... I have learnt to know Manning. ... Have been disappointed in my Dublin Hospital plan. Formed my Paris one. ... Lastly all my admirers are married ... and I stand with all the world before me. ... It has been a baptism of fire this year.'

She determined to go to Paris in February. But Fanny and Parthe were not yet defeated; Fanny discovered she could not bear the idea that Florence was going abroad and suggested a new scheme. Florence had once said she wanted to found a sisterhood. Very well, let her found a sisterhood now, in Aunt Evans's empty house at Cromford Bridge. Everything should be provided – money, furniture, equipment. She declined. Parthe then suggested Forest Lodge, a vacant house on the Embley estate, which Miss Nightingale described as the only place on earth more unsuitable for the purpose than Cromford Bridge House. She declined again.

She had now been at home for some weeks, and Fanny must

have seen for herself how her presence exasperated Parthe. She gave way – partially. Florence might go to Paris for a short time on a visit to Clarkey, but the horrid name of the Sisters of Charity was not to be mentioned. Fanny spoke of dressmakers and wrote to Clarkey enjoining her to make sure that while Florence was in Paris she bought clothes for the coming season. Parthe was furious, possessiveness and jealousy consuming her with a twin flame. On January 29, 1853, she wrote angrily to Clarkey: 'Truth is a good thing and the history of the last year (the others much like it) is one month with the Fowlers in Ireland, three ... in London, three ... at Harrogate and Cromford Bridge, three at the water cure and Grandmammas's ... so that I hope she has passed a very pleasant year, but meantime these eternal poor have been left to the mercies of Mamma and me, both very unwell and whose talkey talkey broth and pudding she holds in very great contempt. ... I believe she has little or none of what is called charity or philanthropy, she is ambitious – very and would like well enough to regenerate the world with a grand *coup de main* or some fine institution, which is a very different thing. Here she has a circle of admirers who cry up everything she does or says as gospel and I think it will do her much good to be with you, who, though you love and admire her, do not believe in the wisdom of all she says *because* SHE says it. I wish she could be brought to see it is the intellectual part which interests her, not the manual. She has no *esprit de conduite* in the practical sense. When she nursed me everything which intellect and kind intention could do was done but she was a shocking nurse. Whereas her influence on people's minds and her curiosity in getting into varieties of minds is insatiable. After she has got inside they generally cease to have any interest for her.'

Yes there was, in spite of the gentleness, the sympathy, the charming intelligence, something about Florence which chilled. Impossible to move her, or to influence her by a personal appeal. She did not know what personal feelings were; in a private note she wrote that never in her life did she recollect being swayed by a personal consideration. She lived on a different plane, out of reach, frighteningly, but also infuriatingly, remote.

On February 4 Miss Nightingale arrived at 120 rue du Bac. She was like a child out of school. Anna von Mohl, a young niece of M. Mohl, described her as being 'so thankful to drop being lady-like'. She would not take cabs but went everywhere in omnibuses.

Anna, fascinated, wrote she was 'on the point of falling in love with Florence'. Her plan was to stay for a month with Clarkey and make a survey of all the hospitals in Paris. She then intended, thanks to Manning, to enter the Maison de la Providence, the hospital of the Sisters of Charity in the rue Oudinot, as a postulante to undergo a training in nursing. She was to wear the convent dress, 'the dress of a nun', and 'render all necessary service to the sick' under the direction of the sisters, but she was to eat and sleep in a separate cell and not enter the dormitories or the refectory of the sisters.

The day of her entry into the Maison de la Providence approached and final arrangements were made. She presented herself to the Reverend Mother, was approved, and an hour was fixed for her admission. She had almost come to her last day in the rue du Bac – when Fate struck again. Her grandmother was taken ill, and she was recalled to England. Once more the arrangements with the Sisters of Charity were cancelled, and she hurried to Tapton, arriving in time to nurse her grandmother through her last days. 'I can never be thankful enough that I came,' she wrote to Hilary Bonham Carter on March 26. 'I was able to make her be moved and changed and to do other little things which perhaps soothed the awful passage and which perhaps would not have been done as well without me.'

After Tapton she went alone to Lea Hurst. She intended the separation from her family to be final, and before she went to Paris she had decided to take a post when her training was completed. Early in April 1853 Liz Herbert wrote that through Lady Canning she had heard of what might prove a suitable opening. The Institution for the Care of Sick Gentlewomen in Distressed Circumstances had got itself into difficulties. It was to be reorganized and moved from its present premises. The committee, of which Lady Canning was chairman, were looking for a Superintendent to undertake the reorganization. Liz Herbert suggested Florence, and Lady Canning, after consulting her committee, wrote describing the post and its requirements. On April 8 Miss Nightingale wrote to Clarkey: ' ... It is no use my telling you the history of the negotiations which are enough to make a comedy in 50 acts. ... I am afraid I *must* live at the place. If I don't, it will be a half and half measure which will satisfy no one. ... I can give you no particulars, dearest friend, because I don't know any. I can only say that, unless I am left a free agent and am to organize the thing myself, and not they, I will have nothing to do with it. ... But there are no Surgeon Students or Improper Patients there at all, which is, of course, a great recommendation in the eyes of the Proper.'

Clarkey advised her to be sure to 'trample on the Committee and ride the Fashionable Asses rough shod round Grosvenor Square.'

On April 18 there was an interview, and Lady Canning wrote the same day to Mrs Herbert: 'I write a line in great haste to say that I was delighted with Miss N's quiet sensible manner. In one short acquaintance I am sure she must be a most remarkable person. It is true that Miss Nightingale looks very young but that need not matter and I hope the old matron or housekeeper will in point of outward appearance supply the young Miss N's deficiencies in years.' Miss Nightingale had suggested she should bring, as her personal attendant, a 'superior elderly respectable person' at her own expense.

When the news was broken to Fanny and Parthe, sickeningly familiar scenes took place. Parthe wept, raged, worked herself into frenzy, collapsed, and had to be put to bed. Fanny stormed, lamented, and had to be given sal volatile. Meals were sent away untouched. Ordinary life was at an end. W. E. N. took refuge in the Athenaeum Club. Among the Verney Nightingale papers are two sheets of Athenaeum Club notepaper scribbled back and front in W. E. N.'s strange difficult hand – he always used a quill and abhorred 'great Iron Spikes':

Memorandum April 20th

I have this day reached the conclusion that Parthe can no more control or moderate the intensity of her interest in Flo's doings than she can change her physical form, and that her life will be sacrificed to the activity of her thoughts, unless she removes herself from the scene immediately – the only question being where to go ...

Having come to the resolution that it is entirely beyond your mental strength to give up interference in your sister's affairs and being equally sure that your health cannot stand the strain we advise you to retire from London and take to your books and country occupations till her proceedings are settled.

23rd April

I doubt my own thoughts.

Retirement might do more harm than good – what then?

Reconsidered.

Matters might be worse if I were alone in mediation – Query then, is the case hopeless?

He did nothing. Parthe was not sent to the country but remained in London, passing from fit to fit of hysterics, while Miss Nightingale conducted her negotiations with the committee. W. E. N. did, however, take one vital step – he decided to allow Miss Nightingale £500 a year. Fanny was extremely angry and insisted that since Florence was independent she should pay her share of the bill at the Burlington. She paid and subsequently discovered that Fanny had charged her more than her fair share.

Negotiations with the committee, 'those Fashionable Asses with their "offs" and "ons" poor fools!' were trying. Their hesitations centred upon her social position. She was a young lady in society – was it not peculiar for a young lady to wish for such a post; could a lady take orders, even from a committee of other ladies; should a lady, even in these days of strange mingling of ranks, nurse one who was not a lady; was it nice for a lady to be present at medical examinations and, worse still, at operations?

The deciding factor in the situation was a family quarrel. The Nicholsons disapproved of Florence. 'Very unkind things' were now said. She was described as 'going into service'; her conduct 'could not be looked on in a charitable light'. It happened that one of the committee knew Marianne, and asked her if Miss Nightingale's parents approved the step she proposed to take. Marianne drew a dramatic picture of Florence in opposition to her family: Parthe prostrated, Fanny in tears. The Committee, horrified, decided to have nothing more to do with Miss Nightingale and a letter was written informing her that negotiations were broken off.

However, Fanny and Parthe now completely changed front. Though they might disapprove, they were not going to have Florence attacked by Marianne, and Parthe rushed to her defence. She said unpleasant things about Marianne. Some of them were repeated to her brother Lothian Nicholson, who lost his temper, wrote angrily to Parthe, and finally wrote to Miss Nightingale suggesting with some justice that Fanny and Parthe had provoked gossip by their own conduct, but because poor

Marianne had the reputation of talking, everything was laid at *her* door.

Miss Nightingale replied with a masterpiece of tact. 'Dear friend,' she began, and apologized for an 'overwhelming quantity of work' which prevented her from replying sooner. She praised Lothian's attachment to his sister; he was quite right to defend her; but she preferred 'not to go into the matter with you. I hope you will come and see me "in service" when you have a day to spare in London. Finally, dear Lothian, one word, our old – and I hope real – friendship encourages me to say it. Do not engage in any paper wars. You will convince nobody and arrive at no satisfaction yourself.' They remained friends and fifteen years later were still meeting.

Despite everything, by the end of April she had successfully completed her negotiations. She was to receive no remuneration, and she was to bear all the expenses of the Matron she brought with her to compensate for her youthful appearance, but she was to be in complete control not only of the management of the Institution but of its finances. Her duties were to begin as soon as new premises could be found. In the interval she proposed to go back to Paris and at last accomplish her training with the Sisters of Charity.

There was an explosion from Fanny and Parthe. What, when Florence was preparing to leave home, would she not devote the few weeks that were left to her mother and sister? She would not, and she went to Paris on May 30.

For the third time she attempted to train at the Maison de la Providence, and for the third time she was defeated, for after a fortnight in the convent she developed measles, *'une rougeole intense'*. 'And, of all my adventures of which I have had many and queer, as will be (never) recorded in the Book of my Wanderings, the dirtiest and queerest I have ever had has been a measles in the cell of a Sœur de la Charité,' she wrote to Clarkey on June 28. 'It is like the Mariage de Mademoiselle; who could have foreseen it? ... For me to come to Paris, to have the measles a second time, is like going to the Grand Desert to die of getting one's feet wet. ...'

Clarkey was in England, but as soon as Miss Nightingale was convalescent M. Mohl, 'in his kind paternity', brought her to the rue du Bac and put her to bed in the back drawing-room, the same room in which she had seen M. Mohl and Fauriel boil the kettle for tea on a January afternoon sixteen years ago. Her convalescence was spent in conversation with him. 'Her gentle manner,' he wrote to W. E. N., 'covers such a depth and strength of mind and thought.' Before she came home, she went to the dressmakers and was fitted for a 'great panjandrum of black velvet'. On July 13 she reached London, but she would not join Fanny and Parthe. She took rooms in Pall Mall.

These rooms caused fresh lamentations. Even Clarkey, who was staying at Embley, was moved to suggest that she might spend her free time at least with her family. 'I have not taken this step Clarkey dear,' wrote Miss Nightingale in August 1853, 'without years of anxious consideration. I mean the step of leaving them. I do not wish to talk about it – and this is the last time I shall ever do so. ... I *have* talked matters over ("made a clean breast" as you express it) with Parthe, *not once but thousands of times*. Years and years have been spent in doing so. It has been, therefore, with the deepest consideration and with the fullest advice that I have taken the step of leaving home, and it is a *fait accompli*. ... So farewell, Clarkey dear, don't let us talk any more about this. It is, as I said before, a *fait accompli*.'

From July 13 to August 12 she was in London with Aunt Mai supervising the alterations to the new premises chosen for the Institution. Fanny refused to give her blessing – 'it would be useless upon what I consider as being an impossible undertaking.' On August 12, 1853, she went into residence in the new premises, number 1 Harley Street.

6

'I AM living in an ideal world of lifts, gas, baths, and double and single wards,' wrote Miss Nightingale to Hilary Bonham Carter in the summer of 1853. She was moving in her natural element. From Paris, first from her cell at the Maison de la Providence, then from her convalescent couch at 120 rue du Bac, she had kept a firm hand on her committee, issuing precise instructions to them in long, enormously detailed letters.

Her requirements were not merely exacting; they were revolutionary. She had a scheme for saving work by having hot water 'piped up to every floor'. She wanted a 'windlass installation', a lift to bring up the patients' food. On June 5, 1853, she wrote to Lady Canning: '... The nurse should never be obliged to quit her floor, except for her own dinner and supper, and her patients' dinner and supper (and even the latter might be avoided by the windlass we have talked about). Without a system of this kind, the nurse is converted into a pair of legs. *Secondly*, that the bells of the patients should all ring in the passage outside the nurse's door *on that story* and should have a valve which flies open when its bell rings, and *remains* open in order that the nurse may see who has rung.'

Her committee became dazed. Forced to answer the innumerable questions raised in her letters, sent out on expeditions to unknown parts of London to view 'windlass installations', and new systems of bells with valves, they had the sensation of having unknowingly released a genie from a bottle. They were, she said, 'children in administration'. In fact, they had never been called on to administer anything before. The Institution had been managed by two committees, a Ladies' Committee and a Gentlemen's Committee. The Gentlemen had transacted all the business and paid the bills. Miss Nightingale returned from Paris to find nothing had been accomplished. On August 20 she wrote to Clarkey: 'I have had to prepare this immense house for patients in ten days – without a bit of help but only hindrance from my committee. I have been "in service" ten days and have had to

86

furnish an entirely empty house in that time. We take in patients this Monday and have not got our workmen out yet. From Committees, Charity and Schism, from the Church of England, from philanthropy and all deceits of the devil, Good Lord deliver us.'

The accounts of the Institution were in confusion. 'I am seriously uneasy about our funds,' she had written to Hilary Bonham Carter on July 24. '... The Committee are wholly regardless of money. £1200 we had in the funds they have taken out for the alteration and furniture of this house, and spent every penny of it.' The administration of the Institution was also in confusion. The two committees quarrelled with each other, and among themselves, the doctors did the same. 'There is as much jealousy in the Committees of one another and among the medical men of one another as ever what's his name had of Marlborough,' she wrote to W. E. N. on December 3, 1853.

During the first week of her appointment Miss Nightingale and her committee had a serious difference. She was determined the Institution should be non-sectarian; the committee was determined it should be Church of England. On August 20 she wrote to Clarkey: 'My Committee refused me to take in *Catholic* patients, whereupon I wished them good morning, unless I might take in Jews and their Rabbis to attend them. So now it is settled, and *in print* that we are to take in all denominations whatever, and allow them to be visited by their respective priests and Muftis, provided *I* will receive (in any case *whatsoever* that is *not* of the Church of England) the obnoxious animal at the door, take him upstairs myself, remain while he is conferring with his patient, make myself *responsible* that he does not speak to, or look at, *anyone else*, and bring him downstairs again in a noose, and out into the street. And to this I have agreed! And this is in print! Amen.'

She gained her point, but there was disapproval. Some members of the committee were shocked, and an opposition formed against her. In October, Liz wrote offering to come up from Wilton for a committee meeting: 'I thought some wicked cats might be there who would set up their backs and if so I would like to set mine up too.'

Miss Nightingale was not what her committee had expected. Her genius was of an unromantic character. She perceived that unorganized devotion, unorganized self-sacrifice were useless. To bring about the installation of a row of bells 'with valves that flew open' when the patient called was more effectual than to turn oneself into a devoted nurse, toiling endlessly up and down stairs because no such bells existed. To put in the best possible kitchen stove, to descend into the coal cellar and rake over the coal to ensure the coal merchant had not delivered an undue proportion of dust, to check stores and linen and provide patients with clean beds and good food were more effectual than to sit through the watches of the night cheering the dying moments of the patient expiring from scurvy and bed sores. But it was not so picturesque.

She gave devotion generously, and she did an immense amount of practical nursing in the Institution herself, but she was always aware that its success was impossible without a balanced expenditure and a proper system of keeping accounts.

She set herself to manage the committee. By December she had learned how to get her own way. 'When I entered "into service" here,' she wrote to W. E. N., 'I determined that, happen what would, I NEVER would intrigue among the Committee. Now I perceive that I do all my business by intrigue. I propose, in private, to A, B, or C, the resolution I think A, B, or C, most capable of carrying in Committee, and then leave it to them, and I always win.'

She went down into the kitchen and herself turned out cupboards and store-rooms, finding, she wrote to Clarkey in August 1853, that while there were no brooms, brushes, or dusters, jam at 1s. a pot had been ordered in '£2 worth at a time'. She had 52 pots of jam made in the kitchen of the Institution at a cost of $3\frac{1}{2}d.$ a pound. The grocer's boy had been calling 'two or three times a day and bringing everything by the ounce'. She gave out contracts to the best firms – one of her contracts was with Fortnum and Mason – and secured wholesale prices.

The bed linen and furniture of the Institution, she told Clarkey, were 'dirty and neglected. Table linen and kitchen linen ragged and filthy. . . . Chairs with covers, not washable, but put on with nails and soaked with dirt. . . . Saucepans deficient.' As the committee had already spent more money than the Institution possessed, she had to make the best of what was at her disposal. She enlisted Fanny's help and 'had odd pieces of linen washed at home and patched together . . . pieced out carpets, contrived bed covers out of old curtains.'

The original staff from Chandos Street did not long survive.

The housekeeper left after a single interview; the house surgeon resigned after a month. She secured a successor who 'dispensed the medicines in the house, saving our bill at the druggist's of £150 per annum.'

She was in tremendous spirits. On October 20 she wrote Fanny a letter, which was bursting with gaiety, asking for a pair of comfortable old boots – she was spending all day and most of the night on her feet: 'Oh my boots! Where are ye, my boots. I never shall see your pretty faces more. My dear I *must* have them boots. ... More flowers, more game, more grapes.'

Within six months opposition had collapsed. On December 3 she wrote to W. E. N.: 'I am now in the heyday of my power. ... Lady — who was my greatest enemy is now, I understand, trumpeting my fame through London.'

It was not an easy life. She had disappointments. She had to struggle with suspicion and inefficiency. 'The chemists,' she had told W. E. N. in December 1853, 'sent me a bottle of ether labelled Spirits of Nitre, which, if I had not smelt it, I should certainly have administered, and should have an inquiry into poisoning.' The builders did not carry out her orders properly. 'The whole flue of a new gas stove came down the second time of using it, which, if I had not caught in my arms would certainly have killed a patient.' Medically there were heartbreaking failures. 'We have had an awful disappointment,' she wrote to W. E. N. in the spring of 1854, 'in a couching for a cataract, which has failed. The eye is lost ... and I am left, after a most anxious watching, with a poor blind woman on my hands, whom we have blinded, and a prospect of insanity. I had rather ten times have killed her.'

Fanny and Parthe continued to disapprove, though Fanny with ineradicable generosity sent regular weekly hampers of flowers, vegetables, game, and fruit from Embley for the patients. Miss Nightingale had her own sitting-room at Harley Street where she gave her friends tea out of special blue cups. Parthe was persuaded to accept an invitation but collapsed in hysterics as she crossed the threshold of the Institution. She nevertheless could write in a private note of January 1, 1854: 'I have never repented nor looked back, not for one moment. And I begin the New Year with more true feeling of a happy New Year than I ever had in my life.'

Yet she found time to go to parties, and in the season of 1854 received invitations from Lady Beresford, Lady Canning, Lady Cranworth, Lady Palmerston. Gay little notes went round by hand to Harley Street from frivolous friends. 'Lock up your young ladies, leave someone else to stir the gruel and cab round to me.' 'Dearest do come to an evening party, it will be rather a squeeze.' 'Don't get too radical, sceptical and querical my Flo.' With the Herberts she was on a footing of closest intimacy. 'Dearest', 'My dearest', 'Dearest old Flo', Liz Herbert wrote.

Her patients worshipped her, writing innumerable adoring letters: 'My dearest kind Miss Nightingale. I send you a few lines of love.' 'I felt so lonely when I saw you going away from me.' 'All your affectionate kindness to me comes before me now and causes me many tears.' 'I am your affectionate, attached and grateful.' 'Thank you, thank you darling Miss Nightingale.' 'You are our sunshine ... were you to give up all would soon fade away and the whole thing would cease to be.'

Her sympathy with impoverished struggling women penetrated into every detail of their lives. She understood their loneliness, their perpetual financial difficulties, the burden of other relatives even poorer than they. She sent a poor governess to Eastbourne at her own expense and arranged that she should be visited and taken for drives. 'I know not how to thank you my dear dear Miss Nightingale. I cannot even express how much indebted I am to you,' wrote the patient. Often she sent financial help. 'How gratefully I accept your offer of defraying my poor aunt's expenses. My mother has unfortunately no means of settling it herself.' She saved women whose resources had been exhausted from going straight from a bed of sickness to a new post. Again and again a letter runs: 'I cannot thank you enough for this extra rest.' At Harley Street her correspondence was very large. The patients wrote; their relatives wrote; poor friendless women who were complete strangers were 'emboldened by your very great kindness to my afflicted friend' to confide their fears of dreaded secret ailments.

In December 1853 Richard Monckton Milnes, after staying at Embley, wrote to his wife: 'They talk quite easily about Florence, but her position does not seem to be very suitable.' She had enjoyed the period of reorganization, but as soon as the Institution was running smoothly she became restless. By January 1854 she was speaking of 'this little mole hill'.

In the spring of that year she began to visit hospitals and collect facts to establish a case for reforming conditions for hospital nurses. On May 29 Liz Herbert wrote: 'Sidney has begged me to write and ask you whether you can give him any facts in writing as to abuses which exist in — Hospital. Sidney says if he could get some authentic information on the subject of the nurses, their bad pay and worse lodging he could get the evil more or less remedied and public attention at any rate turned that way.'

Soon letters were passing almost daily, Miss Nightingale submitting reports and Sidney Herbert asking for 'additional information ... as soon as possible'. Reform was difficult. Within the hospitals there was jobbery. Hospital appointments were often held as the result of bribery or nepotism, and the official who supported reform found his appointment in danger. Outside the hospitals there was indifference; their conditions were accepted as a necessary horror. The number of the enlightened who, like the Herberts, pressed for improved conditions and a better type of nurse was very small. Most people agreed with Lady Palmerston. 'Lady Pam thinks ... the nurses are very good now; perhaps they do drink a little, but ... poor people it must be so tiresome sitting up all night,' wrote Lord Granville.

In any case, where was a better type of nurse to come from? Superior nurses did not exist. In June 1854 a doctor who had met Miss Nightingale in Paris wrote asking her to recommend two reliable skilful nurses to act as matrons in colonial hospitals. She had to reply that she knew none: 'Alas I have no fish of that kind.' It was absurd to create a demand which could not be supplied. Before any scheme of nursing reform was embarked upon, a training school capable of producing a supply of respectable, reliable, qualified nurses must be brought into existence. Her first task must be to produce a new type of nurse.

She confided in Dr Bowman, one of the best-known surgeons of his day, surgeon to the Institution, and her devoted admirer. King's College Hospital, where he held a senior appointment, was being reorganized and rebuilt, and his influence would be sufficient to obtain her the post of Superintendent of Nurses, where she would have scope for training a new type of nurse. Rumours reached Embley, and Fanny and Parthe broke into lamentations: the suggestion of a hospital post struck them with horror as fresh as if it had been a new idea.

But Miss Nightingale was not at home to be reproached in person; she would not go home, she remained in Harley Street.

The eager, susceptible, over-affectionate girl had become the elegant, composed, independent woman of genius. It was now beyond anyone to stop Miss Nightingale in her course. She continued her negotiations with Dr Bowman, and, down at Embley, Fanny's reproaches and Parthe's hysterics sank to ineffectual flutterings. They wrote her letters imploring her to nurse babies, to found a penitentiary; she ignored them, and their voices trailed into silence and were heard no more.

In the summer of 1854, cholera broke out in London, particularly in the miserable, undrained slums round St Giles, to the west of Drury Lane. The hospitals were overcrowded; many nurses died; many, afraid of infection, ran away. In August Miss Nightingale went as a volunteer to the Middlesex Hospital to 'superintend the nursing of cholera patients'. From the Middlesex Hospital she went to Lea Hurst, where Mrs Gaskell was staying. In a long letter to Emily Winkworth Mrs Gaskell repeated Miss Nightingale's account of the epidemic. The authorities at the Middlesex Hospital were 'obliged to send out their usual patients in order to take in the patients brought in every half hour from the Soho district, Broad Street, etc. ... chiefly fallen women of the district. ... The prostitutes came in perpetually – poor creatures staggering off their beat! It took worse hold of them than of any.' Miss Nightingale was 'up day and night, undressing them ... putting on turpentine stupes, etc., herself to as many as she could manage.' The women were filthy and drunken, crazed with terror and pain, and the rate of mortality was very high. All through the night wretched shrieking creatures were being carried in. From Friday afternoon until Sunday afternoon she was never off her feet.

Mr Sam Gaskell, a relative of Mrs Gaskell's, had been prejudiced by what he had heard of Miss Nightingale; he had spoken very contemptuously of her and called her 'your enthusiastic young friend', but when they did meet he was 'carried off his feet'. And Mrs Gaskell herself continued in a letter to Catherine Winkworth dated October 20: 'Oh Katie! I wish you could see her. ... She is tall; very slight and willowy in figure; thick shortish rich brown hair, very delicate colouring; grey eyes which are generally pensive and drooping, but which when they choose can be the merriest eyes I ever saw; and perfect teeth, making her smile the sweetest I ever saw. Put a long piece of soft net, say 1½ yards long and half a yard wide, and tie it round this beautifully shaped head, so as to form a soft white framework for the full oval of her face (for she had the toothache and so wore this little piece of drapery) and dress her up in black silk high up to the long white round throat, and a black shawl on and you may get NEAR an idea of her perfect grace and lovely appearance. . . . She has a great deal of fun and is carried along by that I think. She mimics most capitally, mimics for instance the way of talking of some of the poor Governesses in the Establishment, with their delight at having a man servant, and at having LADY Canning and LADY Mounteagle to do this and that for them.'

And yet a week later Mrs Gaskell was chilled. She too had discovered that Florence was intimidating. Beneath the fascination, the sense of fun, the gentle hesitating manner, the demure wit, there was the hard coldness of steel. On October 27 Mrs Gaskell wrote: 'She has no friend – and she wants none. She stands perfectly alone, half-way between God and his creatures. She used to go a great deal among the villagers here, who dote upon her. One poor woman lost a boy seven years ago of white swelling in his knee, and F. N. went twice a day to dress it. . . . The mother speaks of F. N. – did so to me only yesterday – as of a heavenly angel. Yet the father of this dead child – the husband of this poor woman – died last 5th of September and I was witness to the extreme difficulty with which Parthe induced Florence to go and see this childless widow ONCE whilst she was here; and, though this woman entreated her to come again, she never did. She will not go among the villagers now because her heart and soul are absorbed by her hospital plans, and, as she says, she can only attend to one thing at once. She is so excessively gentle in voice, manner, and movement, that one never feels the unbendableness of her character when one is near her. Her powers are astonishing. . . . She and I had a grand quarrel one day . . . she said if she had influence enough not a mother should bring up a child herself; there should be creches for the rich as well as the poor. If she had twenty children she would send them all to a creche, seeing, of course, that it was a well managed creche. That exactly tells of what seems to me THE want – but then this want of love for individuals becomes a gift and a very rare one, if one takes it in conjunction with her intense love for the RACE; her utter unselfishness in serving and ministering . . . but she is really so extraordinary a creature that anything like a judgement of her must be presumptuous.'

One day Fanny and Mrs Gaskell were alone. Fanny spoke of Florence 'with tears in her eyes', telling Mrs Gaskell, 'We are ducks who have hatched a wild swan'. But it was not a swan they had hatched: in the famous phrase of Lytton Strachey's essay – it was an eagle.

The summer of 1854 marked the end of a chapter. The long agonizing apprenticeship was over, and the instrument uniquely fitted for its purpose was forged. In the world outside Harley Street a catastrophe was taking shape. In March 1854 England and France had declared war on Russia. In September the Allied armies landed in the Crimea. Harley Street, with its unreasonable committee, its 'deficient' saucepans, its ragged linen, had been a dress rehearsal. Now the curtain was about to go up on the play.

7

To the British people the invincibility of the British Army was an article of faith. Waterloo was a recent memory, and it was taken for granted that the nation which had beaten Napoleon could not be defeated. But since Waterloo forty years of economy had run their course, and the army which had won Wellington's victories had ceased to exist. In 1852 the artillery of the British Army consisted of forty field-pieces, many officially described as defective. In 1854, when the army was mobilizing for the Crimea, volunteers had to be drafted into the battalions selected for active service to raise their numbers to the regulation 850. The staff of the supply departments had been reduced to a few clerks, who were overwhelmed by the demands of mobilization. Before the Army sailed, the processes by which the troops were to receive food and clothing, to be maintained in health and cared for when wounded or sick, had already fallen into confusion.

An enormous amount of information exists on the Crimean War. While it was in progress, four Parliamentary Commissions of Inquiry investigated its disasters. Three of them went to the Crimea; the fourth, which sat in London, examined civil servants and officials in Government service as well as witnesses from the seat of war. The resulting mass of evidence fills a shelf of Blue Books in whose innumerable pages, from which the stench of misery and filth and despair seems palpably to rise, the Crimean War lies embalmed.

But in the spring of 1854, confidence was complete. The Guards were a magnificent body of fighting men as they marched through London to embark. The crowds which cheered them did not know that behind these splendid troops, the flower of the British Army, were no reserves. They were doomed to perish, and when they perished, their ranks were filled with raw recruits made 'pretty perfect in drill in sixty days'.

The first operation was not to be in the Crimea. The British Army was to relieve Silistria, in Roumania, then a Turkish province, where the Russians were besieging the Turks. A base was established at Scutari, a large village on the Asian shore of the Bosphorus, and in June 1854 the British Army disembarked at Varna, in Bulgaria. Nothing was accomplished. A cholera epidemic broke out; the army became an army of invalids, and the Turks raised the siege of Silistria on their own account. The Allies then proceeded to the true objective of the war, the destruction of the great naval base recently constructed by the Russians at Sebastopol.

Though the plan of a descent on Sebastopol was an open secret and had been discussed in the Press, it had never been officially intimated to the supply departments; consequently no preparations had been made. When the British Army embarked at Varna for the Crimea, there were not enough transports to take both the army and its equipment across the Black Sea. Thirty thousand men were crammed in, but pack animals, tents, cooking equipment, hospital marquees, regimental medicine chests, bedding, and stores had all to be left behind. Twenty-one wagons only were brought for 30,000 men going into action. On September 14 the army disembarked at a cove with the sinister name of Calamita Bay. 'My God,' exclaimed Dr Alexander, 1st class Staff Surgeon of the Light Division, 'they have landed this army without any kind of hospital transport, litters or carts or anything.' Cholera still raged, and over 1,000 cholera cases were sent back to Scutari.

A week later, the British and the French won the hard-fought battle of the Alma, and the wounded paid the price of the abandonment of the army's hospital equipment. There were no bandages, no splints, no chloroform, no morphia. The wounded lay on the ground or on straw mixed with manure in a farmyard. Amputations were performed without anaesthetics; the victims sat on tubs or lay on old doors; the surgeons worked by moonlight because there were no candles or lamps. And another 1,000 cholera cases were sent back to Scutari.

Of this the British public knew nothing. Nor did they know what awaited the wounded and the sick when they reached the base at Scutari. At Scutari were enormous barracks, the headquarters of the Turkish artillery. These barracks and the hospital attached had been handed over to the British, and the British authorities assumed that the hospital, known as the General Hospital, would be adequate. The unexpected disaster of the cholera epidemic produced total disorganization. The first 1,000 cholera cases sent back after the landing at Calamita Bay filled the hospital to overflowing; drugs, sanitary conveniences, bedding, doctors were insufficient. While Dr Menzies, senior Medical Officer, was struggling with the crisis, he was notified that many hundreds of battle casualties from the Alma and another 1,000 cholera cases were on their way. Since the General Hospital was filled, he was ordered to convert the artillery barracks into a hospital. It was an impossible task. The vast building was bare, filthy, and dilapidated. There was no labour to clean it; there was no hospital equipment to put in it.

Meanwhile the sick and wounded were enduring a ghastly journey across the Black Sea. They were conveyed in 'hospital ships' which figured well on paper but in fact were ordinary transports equipped 'with some medicines and medical appliances'. They were packed far beyond their capacity. One, the *Kangaroo*, fitted to receive 250 sick, received between 1,200 and 1,500. Cholera cases, battle casualties, were crammed in together. Too weak to move, too weak to reach the sanitary conveniences, they fell on each other as the ship rolled and were soon lying in heaps of filth. Men with amputations were flung about the deck screaming with pain.

When the men arrived at the Barrack Hospital, there were no beds. They lay on the floor wrapped in the blankets saturated with blood and ordure in which they had been lying since they left the battlefield. No food could be given them because there was no kitchen. No one could attend to them because there were not sufficient doctors. Some of them lay without even a drink of water all that night and through the next day. There were no cups or buckets to bring water in. There were no chairs or tables. There was not an operating table. The men, half naked, lay in long lines on the bare filthy floors of the huge dilapidated rooms.

Such scenes of horror were nothing new in Britain's military annals: similar miseries had been endured by the British Army many times before. During the winter of 1759, outside Quebec, outside Havana in 1762, during the retreat to the Ems in 1797, worse miseries were endured than in the Crimea. In the disastrous Walcheren expedition of 1809, a whole army was lost through sickness. Men died in thousands in the general hospitals of the Peninsula; the Guards were so reduced by sickness that they had had to fall out of the campaign from November 1812 to June 1813.

But these horrors had remained unknown. England rang with the story of Scutari because with the British Army was the first war correspondent, William Howard Russell of *The Times*.

'By God, Sir, I'd as soon see the devil,' said General Pennefather to Russell when they met in the Crimea; but Pennefather did not order Russell home. The Crimea was a casual war. Numbers of tourists, known to the army as 'T.G.'s', 'Travelling Gentlemen', camped with the troops. Private philanthropists came out at their own expense. Though Russell and his paper were abominated by the army authorities – *The Times* under the editorship of Delane was a Radical newspaper – he was never obstructed. The military aristocrats of the high command were content to ignore him. 'Lord Raglan,' wrote Russell, 'never spoke to me in his life. ... I was regarded as a mere camp follower, whom it would be impossible to take more notice of than you would of a crossing sweeper, without the gratuitous penny.'

Russell was an Irishman with an Irishman's capacity for indignation, and in dispatches published on October 9, 12, and 13 he furiously described the sufferings of the sick and wounded. 'It is with feelings of surprise and anger that the public will learn that no sufficient preparations have been made for the care of the wounded. Not only are there not sufficient surgeons ... not only are there no dressers and nurses ... there is not even linen to make bandages. ... Can it be said that the battle of the Alma has been an event to take the world by surprise? Yet ... there is no preparation for the commonest surgical operations! Not only are the men kept, in some cases for a week, without the hand of a medical man coming near their wounds, ... but now ... it is found that the commonest appliances of a workhouse sick ward are wanting, and that the men must die through the medical staff of the British Army having forgotten that old rags are necessary for the dressing of wounds.'

The revelation burst on the nation like a thunderclap, and on October 13 Sir Robert Peel, the third baronet, opened '*The Times* Fund' for supplying the sick and wounded with comforts. The same day *The Times* published another dispatch from Russell. 'The manner in which the sick and wounded are treated is worthy only of the savages of Dahomey. ... There are no dressers or nurses to carry out the surgeons' directions, and to attend on the sick during the intervals between his visits. Here the French are greatly our superiors. Their medical arrangements are extremely

98

good, their surgeons more numerous and they have also the help of the Sisters of Charity ... these devoted women are excellent nurses.'

The country seethed with rage. Russell's statement that British arrangements compared unfavourably with those of the French was intolerable, and the next day a letter in *The Times* demanded angrily, 'Why have we no Sisters of Charity?'

It was read by Sidney Herbert, who in December 1852 had been appointed Secretary at War, and was now responsible for the treatment of the sick and wounded. The administration of the British Army was then divided between two Ministers, the Secretary for War and the Secretary at War. The Secretary at War was responsible for the financial administration of the army, and since the cheese-paring, the callous economies, the criminally inadequate arrangements had been executed in his name, the blame must lie at the door of Sidney Herbert. His political position was now extremely delicate. His mother had been Russian, daughter of the Russian Ambassador, and the famous Woronzoff road which was to be of such overwhelming importance to the British Army in the Crimea led to the Woronzoff palace at Yalta which belonged to his uncle. Suspicion was inevitable; and a storm of national fury burst on his head. The military authorities, enraged by the interference of *The Times*, refused to admit that anything was wrong. Sidney Herbert was not convinced and acted on his own responsibility. He wrote to the British Ambassador at Constantinople, Lord Stratford de Redcliffe, giving him *carte blanche* to purchase anything he considered necessary for the hospitals, and on October 15 he wrote to Miss Nightingale inviting her to go to Scutari in command of a party of nurses. She would go with the Government's sanction and at the Government's expense.

She had already acted on her own account and, without consulting the Herberts, had arranged to sail for Constantinople with a party of nurses in three days' time. She had hesitated to approach them, embarrassed by the attacks being made on Sidney Herbert; but when her plans were completed, she called at 49 Belgrave Square on the morning of Saturday, October 14. The Herberts had gone to Bournemouth for the week-end.

On Saturday afternoon Miss Nightingale wrote to Liz Herbert: 'My dearest I went to Belgrave Square this morning for the chance of catching you, or Mr Herbert even, had he been in Town. A small private expedition of nurses has been organized for Scutari and I have been asked to command it. I take myself out and one nurse. . . . I do not mean that I believe *The Times* accounts, but I do believe we may be of use to the poor wounded wretches.' She asked Liz to negotiate her release from her engagement with the Harley Street committee – unless the committee thoroughly approved she could not honourably break her engagement; and, would Sidney approve? 'What does Mr Herbert say to the scheme itself? Does he think it would be objected to by the authorities? Would he give us any advice or letters of recommendation? And are there any stores for the Hospital he would advise us to take out.' Finally, would Liz write to the Ambassadress, Lady Stratford de Redcliffe, and say: 'This is not a Lady but a real Hospital Nurse . . . and she has had experience.'

This letter crossed one written by Sidney Herbert at Bournemouth on the Sunday, in which he formally asked her to take charge of an official scheme for introducing female nurses into the hospitals of the British Army:

DEAR MISS NIGHTINGALE,

You will have seen in the papers that there is a great deficiency of nurses at the Hospital at Scutari.

The other alleged deficiencies, namely of medical men, lint, sheets, etc., must, if they have really ever existed, have been remedied ere this, as the number of medical officers with the Army amounted to one to every 95 men in the whole force, being nearly double what we have ever had before, and 30 more surgeons went out 3 weeks ago, and would by this time, therefore, be at Constantinople. A further supply went on Thursday, and a fresh batch sail next week.

As to medical stores, they have been sent out in profusion; lint by the *ton* weight, 15,000 pairs of sheets, medicine, wine, arrowroot in the same proportion; and the only way of accounting for the deficiency at Scutari, if it exists, is that the mass of stores went to Varna, and was not sent back when the Army left for the Crimea; but four days would have remedied this. In the meanwhile fresh stores are arriving.

But the deficiency of female nurses is undoubted, none but male nurses having ever been admitted to military hospitals.

It would be impossible to carry about a large staff of female nurses with the Army in the field. But at Scutari, having now a fixed hospital, no military reason exists against their introduction, and I am confident they might be introduced with great benefit, for hospital orderlies must be very rough hands, and most of them, on such an occasion as this, very inexperienced ones.

I receive numbers of offers from ladies to go out, but they are ladies who have no conception of what an hospital is, nor of the nature of its duties; and they would, when the time came, either recoil from the work or be entirely useless, and consequently – what is worse – entirely in the way. Nor would these ladies probably ever understand the necessity, especially in a military hospital, of strict obedience to rule ...

There is but one person in England that I know of who would be capable of organizing and superintending such a scheme; and I have been several times on the point of asking you hypothetically if, supposing the attempt were made, you would undertake to direct it.

The selection of the rank and file of nurses will be very difficult: no one knows it better than yourself. The difficulty of finding women equal to a task, after all, full of horrors, and requiring, besides knowledge and goodwill, great energy and great courage, will be great. The task of ruling them and introducing system among them, great; and not the least will be the difficulty of making the whole work smoothly with the medical and military authorities out there. This it is which makes it so important that the experiment should be carried out by one with a capacity for administration and experience. A number of sentimental enthusiastic ladies turned loose into the Hospital at Scutari would probably, after a few days, be *mises à la porte* by those whose business they would interrupt, and whose authority they would dispute.

My question simply is, Would you listen to the request to go and superintend the whole thing? You would of course have plenary authority over all the nurses, and I think I could secure you the fullest assistance and co-operation from the medical staff, and you would also have an unlimited power of drawing on the Government for whatever you thought requisite for the success of your mission. On this part of the subject the details are too many for a letter, and I reserve it for our meeting; for whatever decision you take, I know you will give me every assistance and advice.

I do not say one word to press you. You are the only person who can judge for yourself which of conflicting or incompatible duties is the first, or the highest; but I must not conceal from you that I think upon your decision will depend the ultimate success or failure of the plan. Your own personal qualities, your knowledge and your power of administration, and among greater things your rank and position in Society give you advantages in such a work which no other person possesses.

If this succeeds, an enormous amount of good will be done now, and to persons deserving everything at our hands; and a prejudice will have been broken through, and a precedent established, which will multiply the good to all time.

I hardly like to be sanguine as to your answer. If it were 'yes', I am certain the Bracebridges would go with you and give you all the comfort you would require, and which their society and sympathy only could give you. I have written very long, for the subject is very near to my heart. Liz is writing to Mrs Bracebridge to tell her what I am doing. I go back to town tomorrow morning. Shall I come to you between 3 and 5? Will you let me have a line at the War Office to let me know?

There is one point which I have hardly a right to touch upon, but I know you will pardon me. If you were inclined to undertake this great work, would Mr and Mrs Nightingale give their consent? The work would be so national, and the request made to you proceeding from the Government who represent the nation comes at such a moment, that I do not despair of their consent. Deriving your authority from the Government, your position would secure the respect and consideration of every one, especially in a service where official rank carries so much weight. This would secure to you every attention and comfort on your way and there, together with a complete submission to your orders. I know these things are a matter of indifference to you except so far as they may further the great objects you have in view; but they are of importance in themselves, and of every importance to those who have a right to take an interest in your personal position and comfort.

I know you will come to a wise decision. God grant it may be in accordance with my hopes!

Believe me, dear Miss Nightingale,

ever yours,
SIDNEY HERBERT.

The terms of this letter were accepted by Miss Nightingale and considered by her to be her charter. They made it clear that from the inception of her mission she was to be an administrator.

It was not as an angel of mercy that she was asked to go to Scutari – relieving the sufferings of the troops was scarcely mentioned. The consideration of overwhelming importance was the opportunity offered to advance the cause of nursing. Were nurses capable of being employed with success to nurse men under such conditions? The eyes of the nation were fixed on Scutari. If the nurses acquitted themselves creditably, never again would they be despised. 'If this succeeds,' Sidney Herbert had written, 'an enormous amount of good will have been done now ... a prejudice will have been broken through and a precedent established which will multiply the good to all time.'

Before she had time to write a reply, he had received her letter, and on the Monday afternoon he called on her at Harley Street, bringing a letter from Liz: 'My own dearest noblest Flo. I *knew* you would do it. . . . God be thanked. Sid longed to go to you last week .. I will write a "cut and dry" letter to the committee in Harley Street and bear all the blame if any can possibly attach itself to such a work! Go then at once, and God prosper it and you. Your own loving E. H. Would that I could come to town to you at once. But my nurse is ill and away and I cannot leave my children. . . .'

Sidney Herbert warned Miss Nightingale that he was by no means satisfied with the assurances he was receiving from the army authorities, and that he was sending out immediately a 'Commission of Enquiry into the State of the Hospitals and the Condition of the Sick and Wounded.' The Commission had three members, a well-known barrister, Mr Benson Maxwell, and two doctors, Dr Cumming and Dr Spence. Its purpose was to establish the facts, but it was not empowered to take action and could not alter existing arrangements. She was to work with the Hospitals Commission, send in official reports, and in addition write privately to him telling him confidentially what she could not write officially.

The number of nurses in the party was fixed at forty. She was doubtful of her ability to control more than twenty, but Sidney Herbert insisted that twenty would not be a sufficiently large number to make the experiment impressive. He would have preferred an even larger number than forty. On Wednesday, October 18, Sidney Herbert, supported by the Duke of Newcastle, placed Miss Nightingale's appointment before the Cabinet. The appointment was unanimously approved, and next day she received a formal confirmation written and signed by Sidney Herbert as Secretary at War. She was appointed 'Superintendent of the Female Nursing Establishment of the English General Hospitals in Turkey,' and her authority was defined: 'Everything relating to the distribution of the nurses, the hours of their attendance, their allotment to particular duties is placed in your hands, subject of course to the sanction and approval of the chief medical officer; but the selection of the nurses in the first instance is placed solely under your control.' Precise as these instructions appeared, they contained a flaw. The words 'Superintendent of the Female Nursing Establishment of the English General Military Hospitals in *Turkey*' were subsequently contended to limit her authority to Turkey and to exclude her from the Crimea.

Her appointment caused a sensation. The story of the Cabinet meeting, the official instructions, the letter to the Commander-in-Chief flew from mouth to mouth. No woman had ever been so distinguished before, and Fanny and Parthe were ecstatic. Forgetting they had brought her to the verge of insanity by their opposition, they congratulated themselves on the scope of the experience which qualified her for her mission. 'It is a great and noble work,' wrote Parthe to a favourite cousin. 'One cannot but believe she was intended for it. None of her previous life has been wasted, her experience all tells, all the gathered stores of so many years, her Kaiserswerth, her sympathy with the R. Catholic system of work, her travels, her search into the hospital question, her knowledge of so many different minds and classes. . . .'

Parthe and Fanny hastened from Embley to London to share in the excitement, and, in the haste of packing, the owl Athena was left shut in an attic, where she was found later dead; she required constant attention and was subject to fits. When the lifeless body was put into Miss Nightingale's hands, she burst into tears. 'Poor little beastie,' she said, 'it was odd how much I loved you.' It was the only sign of emotion she showed on the eve of departure. Otherwise she was, wrote Parthe, 'as calm and composed as if she was going for a walk.'

She had made up her mind to start on Saturday, October 21, four clear days only after she had received Sidney Herbert's letter. Nurses had to be engaged; she was determined they should wear uniform, which must be made; tickets and berths must be reserved. Σ and her husband had agreed to accompany her, and Mr Bracebridge took over the finances of the expedition and made the travelling arrangements, adding to the prevailing excitement by hiring one of the new fast hansom cabs and driving about London at ten miles an hour.

The headquarters of the expedition were at the Herberts' London house, 49 Belgrave Square. Mary Stanley, Mrs Bracebridge, Lady Canning, and Lady Cranworth sat all day in the dining-room prepared to receive a rush of applicants – but few came. It had been intended to engage forty, but in the end only thirty-eight women who could conceivably be considered suitable presented themselves. 'I wish people who may hereafter complain of the women selected could have seen the set we had to choose from,' wrote Mary Stanley to Liz Herbert in October 1854. 'All London was scoured for them. ... We felt ashamed to have in the house such women as came. One alone expressed a wish to go from a good motive. Money was the only inducement.' 'As to that stuff about the "enthusiasm" of the nursing in the Crimean campaign – that is all bosh,' wrote Miss Nightingale to Sir John McNeil in 1867; 'we had, unfortunately for us, scarcely one woman sent out who was even up to the level of a head nurse.' The nurses were to receive 12s. to 14s. a week with board, lodging, and uniform. After three months' good conduct they received 16s. to 18s., and after a year's good conduct 18s. to 20s. The average wage of a nurse in a London hospital was 7s. to 10s. Each nurse signed an agreement submitting herself absolutely to Miss Nightingale's orders. Misconduct with the troops was to be punished by instant dismissal. A nurse invalided home was to have her expenses paid first class, but one sent home for misconduct must travel third class on salt rations. No young women were accepted, the majority being stout elderly old bodies. Miss Nightingale wrote later from Scutari that in future 'fat drunken old dames of fourteen stone and over must be barred, the provision of bedsteads is not strong enough.' A uniform dress was provided, but each nurse brought with her underclothing, four cotton nightcaps, one cotton umbrella and a carpet bag. No coloured ribbons or flowers were allowed. No nurse in any circumstances was to go out alone or with only one other nurse. She must either be with the housekeeper from Harley Street, Mrs Clark, or with three other nurses. In no circumstances was any nurse to go out without leave. Strong liquor was permitted in moderate quantities. At dinner each nurse was to be allowed one pint of porter or one pint of ale, at supper half a pint of porter, or half a pint of ale, or one-glass of Marsala or one ounce of brandy.

Lady Canning and Lady Cranworth kept a 'large and melancholy' book in which they recorded the particulars of each applicant. Candidates came from the humblest class – 'Maid-of-all work', 'Very poor', 'Has been for a few days in St George's Hospital'. Subordinate clerks in Government service signed testimonials recklessly. 'Many,' wrote Miss Nightingale, 'were (undisguisedly) sent out as paupers to be provided for, who could not otherwise gain their living.'

Fourteen professional nurses who had experience of serving in hospitals were engaged; the remaining twenty-four were all members of religious institutions. The party was non-sectarian; nurses, insisted Miss Nightingale, were to be selected 'with a view to fitness and without any reference to religious creed whether Roman Catholic nuns, Dissenting Deaconesses, Protestant Hospital nurses or Anglican sisters.'

With the assistance of Manning it was arranged that ten Roman Catholic nuns, five from a convent in Bermondsey and five from an orphanage in Norwood, should join the party, and it was conceded that they should be completely under Miss Nightingale's control. If she were to weld this heterogeneous, undisciplined collection of women into an efficient instrument, she must have absolute and unquestioned authority; her word must be law; a nun or a sister nursing for Miss Nightingale must take her nursing orders from Miss Nightingale and not from her mother superior; and the mother superior must take her nursing orders from Miss Nightingale and not from the bishop.

It was an extraordinary concession for Manning to have obtained, and, as far as the original nuns were concerned, it worked with perfect smoothness. The five Norwood nuns, though amiable, proved inexperienced, but the five from Bermondsey were very nearly the most valuable members of the party. Their superior, known as Rev. Mother Bermondsey, became one of Miss Nightingale's dearest friends.

Three other religious bodies were approached for nurses. St John's House, a High Church sisterhood in Blandford Square, Miss Sellon's Anglican sisterhood, known as the Sellonites, in Devonport, and an Evangelical body, the Protestant Institution for Nurses, in Devonshire Square.

The Sellonites agreed to accept Miss Nightingale's authority and sent eight sisters who were especially valuable as they had had experience in nursing cholera in the slums of Plymouth and Devonport during the cholera epidemic of 1853. The authorities of St John's House demurred, but after being visited first by Sidney Herbert and then by the Chaplain-General of the Forces allowed themselves to be persuaded and sent six sisters. The Protestant Institution flatly refused – their nurses were to be controlled only by their own committee. The refusal was unfortunate. As a result, the party contained a preponderance of Roman Catholics and members of the High Church. Out of the thirty-eight nurses, twenty-four were either professed nuns or Anglican sisters. The remaining fourteen, the hospital nurses, were, as Clarkey observed, of no particular religion unless the worship of Bacchus should be revived.

But religious differences were not the only difficulty. Amongst women who were prepared to devote themselves to the sick, there were two totally different conceptions of the functions of a nurse. The hospital nurse, drunken, promiscuous, and troublesome, considered that her function was to tend her patient's sick body and restore him to physical health by carrying out the doctor's orders. The religious orders, sisters and nuns, were neither drunken nor promiscuous, but were apt to be more concerned with the souls of their patients than with their bodies. Since the middle of the eighteenth century the great medieval tradition of nursing among religious orders had decayed. Physical and spiritual were thought incompatible; at one point the sisters of St Vincent de Paul had been forbidden to put diapers on boy babies. Lofty sentiments were encouraged but cleanliness was ignored. 'Excellent self devoted women,' wrote Miss Nightingale of certain nuns, 'fit more for heaven than a hospital, they flit about like angels without hands among the patients and soothe their souls while they leave their bodies dirty and neglected.' This conception was not held only by religious orders. It was shared by a number of educated women who spent much of their time among the sick, but described themselves not as nurses but 'ladies'.

Miss Nightingale refused to admit 'ladies', as such, into her party. All must be nurses; all must eat the same food, have the same accommodation, wear the same uniform, except the nuns and sisters, who were allowed to wear their habits. And the uniform was extremely ugly. It consisted of a grey tweed dress, called a 'wrapper', a grey worsted jacket, a plain white cap, and a short woollen cloak. Over the shoulders was worn a holland scarf described as 'frightful', on which was embroidered in red the words 'Scutari Hospital'. There was no time to fit individual wearers; various sizes were made up and issued as they came in, with unhappy results. Small women got large sizes; tall women got small. That a 'lady' could be induced to appear in such a get-up was certainly a triumph of grace over nature, wrote one of the nuns. The uniform had not been designed to make the wearer look attractive. Scutari was a disorderly camp, teeming with drink-shops, prostitutes, and idle troops, and a distinguishing dress was necessary for the nurses' protection. A Crimean veteran told Sir Edward Cook that he saw a nurse seized by a soldier in the street of Scutari, but the man's mate recognized the uniform. 'Let her alone,' he said, 'don't you see she's one of Miss Nightingale's women.'

Before Miss Nightingale left England, she called again on Dr Andrew Smith. He was jocose. The ladies, he assured her, would undoubtedly be a comfort to the men. Ladies had finer instincts; they might, for instance, see a spot on a sheet where a mere man might easily overlook it. As for medical duties – well, he did not think Miss Nightingale and her nurses could possibly go wrong in administering a nice soothing drink of capillary syrup to any man who seemed uncomfortable. She contemplated taking a quantity of stores, but he assured her stores were unnecessary. There was now a positive profusion of every kind of medical comfort at Scutari.

On Saturday morning, October 21, 1854, the party left London Bridge to travel *via* Boulogne to Paris. One night was to be spent in Paris and four nights in Marseilles, where in spite of the assurances of Sidney Herbert and Dr Andrew Smith Miss Nightingale intended to buy a large quantity of miscellaneous provisions and stores. Uncle Sam was to go as far as Marseilles to assist her. From Marseilles the party were to proceed to Constantinople in a fast mail boat, the *Vectis*.

Among the Nightingale papers is preserved a small oblong black notebook, fastened with an elastic band and covered with American cloth. It contains three letters, the only personal papers Miss Nightingale took with her to Scutari. One from Fanny bestowed on her the maternal blessing she had so long sought in vain; one from Manning commended her to the Protection, Worship, and Imitation of the Sacred Heart; the third was from Richard Monckton Milnes – 'I hear you are going to the East,' he wrote. ' ... you can undertake that, when you could not undertake me.'

The party reached Boulogne at dinner-time and were given an ovation. Many of the fisherwives of Boulogne had sons and brothers in the French Army, and they seized and shouldered the baggage and carried it in triumph to the hotel, refusing to accept payment. The landlord placed his establishment at the disposal of the party, desired them to order what they would for dinner, and refused to be paid.

The ladies would not sit with the nurses at dinner, though the nurses were in difficulty, as they knew no French. Miss Nightingale waited on the nurses and ate with them herself. 'We never had so much care taken of our comforts before,' one of them said to her. 'It is not people's way with us.'

The party arrived at the Gare du Nord, Paris, at 10 P.M., and was welcomed by an enthusiastic crowd and cheered on the way to the hotel where M. Mohl had arranged rooms and supper. Uncle Sam, writing to Embley, described Florence and Mrs Bracebridge going from room to room trying to fit the party in, followed by Mr Bracebridge, who, carrying a large box with all the cash in it under his arm, was highly excited, constantly interrupting Florence with exclamations and irrelevant reminiscences, and reproaching her for being so confoundedly silent. Mr Bracebridge was followed in turn by M. Mohl, who implored him to come downstairs and eat his supper like a good boy.

Miss Nightingale had hoped to add to her party some Sisters of Charity of British nationality from the convent of St Vincent de Paul, but permission was refused.

The next day the party left for Marseilles. In Marseilles Miss Nightingale set about purchasing stores. In her bedroom, but NOT, explained Uncle Sam, at bedtime, she received a motley crowd of merchants, shopkeepers, dealers, officials from the French Government and the British Consulate, army officers, *The Times* correspondent, and a Queen's Messenger 'with the same serenity as in a drawing-room'. She was looking handsomer than ever, he noted, and the impression she created was extraordinary.

On October 27 the party sailed in the *Vectis*. She was a horrible ship, built for carrying fast mails from Marseilles to Malta, infested with huge cockroaches and so notorious for her discomfort that the Government had difficulty in manning her. Miss Nightingale, a wretchedly bad sailor, was prostrated by sea-sickness. On the second day out the *Vectis* ran into a gale. The guns with which she was armed had to be jettisoned; the steward's cabin and the galley were washed overboard. Miss Nightingale suffered so severely that when Malta was reached she was too weak to go ashore.

The rest of the party went sightseeing in the charge of a major of militia. The party was made up partly of Anglican sisters in black serge habits, partly of Roman Catholic nuns in white habits, and partly of hospital nurses. The hospital nurses were placed in the middle where they would have no chance to misbehave, and the major marched the party from point to point in military formation. The major would shout, 'Forward black sisters,' and the Anglican sisters in their black serge habits got into motion; but the white nuns would straggle, and there came a shout, 'Halt! Those damned white sisters have gone again.' Malta was full of idle troops, and soon the party was followed by a crowd of soldiers. One of the Anglican sisters heard a sergeant remark that he should think 'them ancient Amazons we read about took a deal of drilling.'

On November 3, still in atrocious weather, the *Vectis*, 'blustering, storming, shrieking,' wrote Miss Nightingale, rushed up the Bosphorus, and anchored off Seraglio Point next day. Constantinople, in the pouring rain, looked like a washed-out daguerreotype. On the opposite shore stood the enormous Barrack Hospital. Everyone was on deck eager to see their goal. 'Oh, Miss Nightingale,' said one of the party, 'when we land don't let there be any red tape delays, let us get straight to nursing the poor fellows!' Miss Nightingale, gazing at the gigantic pile, replied: 'The strongest will be wanted at the wash tub.'

At breakfast-time the *Vectis* anchored, and during the morning Lord Stratford, the British Ambassador at Constantinople, sent across Lord Napier, the Secretary of the Embassy. Lord Napier found Miss Nightingale, exhausted from the effects of prolonged sea-sickness, stretched on a sofa. Fourteen years later he recalled their first meeting: ' ... I was sent by Lord Stratford to salute and welcome you on your first arrival at Scutari ... and found you stretched on the sofa where I believe you never lay down again. I thought *then* that it would be a great happiness to serve you.'

The nurses were to go to the hospital at once, for wounded were expected from the battle of Balaclava, fought on October 25. Painted caïques, the gondola-like boats of the Bosphorus, were procured, the nurses were lowered into them with their carpetbags and umbrellas, and the party was rowed across to Scutari.

The rain having ceased, a few fitful gleams of sunshine lit up the Asian shore, which, as it grew clearer, lost its beauty. The steep slopes to the Barrack Hospital were a sea of mud littered with refuse; there was no firm road, merely a rutted, neglected track. As the caïques approached a rickety landing-stage, the nurses shrank at the sight of the bloated carcass of a large grey horse, washing backward and forward on the tide and pursued by a pack of starving dogs, who howled and fought among themselves. A few men, limping and ragged, were helping each other up the steep slope to the hospital, and groups of soldiers stood listlessly watching the dead horse and the starving dogs. A cold wind blew. Some wretched-looking women shivered in tawdry finery.

The nurses disembarked, climbed the slope, and passed through the enormous gateway of the Barrack Hospital, that gateway over which Miss Nightingale said should have been written 'Abandon hope all ye who enter here.' Dr Menzies and Major Sillery, the Military Commandant, were waiting to receive them. That night Lord Stratford wrote to the Duke of Newcastle: 'Miss Nightingale and her brigade of nurses are actually established at Scutari under the same roof with the gallant and suffering objects of their compassion.'

8

From the European shore of the Bosphorus, from the magnificent house where the British Ambassador lived, the great quadrangle of the Turkish Barracks glimmered golden, magnificent as a giant's palace, but at close quarters romance vanished. Vast echoing corridors with floors of broken tiles and walls streaming damp, empty of any kind of furniture, stretched for miles. Later Miss Nightingale calculated there were four miles of beds. Everything was filthy; everything was dilapidated. The form of the building was a hollow square with towers at each corner. One side had been gutted in a fire and could not be used. The courtyard in the centre was a sea of mud littered with refuse. Within the vast ramifications of the barracks were a depot for troops, a canteen where spirits were sold, and a stable for cavalry horses. Deep in the cellars were dark and noisome dens where more than 200 women, who had been allowed by an oversight to accompany the army, drank, starved, gave birth to infants, carried on their trade as prostitutes, and died of cholera. 'But it is not a building, it's a town!' exclaimed a new arrival.

To reach the Barrack Hospital meant martyrdom for wounded men. There was no pier, and the rickety landing-stage could only be used by small boats. The men were taken out of the sick transports and lowered into caïques or rowing-boats; after landing they were jolted on stretchers over rough ground up a precipitous slope.

Although so near Constantinople the situation was isolated. The only communication with Constantinople was by boat, and the Bosphorus was swept by sudden storms which cut off all communication for three or four days at a time. At Scutari were the principal cemeteries of Constantinople, but no markets or shops, only a 'profusion of tombs, fountains and weeping willows' – and ample opportunities for drunkenness and vice. As soon as the British Army occupied Scutari, a horde of Jews, Greeks, and Armenians descended. Tents, booths, ramshackle sheds used as drinking-shops and brothels sprang up round the barracks, and spirits of the worst quality were drunk by the troops in enormous quantities. Regiments sent to Scutari rapidly deteriorated, and on one night, out of 2,400 troops stationed in the barracks, 1,400 were reported drunk.

These were obvious drawbacks, but the vast building hid a more fatal secret. Sanitary defects made it a pest house, and the majority of the men who died there died not of the wounds or sickness with which they arrived, but of disease they contracted as a result of being in the hospital.

The catastrophe which destroyed the British Army was a catastrophe of sickness, not of losses in battle. There were two different sicknesses. The troops on the heights before Sebastopol fell sick of diseases resulting from starvation and exposure. When they were brought down to Scutari and entered the Barrack Hospital, they died of fevers resulting from the unsanitary construction of the Barrack Hospital assisted by insufficient food, filth, and overcrowding. The second sickness was the more fatal. When the war was over, it was found that the mortality in each regiment depended on the number of men which that regiment had been able to send to Scutari.

When Miss Nightingale entered the Barrack Hospital on 5 November, 1854, there were ominous signs of approaching disaster, but the catastrophe had not yet occurred. Food, drugs, medical necessities had already run short, the Barrack Hospital was without equipment, and in the Crimea supply was breaking down. Winter was swiftly advancing, and each week the number of sick sent to Scutari steadily increased.

There were men in the Crimea, there were men in Scutari, there were men at home in England who saw the tragedy approach. They were powerless. The system under which the health of the British Army was administered defeated them. The exactions, the imbecilities of the system killed energy and efficiency, crushed initiative, removed responsibility, and were the death of common sense.

Three departments were responsible for maintaining the health of the British Army and for the organization of its hospitals. The Commissariat, the Purveyor's Department, and the Medical Department. They were departments which during forty years of economy had been cut down nearer and still nearer the bone. In 1853, Dr Andrew Smith, the Director-General of the British Army Medical Service, received 1,200 pounds sterling a year and had only twelve clerks to execute the entire administration of his department. The Purveyor's Department had been reduced to a staff of four, and at the outbreak of war it was extremely difficult to find anyone with sufficient experience to send out as a Purveyor-in-Chief. Mr Ward, 'poor old Ward', the Purveyor at Scutari, was over seventy years of age, a veteran not only of the Peninsula but of Walcheren. His staff consisted of two inexperienced clerks and three boys who also acted as messengers. Mr Filder, the Commissary-General, a Peninsula veteran, complained in 1855 that he was expected with three incompetent clerks to conduct and record the whole business of supplying the British Army in the Crimea.

These departments had no standing. Dr Andrew Smith told the Roebuck Committee that it would have been considered impertinence on his part to approach the Commander-in-Chief with suggestions as to the health of the army. A commissary officer did not rank as a gentleman, while the purveyor was despised even by the commissary. Ill-paid, despised, not highly qualified and painfully anxious for promotion, their fear of their superior officers, especially of the military authorities, was abject. It was something, wrote Miss Nightingale, which no one outside the army had any idea of; it was absolutely Chinese. Men of courage, of determination, and of character might have risen above the system, as Dr Alexander, the ablest man in the medical service, rose above it, but such men did not usually choose to become commissariat officers, purveyors, or army surgeons.

The method by which the hospitals were supplied was confused. The Commissariat were the caterers, bankers, carriers, and store-keepers of the army. They bought and delivered the standard daily rations of the men whether they were on duty or in hospital. The bread and the meat used in the hospitals, the fuel burned there were supplied by the Commissariat. But the Commissariat did not supply food for men too ill to eat their normal rations. At this point the Purveyor stepped in. All invalid foods, known as 'medical comforts', sago, rice, milk, arrowroot, port wine, were supplied by the Purveyor. But though these comforts were supplied to the hospital by the Purveyor, he did not obtain them: all the Purveyor's contracts were made by the Commissariat. The Purveyor never dealt directly with his merchant and had no power over him. If goods were unsatisfactory, the Purveyor could only complain to the Commissariat. Though the standard daily rations of the men in hospital were bought and delivered by the Commissariat, it was the Purveyor who cooked and distributed them. Yet he had no authority over their price, suitability, or quality, having to accept what the Commissariat sent unless he could claim the consignment was unfit for human consumption. Mr Filder, the Commissary-General, cross-examined by the Roebuck Committee, said with heat that for his part he never had understood where the duties of the Commissariat ended and the duties of the Purveyor began. Mr Benson Maxwell, an eminent lawyer and a member of the Hospitals Commission, declared that though he had spent some weeks in the hospitals he was perfectly unable to disentangle the respective duties of Commissariat and Purveyor.

Relations between the doctors and the Purveyor were even more obscure. A doctor might order a man a special diet, but it depended on the Purveyor whether the patient received it or not. Having made a requisition on the Purveyor, the doctor was powerless. Dr Andrew Smith stated before the Roebuck Committee that he could not say what his position was with regard to the Purveying Department. If he made a complaint, the Purveyor told him it was not his province. 'Then,' said Dr Smith, 'I must go to the War Office and get them to carry out what I ought to have the power to carry out.' He was asked: 'Has this uncertainty with regard to the power of providing necessaries and comforts in the hospitals been in existence ever since you have been at the Medical Board?' 'Yes, and long before that.'

Though the system placed executive power in the hands of the Commissariat and the Purveyor, it was only a limited power. Certain goods only might be supplied. Each department had a series of 'warrants' naming definite articles. 'The Purveyor,' wrote Miss Nightingale, 'only gives such amounts of articles as are justifiable under his "warrants", by which he is governed, and is not responsible for those wants of the soldier in hospital which are in excess of the warrants, whatever may be the evidence before him, either in the requisition of the medical officer or the personal observation which, it would appear, he was bound to make of what was close under his eyes.'

The result was the extraordinary shortages. When the sick and wounded came down to Scutari from the Crimea, they were in the majority of cases without forks, spoons, knives, or shirts. The regulations of the British Army laid down that each soldier should bring his pack into hospital with him, and his pack contained a change of clothing and utensils for eating. These articles were consequently not on the Purveyor's warrant. But most of the men who came down to Scutari had abandoned their packs after Calamita Bay, or on the march from the Alma to Balaclava, at the orders of their officers. Nevertheless, the Purveyor refused to consider any requisitions on him for these articles.

Officials were trained not to make trouble, not to spend money, never to risk responsibility; and at Scutari, grossly overworked as they were, they were placed in a situation which demanded courage and resource. The system, while it discouraged action, was enormously prolific of forms, requisitions, dockets, cross-checks, authorizations, and reports. In the hospitals at Scutari every requisition, however trifling, had to be checked and counter-signed by two doctors, one of them a senior officer. No medical officer was permitted to use his discretion. The surgeon on duty had to make as many as six different daily records of the 'Diet Roll', the particulars of food and comforts to be consumed by each patient. As soon as a man attained proficiency in his profession and became a first-class surgeon, he spent so much time filling in forms and drawing up reports that the care of the patients was left to inexperienced juniors. Dr Menzies, Senior Medical Officer at the Barrack Hospital, stated that he was so inundated with office work that he had no time to go into the wards. 'It must be admitted,' the Roebuck Committee agreed, 'that he had no time left for what should have been his principal duty, the proper superintendence of these hospitals.'

The Barrack Hospital was the fatal fruit of the system. When the General Hospital was unexpectedly filled with cholera cases and Dr Menzies was abruptly notified that the casualties from the Alma and a further large number of cholera cases were on their way, he was instructed to turn the Turkish Barracks into a hospital. The preparation and equipment of a hospital formed no part of his duties, his task being to instruct the Purveyor. He sent for 'poor old Ward' and told him to prepare the Turkish Barracks for the reception of wounded. He had then, in accordance with the rules of the service, performed his duty. How Mr Ward was to conjure hospital equipment at a moment's notice out of the drink-shops, brothels, and tombs of Scutari, how he was to collect labour to clean the vast filthy building when no labour existed nearer than Constantinople, was not Dr Menzies' concern.

Mr Ward also knew the correct procedure. He had no authority to expend sums of money in purchasing goods in the open market, and in any case many of the articles required were not on his warrant. He requisitioned the Commissariat on the proper forms, the Commissariat wrote on the forms 'None in store', and the matter was closed. The wounded arrived and were placed in the building without food, bedding, or medical attention. At a later date Dr Menzies instructed the Purveyor to issue the men shirts. This was not done, and the men continued to lie naked. Dr Menzies was asked by the Roebuck Committee why he had not seen to it that his order was carried out. He replied that it was no part of his duty to see that an order was executed. Having issued the instruction correctly and placed it on record, his duty was done. 'Their heads,' wrote Miss Nightingale in 1855, 'are so flattened between the boards of Army discipline that they remain old children all their lives.'

The destruction of the British Army in the Crimean campaign was materially assisted by the attitude of his officer to the private soldier. Savage physical suffering was endured by officers and men alike, and the officers were courageous, stoical, physically tough – Sir George Brown, who commanded the Light Division, had had his arm cut off in the Peninsula and had been thrown on some straw in the bottom of a cart; Lord Raglan had had his arm amputated without an anaesthetic after Waterloo and had called out: 'Here, bring that arm back, there is a ring my wife gave me on the finger.' But officers regarded the men they commanded as denizens of a different world.

The private soldier of 1854 did not bear a good character. The young man who was the disgrace of his village, the black sheep of the family, enlisted. The Duke of Wellington described his army, the army which won the victories of the Peninsula and Waterloo, as 'the scum of the earth enlisted for drink'. Officers had no feeling of responsibility toward their men. 'During the time I have been in the Crimea, that is since the landing ... no general officer has visited my hospital nor, to my knowledge, in any way interested himself about the sick,' Dr Brush of the Scots Greys wrote to the Hospitals Commission. When it became evident that the army would have to winter before Sebastopol under conditions of appalling severity, a large number of officers threw up their commissions and went home. Many of these, like Lord George Paget, who had taken part in the Charge of the Light

Brigade, were men of unquestioned personal courage. They were astounded when they were cut in their clubs.

Miss Nightingale was told: 'You will spoil the brutes'; she heard the troops described by their officers as 'animals', 'black-guards', 'scum'. And the medical authorities were enraged by what they considered unreasonable demands – clean bedding, soup, hospital clothing were 'preposterous luxuries'. 'Poor old Ward', cross-examined as to the state of the Barrack Hospital by the Hospitals Commission in December 1854, said: 'I served through the whole of the Peninsula War. The patients never were nearly so comfortable as they are here. ... In general the men were without bedsteads. Even when we returned to our own country from Walcheren and Corunna the comforts they got were by no means equal to what they have here.'

The doctors at Scutari received the news of Miss Nightingale's appointment with disgust. They were understaffed, overworked; it was the last straw that a youngish Society lady should be foisted on them with a pack of nurses. Of all Government follies, this was the worst. However, they had no choice but to submit; open opposition would be dangerous, for Miss Nightingale was known to have powerful backing, to be the intimate friend of Sidney Herbert and on friendly terms with half the Cabinet. Opinion was divided as to whether she would turn out a well-meaning, well-bred nuisance or a Government spy. For their part, regimental officers received the news with an indulgent smile. Colonel Anthony Sterling, attached to the Highland Brigade, wrote in November 1854: 'The ladies seem to be on a new scheme, bless their hearts. ... I do not wish to see, neither do I approve of, ladies doing the drudgery of nursing.'

However, on November 5 Miss Nightingale and her party were welcomed into the Barrack Hospital with every appearance of flattering attention and escorted into the hospital with compliments and expressions of goodwill. When they saw their quarters, the picture abruptly changed. Six rooms, one of which was a kitchen and another a closet ten feet square, had been allotted to a party of forty persons. The same space had previously been allotted to three doctors and, in another part of the hospital, was occupied solely by a major. The rooms were damp, filthy, and unfurnished, except for a few chairs. There were no tables; there was no food. Miss Nightingale made no comment, and the officials withdrew. It was a warning, a caution against placing reliance on the flowery promises, the resounding compliments of Stratford Canning, first Viscount Stratford de Redcliffe.

Lord Stratford had been British Ambassador to Constantinople three times and associated with Turkey since 1807. His influence was immense; he was virtually a dictator; his latest 'reign' at Constantinople had lasted, with a two years' intermission, for sixteen years. The Turks called him 'the great Elchi', the great ambassador. Physically, he was extremely handsome, and he prided himself on his presence – 'the thin rigid lips, the majesty of brow of a Canning'. He lived magnificently and travelled with twenty-five servants and seventy tons of plate.

Miss Nightingale described Lord Stratford as bad-tempered, heartless, pompous, and lazy. He loved to consort with kings and emperors; he loved to write bad poems in majestic rhythms and keep his attachés up until the small hours while he read them aloud. He was jealous of his inferiors. 'The Elchi,' wrote Lord Napier, 'would never employ anyone on serious work who was at all near himself, so I spent the best years of my life at a momentous crisis doing nothing.' He was not the man to interest himself in a hospital for common soldiers. In his magnificent palace on the Bosphorus he lived for two years with, said Miss Nightingale, 'the British Army perishing within sight of his windows', and during those two years he visited the hospitals only once, when she 'dragged' him there for a visit of only one and a half hours.

After receiving Sidney Herbert's letter, Lord Stratford informed Dr Menzies that if anything was required for the hospitals both *The Times* fund and public money were available. Dr Menzies was thoroughly alarmed; the suggestion that civilian funds should be used to make good deficiencies in army administration struck him with horror. He refused to admit anything was wrong: as far as present wants extended, the hospitals were satisfactorily supplied, and as for future needs he referred once more to the stores expected from Varna. The Ambassador accepted this assurance. He did not go across and see the hospitals for himself, nor did he send anyone else to inspect them or ask for details. He wrote to Sidney Herbert that there did not appear to be anything required and passed on to a project very near his heart: the subscriptions to *The Times* fund would be difficult to return, and he pressed that they should be devoted to the building of a Protestant Church in Constantinople. Though he was strongly Protestant in sympathies, the project was by no means a

119

religious one. It would be a diplomatic triumph. To have procured permission from the Sultan to build a church of a rival
religion in Constantinople, a Mohammedan city, was a mark of
extraordinary favour, and the building of the church would
immensely increase British prestige.

That night, as Miss Nightingale was calculating how she could
cram her party of forty into five small rooms and a kitchen, Lord
Stratford wrote a flowery letter to the Duke of Newcastle com-
plimenting her on the 'accomplishments' she brought into the
field of charity and venturing to hope that 'much comfort may be
derived by the sick and wounded from that attractive source'.

Fourteen nurses were to sleep in one room, ten nuns in another;
Miss Nightingale and Mrs Bracebridge shared the closet; Mr
Bracebridge and the courier-interpreter slept in the office; Mrs
Clark, who was to be cook, and her assistant must go to bed in
the kitchen. There was one more room upstairs, and the eight
Sellonites must sleep there. They went upstairs, and hurried back.
The room was still occupied – by the dead body of a Russian
general. Mr Bracebridge fetched two men to remove the corpse
while the sisters waited. The room was not cleaned, and there was
nothing to clean it with; it was days before they could get a broom,
and meanwhile the deceased general's white hairs littered the
floor. There was no furniture, no food, no means of cooking food,
no beds. Most of the party prepared to sleep on so-called Turkish
'divans', raised wooden platforms running round the rooms on
which the Turks placed bedding; there was, however, no bedding.
While the nurses and sisters unpacked, Miss Nightingale went
down into the hospital and managed to procure tin basins of
milkless tea. As the party drank it, she told them what she had
discovered.

The hospital was totally lacking in equipment. It was hopeless
to ask for furniture. There was no furniture. There was not even
an operating table. There were no medical supplies. There were
not even the ordinary necessities of life. For the present the
nurses must use their tin basins for everything, washing, eating,
and drinking.

They must be prepared to go short of water. The allowance was
limited to a pint a head a day for washing and drinking, including
tea, and it was necessary to line up in one of the corridors where
there was a fountain to obtain it. Tomorrow the situation would
become worse; a battle at Balaclava had been fought on October

25, and transports loaded with sick and wounded were expected.

The party had to go to bed in darkness, for the shortage of lamps and candles was acute. Sisters and nurses lying on the hard divans tried to console themselves by thinking how much greater were the sufferings of the wounded in the sick transports. The rooms were alive with fleas, and rats scurried beneath the divans all night long. The spirits of all, wrote Sister Margaret Goodman, one of the Sellonites, sank.

The doctors ignored Miss Nightingale. She was to be frozen out, and only one doctor would use her nurses and her supplies. Mr Macdonald told the Hospitals Commission: 'Nurses were offered by Miss Nightingale and not accepted'; and he experienced similar difficulty himself. He had *The Times* fund to spend; the urgency of the need for supplies was tragically evident, but he had the greatest difficulty in 'squeezing out' of the doctors an admission of what was needed. The medical authorities drew together in a close defensive phalanx. Admit failure! Accept help for the army from civilians, from *The Times* under whose attacks the army authorities were smarting! From a high Society miss who happened to be on dining terms with the Cabinet! Their experience of army methods, of confidential reports, told them that the man who consorted with Miss Nightingale or who supplied his wards through *The Times* fund would be a marked man.

She realized that before she could accomplish anything she must win the confidence of the doctors. She determined not to offer her nurses and her stores again, but to wait until the doctors asked her for help. She would demonstrate that she and her party wished neither to interfere nor attract attention, that they were prepared to be completely subservient to the authority of the doctors.

It was a policy which demanded self-control; the party were to stand by, see troops suffer, and do nothing until officially instructed. Though Miss Nightingale could accept the hard fact that the experiment on which she had embarked could never succeed against official opposition, yet she inevitably came into conflict with her nurses.

A day passed, and some stores arrived. She made them sort old linen, count packages of provisions. The hardships of life continued. They stood in the corridor to get their pint of water. They ate out of the tin bowls, wiped them with paper, washed their faces and hands in them, wiped them again and drank tea from them. Discomfort would have been ignored if the sufferings of the wounded had been relieved, but they were not relieved. The cries of the men were unanswered while old linen was counted and mended – this was not what they had left England to accomplish. They blamed Miss Nightingale.

On Sunday, November 6, the ships bringing the wounded from Balaclava began to unload at Scutari. As on other occasions the arrangements were inadequate, and the men suffered frightfully; they were brought up to the hospital on stretchers carried by Turks, who rolled their bleeding burdens about, put the stretchers down with a bump when they needed a rest, and on several occasions threw the patient off. Screams of pain were the accompaniment to the unhappy procession, and Sister Margaret Goodman recorded the case of a soldier who died as a result.

Still Miss Nightingale would not allow her nurses to throw themselves into the work of attending on these miserable victims. She allocated twenty-eight nurses to the Barrack Hospital and ten to the General Hospital a quarter of a mile away. All were to sleep in the Barrack Hospital, and all were to wait. No nurse was to enter a ward except at the invitation of a doctor. However piteous the state of the wounded, the doctor must give the order for attention. She sent her nurses to church to sit through an admirable sermon by the chief Chaplain, Mr Sabin. If the doctors did not choose to employ the nurses, then the nurses must remain idle.

She was also determined to send no nurse into the wards until she knew that nurse could be relied on. The reliability of the nurse was as important to the success of the experiment as the cooperation of the doctors, and for nearly a week the party were kept shut up in their detestable quarters making shirts, pillows, stump-rests, and slings, and being observed by her penetrating eye. The time, sighed one of the English Sisters of Mercy, seemed extremely long.

In any case, no directions had been issued governing the employment of nurses. They were entirely in the hands of the doctors. 'No general order,' wrote Miss Nightingale in 1856, 'ever existed defining the duties of the nurses in the various hospitals to which they were respectively attached. ... The number admitted into each division depended on the medical officer of that Division, who sometimes accepted them, sometimes refused them, sometimes accepted them after they had been refused.'

Miss Nightingale herself rigidly obeyed regulations. On a later occasion she was sitting by the bedside of a man critically ill and found his feet stone cold. She told an orderly to fetch a hot-water bottle. The man refused, saying he had been told to do nothing for a patient without directions from a medical officer. She accepted the correction, found a doctor, and obtained a requisition in proper form.

For weeks she stood by in silence while the skill of highly efficient nurses was wasted. 'Our senior medical officer here,' she wrote to Sidney Herbert in January 1855, 'volunteered to say that my best nurse, Mrs Roberts, dressed wounds and fractures more skilfully than any of the dressers or assistant surgeons. But that it was not a question of efficiency, nor of the comfort of the patients, but of the "regulations of the service".'

She was first able to get a footing in the hospital through the kitchen. A state of starvation existed in the Barrack Hospital. According to regulations a private soldier in hospital was placed on what was known as a whole diet, a half-diet, or a spoon diet, the first representing the man's ordinary rations cooked for him by the hospital, the second about half his rations, and the third liquid food. In addition he was supposed to receive 'extra diet', wine, milk, butter, arrowroot, jelly, milk puddings, eggs, etc., as prescribed by the surgeon attending him and procured through the Purveyor.

But to cook anything at the Barrack Hospital was practically impossible. The sole provision for cooking was thirteen Turkish coppers each holding about 450 pints. There was only one kitchen. There were no kettles, no saucepans; the only fuel was green wood. The tea was made in the coppers in which the meat had just been boiled, water was short, the coppers were not cleaned, and the tea was undrinkable. The meat for each ward was issued to the orderly for the ward, who stood in line to receive it from the Purveyor's Department. The Purveyor was understaffed, and when the hospital had 2,500 patients one clerk did all the issues, and the orderlies had to wait an hour or more. When the orderly had the meat, he tied it up, put some distinguishing marks on it, and dropped it into the pot. Some of the articles used by the orderlies to distinguish their meat included red rags, buttons, old nails, reeking pairs of surgical scissors, and odd bits of uniform. The water did not generally boil; the fires smoked abominably. When the cook considered that sufficient time had been taken up in cooking, the orderlies threw buckets of water on the fires to put them out, and the contents of the coppers were distributed, the cook standing by to see that each man got his own joint; the joints which had been dropped in last were sometimes almost raw. The orderly then carried the meat into the ward and divided it up, usually on his bed, and never less than twenty minutes could elapse between taking it out of the pot and serving it. Not only were the dinners always cold, but the meat was issued with bone and gristle weighed in, and some men got portions which were all bone. Those who could eat meat usually tore it with their fingers – there were almost no forks, spoons, or knives. Men on a spoon diet got the water in which the meat had been cooked, as soup. There were no vegetables: only, sometimes, dried peas.

Orderlies cooked extras over fires of sticks in the wards and the courtyard. One of them, Edward Jennings, told the Hospitals Commission on December 14, 1854: 'I boil chickens in an old tin in the ward. I also cook the sago and other things as well as I can ... the doctor does not give me any directions. I cook all the extras and give them to the man at once and he can do what he likes with them. ... I never did anything in the way of cooking until I became an orderly.' The administration of medicines was left to the orderlies, and it was their practice to give the day's medicine in one draught. When wine was ordered, the orderlies drank it themselves. They also ate the rations of men who were ill or asleep. One of the Sellonite sisters saw a young orderly eat up eight dinners.

The food was almost uneatable by men in rude health; as a diet for cholera and dysentery cases it produced agonies. The torture endured by the men when the pangs of hunger were super-imposed on diarrhoea was frightful. 'I have never seen suffering greater,' wrote one observer.

The day after Miss Nightingale arrived she began to cook 'extras'. She had bought arrowroot, wine and beef essences, and portable stoves in Marseilles. On November 6, with the doctors' permission, she provided pails of hot arrowroot and port wine for the Balaclava survivors, and within a week the kitchen belonging to her quarters had become an extra diet kitchen, where food from her own stores was cooked. For five months this kitchen was the only means of supplying invalid food in the Barrack Hospital. She strictly observed official routine, nothing being supplied from the kitchen without a requisition signed by a doctor. No nurse was permitted to give a patient any nourishment without a doctor's written directions.

Cooking was all she had managed to accomplish when, on November 9, the situation completely changed. A flood of sick poured into Scutari on such a scale that a crisis of terrible urgency arose, and prejudices and resentments were for the moment forgotten.

9

It was the opening of the catastrophe. The destruction of the British Army had begun. These were the first of the stream of men suffering from dysentery, from scurvy, from starvation and exposure, who were to pour down on Scutari all through the terrible winter. Over in the Crimea on the heights above Sebastopol the army was marooned, as completely as if on a lighthouse. Thousands of men possessed only what they stood up in. After the landing at Calamita Bay and after the battle of the Alma, when the troops were riddled with cholera and the heat was intense, the men had, by their officers' orders, abandoned their packs.

Seven miles below the heights lay Balaclava, the British base. There had been one good road, the Woronzoff road, but the Russians had gained possession of it in the battle of Balaclava on October 25. There remained a rough track. The weather was still moderately good, but the track was not metalled and put into order before the winter. Men to carry out the work were non-existent. There was no native labour to be hired in this deserted spot. There were no tools. Above all, there was no transport. The army was still without wagons or pack animals.

Balaclava had become a nightmare of filth. Lord Raglan had been attracted by its extraordinary harbour, a land-locked lagoon, calm, clear, and almost tideless, so deep that a large vessel could anchor close inshore. But Balaclava was a fishing village of only 500 inhabitants, a single street of white vine-wreathed houses clinging to a precipitous ravine. No steps were taken to inspect Balaclava before it was occupied or to keep it in a sanitary condition. The army which marched in was stricken with cholera, and within a few days the narrow street had become a disgusting quagmire. Piles of arms and legs amputated after the battle of Balaclava, with the sleeves and trousers still on them, had been thrown into the harbour and could be seen dimly through the water. Bodies of dead men rose suddenly and horribly out of the mud to the surface. Anchor chains and cables were fouled by limbs and trunks. The surface of the once translucent water was covered with brightly coloured scum, and the whole village smelled of sulphuretted hydrogen.

On November 5 the Russians had attacked at Inkerman, on the heights above Sebastopol. In a grim battle fought in swirling fog the British were victorious. But victory was not reassuring. The British troops were exhausted; their commanders were shaken by the revelation of Russian strength. It was evident that Sebastopol would not fall until the spring.

And now an ominous incident occurred. A Mr Cattley was attached to the British Army as chief interpreter. Mr Cattley knew the Crimea well, and he sent in his resignation. He saw a great disaster ahead. The British Army was going to winter on the heights before Sebastopol, and the British Army was not only totally destitute of supplies but without the means of being able to transport supplies should they ultimately arrive. Moistened by the dews of autumn, and churned by the wheels of heavy guns, the rough track from Balaclava to the camp had become impassable. Mr Cattley wrote to Lord Raglan warning him that winter was near, that the climate of the Crimea was subject to sudden and terrifying changes, and tendering his resignation. Lord Raglan made light of the warning and besought Mr Cattley to withdraw his resignation. He did so and stayed to die in 1855.

The weather changed rapidly, icy winds blew – and the troops on the heights above Sebastopol had no fuel. Every bush, every stunted tree was consumed, and the men clawed roots out of the sodden earth to gain a little warmth. As it grew colder, they had to live without shelter, without clothing, drenched by incessant driving rain, to sleep in mud, to eat hard dried peas and raw salt meat. The percentage of sickness rose and rose, and the miserable victims began to pour down on Scutari. The authorities were overwhelmed. The first transports were not even expected. Through an oversight, notification that they had sailed was received only half an hour before the sick and wounded began to land. Utter confusion resulted, official barriers were swept away, and everyone was pressed into service. The Hon. and Rev. Sidney Godolphin Osborne, a personal friend of Sidney Herbert, had come out as a volunteer to act as chaplain to the troops in hospital, and had been cold-shouldered by the authorities; now he found himself assisting at operations. Mr Augustus Stafford, M.P., who had come to Scutari to investigate the hospitals privately, and had had difficulty even in obtaining admission, had a saucepan thrust into his hand and was asked to go down to the wretched pier to pour some kind of warm stimulant down the throats of men writhing in agony. 'Everyone helped,' he told the Roebuck Committee, 'the official people were assisting as much as possible but the number of official people was too small and the arrival was so great, a flood of sick came upon them, bursting in so suddenly that the means of the hospital were not able to meet it.'

It was Miss Nightingale's opportunity – at last the doctors turned to her. Her nurses dropped their sorting of linen and began with desperate haste to seam up great bags and stuff them with straw. These were laid down not only in the wards but in the corridors, a line of stuffed sacks on each side with just room to pass between them.

Day after day the sick poured in until the enormous building was entirely filled. The wards were full; the corridors were lined with men lying on the bare boards because the supply of bags stuffed with straw had given out. Chaos reigned. The doctors were unable even to examine each man. Mr Sabin, the head Chaplain, was told that men were a fortnight in the Barrack Hospital without seeing a surgeon. Yet the doctors, especially the older men, worked 'like lions' and were frequently on their feet for twenty-four hours at a time. 'We are lucky in our Medical Heads,' Miss Nightingale wrote to Dr Bowman on November 14. 'Two of them are brutes and four are angels – for this is a work which makes angels or devils of men. .... As for the Assistants, they are all cubs and will, while a man is breathing his last under the knife, lament the "annoyance of being called up from their dinners by such a fresh influx of wounded". But unlicked cubs grow up into good old Bears, tho' I don't know how, for certain it is the old Bears are good.'

The filth became indescribable. The men in the corridors lay on unwashed rotten floors crawling with vermin. As the Rev. Sidney Godolphin Osborne knelt to take down dying messages, his paper became covered thickly with lice. There were no pillows, no blankets; the men lay, with their heads on their boots, wrapped in the blanket or greatcoat stiff with blood and filth which had been their sole covering perhaps for more than a week. There were no screens or operating tables. Amputations had to be performed in the wards in full sight of the patients. Mr Osborne describes the amputation of a thigh 'done upon boards put on two trestles. I assisted ... during the latter part of the operation the man's position became such from want of a table he was supported by my arm underneath, a surgeon on the other side grasping my wrist.' One of Miss Nightingale's first acts was to procure a screen from Constantinople so that men might be spared the sight of the suffering they themselves were doomed to undergo.

She estimated that in the hospital at this time there were more than 1,000 men suffering from acute diarrhoea and only twenty chamber pots. The privies in the towers of the Barrack Hospital had been allowed to become useless; the water pipes which flushed them had been stopped up when the Barracks were used for troops, and when the building was converted into a hospital they had never been unstopped. Mr Augustus Stafford said there was liquid filth which floated over the floor an inch deep and came out of the privy itself into the ante-room. He told the Roebuck Committee: 'The majority of the cases at the Barrack Hospital were suffering from diarrhoea, they had no slippers and no shoes, and they had to go into this filth so that gradually they did not trouble to go into the lavatory chamber itself.' Huge wooden tubs stood in the wards and corridors for the men to use. The orderlies disliked the unpleasant task of emptying these, and they were left unemptied for twenty-four hours on end. In this filth lay the men's food – Miss Nightingale saw the skinned carcase of a sheep lie in a ward all night. 'We have Erysipelas, fever and gangrene,' she wrote, '... the dysentery cases have died at the rate of one in two ... the mortality of the operations is frightful. ... This is only the beginning of things.' By the end of the second week in November the atmosphere in the Barrack Hospital was so frightful that it gave Mr Stafford the prevailing disease of diarrhoea in five minutes. The stench from the hospital could be smelled *outside* the walls.

A change came over the men, said Mr Macdonald. The classification between wounded and sick was broken down. The wounded who had been well before began to catch fevers, 'gradually all signs of cheerfulness disappeared, they drew their blankets over their heads and were buried in silence.'

Fate had worse in store. On the night of November 14 it was noticed that the sea in the Bosphorus was running abnormally high, and there was a strange thrumming wind. Within a few days news came that the Crimea had been devastated by the worst hurricane within the memory of man. Tents were reduced to shreds, horses blown helplessly for miles, buildings destroyed, trees uprooted. The marquees which formed the regimental field hospitals vanished, and men were left half buried in mud without coverings of any kind. Most serious of all, every vessel in Bala-clava harbour was destroyed, amongst them a large ship, the *Prince*, which had entered the harbour the previous day loaded with warm winter clothing and stores for the troops.

The hurricane rendered the situation of the army desperate. Such few stores and such little forage as it possessed were destroyed. Winter began in earnest, with storms of sleet and winds that cut like a knife as they howled across the bleak plateau. Dysentery, diarrhoea, rheumatic fever increased by leaps and bounds. More and more shiploads of sick inundated Scutari. The men came down starved and in rags. 'They were without their shoes and their shirts had been thrown away in utter disgust at their filthiness or torn to shreds ... they were swarming with vermin; their trousers were all torn; their coats ragged ... some-times they came down without any coats at all,' said Mr Mac-donald in his evidence before the Roebuck Committee. The men told the nurses to keep away because they were so filthy. 'My own mother could not touch me,' said one man to Sister Margaret Goodman. By the end of November the administration of the hospital had collapsed.

'In the confusion at Scutari,' Mr Augustus Stafford told the Roebuck Committee, 'I was never able to distinguish where one department began and the other ended. ... Whenever I had any-thing to do with the authorities at Scutari, I never met with any-thing but personal courtesy and a wish to reform the evils ... but through all the departments there was a kind of paralysis, a fear of incurring any responsibility, and a fear of going beyond their instructions.'

For instance, Mr Stafford determined to get the lavatories cleaned. He approached Dr Menzies, who said it was none of his business. 'If he had got in 12 or 13 men to clean out the lava-tories he would immediately have been pounced on by another department and told that it belonged to that department.' Mr Stafford then went to Major Sillery, Military Commandant of the Hospital, who freely admitted the urgent necessity of the work, but asked where the money was to come from. He was 'very nervous and anxious, very much distressed and perplexed'. He had no instructions to execute the work, the money would have to be advanced, and he had no security for repayment. Mr Stafford offered to pay himself. Major Sillery was horror-struck and refused. Mr Stafford declared that if the lavatories were not cleaned he would write a letter and have it read aloud in the House of Commons. He then retired to bed with diarrhoea.

In his evidence before the Roebuck Committee Mr Stafford made it clear that he did not blame Major Sillery. 'He was most anxious to do all he could for the improvement and amelioration of the hospitals, but he had no money. He did not consider, neither did I, that he was called on to risk the money which he, as a man deriving his support from his profession would have had to do, if he had advanced this money for the payment of 16 Turks to cleanse the lavatories.'

And then in the misery, the confusion, a light began to break. Gradually it dawned on harassed doctors and overworked officials that there was one person in Scutari who could take action – who had money and the authority to spend it – Miss Nightingale.

She had a very large sum at her disposal derived from various sources and amounting to over £30,000, of which £7,000 had been collected by her personally; and Constantinople was one of the great markets of the world. During the first horrors of November, the gathering catastrophe of December, it became known that whatever was wanted, from a milk pudding to a water-bed, the thing to do was to 'go to Miss Nightingale'.

Each day she ascertained what comforts were lacking in the Purveyor's Store, what articles supply was short of, what requisitions had been made which had not been met. Mr Macdonald then went into Constantinople and bought the goods, which were placed in her store and issued by her upon requisition in the official form by a medical officer. Nothing, with the exception of letter-paper and pencils, was ever given out without an official requisition duly signed. Gradually, Mr Macdonald told the Roebuck Committee, the doctors ceased to be suspicious and their jealousy disappeared.

In one urgent work she met no opposition. Just as it was no one's business to clean the lavatories, so it was no one's business to clean the wards. The first commission Mr Macdonald executed for Miss Nightingale was the purchase of 200 hard scrubbing-brushes and sacking for washing the floors. She insisted on the huge wooden tubs in the wards being emptied, standing quietly and obstinately by the side of each one, sometimes for an hour

131

at a time, never scolding or raising her voice, until the orderlies gave way and the tub was emptied.

Her next step was to wash the men's clothes. Mr Macdonald stated that for five weeks after he arrived at Scutari no washing was done at all. The Purveyor had been instructed to make a laundry contract and had done so with a Greek, who was quite unable to fulfil his obligations; he either failed to wash at all or washed in cold water, and shirts came back as filthy as they were sent, still crawling with lice. The men said they preferred their own lice to other people's and refused to part with their shirts, stuffing them, filthy and vermin-ridden, under their blankets. The total amount of washing satisfactorily accomplished for the vast hospital was seven shirts. Miss Nightingale made arrangements to rent a house outside the barracks and have the washing done by soldiers' wives. She consulted Dr Menzies, telling him she wished to have boilers put in by the Engineers Corps. 'Oh, but that is putting you to a great deal of trouble,' said Dr Menzies. 'I should think the Purveyor would be able to make arrangements.' The boilers were installed and the cost paid out of *The Times* fund.

Within the hospital her principal ally was Dr McGrigor, 1st class Staff Surgeon, a young, energetic man not, she said, wedded in everything to what had been done in the Peninsula. He accepted her nurses and made full use of them.

For a time she tried to work with Lady Stratford. On November 7, two days after her arrival, she wrote to Lord Stratford asking for sheets, shirts, and portable stoves for cooking 'extras'. He sent her Lady Stratford instead. Lady Stratford would not come across to Scutari (she had been in the Barrack Hospital once and the stench had made her sick), nor did she send linen and stoves, but she offered to get anything that was required in Constantinople. Miss Nightingale asked her to obtain twelve wagons to bring heavy goods up to the Barrack Hospital. Next day she looked out and saw drawn up before her quarters seven glass and gilt coaches and five other vehicles, which she had to pay off out of her own private funds. 'This lark of the Ambassadress's,' she wrote, 'cost Miss Nightingale 500 piastres.'

By the end of December Miss Nightingale was in fact purveying the hospital. During a period of two months she supplied, on requisition of Medical Officers, about 6,000 shirts, 2,000 socks, and 500 pairs of drawers. She supplied nightcaps, slippers, plates, tin cups, knives, forks, spoons 'in proportion'. She procured trays, tables, forms, clocks, operating tables, scrubbers, towels, soap, and screens. She caused an entire regiment which had only tropical clothing to be re-fitted with warm clothing purchased by Mr Macdonald in the markets of Constantinople when Supply had declared such clothing unprocurable in the time – Supply was compelled to get all its goods from England. 'I am a kind of General Dealer,' she wrote to Sidney Herbert on January 4, 1855, 'in socks, shirts, knives and forks, wooden spoons, tin baths, tables and forms, cabbages and carrots, operating tables, towels and soap, small tooth combs, precipitate for destroying lice, scissors, bed pans, and stump pillows.'

Before Sebastopol conditions grew steadily worse. The stores lost in the hurricane were not replaced. Men, sick or well, lay in a foot of water in the mud covered only by a single blanket. Every root had been burned, and the men had to eat their food raw; meat stiff with salt and dried peas. Tea was withdrawn and green coffee, needing roasting and pounding, was issued instead, because good results had been obtained from the use of green coffee in the Caffre War. There was no bread. As the percentage of sick climbed and climbed, double turns of duty were thrown on the survivors. Men were in the trenches before Sebastopol for thirty-six hours at a stretch, never dry, never warmed, never fed. The sick were brought down to Balaclava strapped to mule-litters lent by the French – there was no British transport of any kind – naked, emaciated, and filthy. They were universally suffering from diarrhoea, and strapped to the mules they could not relieve themselves. After waiting hours without food or shelter in the icy wind or driving sleet at Balaclava, they were piled on to the decks of the sick transports and brought down to Scutari. And the catastrophe had not yet reached its height.

At the beginning of December, when the Barrack Hospital was filled to overflowing, a letter from Lord Raglan announced the arrival of a further 500 sick and wounded. It was impossible to cram any additional cases into the existing wards and corridors, and Miss Nightingale, supported by Dr McGrigor, pressed to have put in order the wing of the hospital which had been damaged by fire before the British occupation; it consisted of two wards and a corridor and would accommodate nearly 1,000 extra cases. But the cost would be considerable, and no one in the hospital had the necessary authority to put the work in hand. She

133

had been repeatedly assured by Sidney Herbert that Lord Stratford had *carte blanche;* now she applied to him, and Lady Stratford came across to Scutari escorted by a couple of *attachés.* Preferring not to come inside the hospital, she held conferences with the Purveyor and Major Sillery in the courtyard, and 125 Turkish workmen were engaged to repair the wards. After a few days a dispute about the rate of wages arose, and the Turkish workmen struck. Miss Nightingale wrote to Lord Stratford, who denied the slightest knowledge of the business; Lady Stratford withdrew; worried Major Sillery had neither money nor authority. On this Miss Nightingale took matters into her own hands. She engaged on her own responsibility not 125 but 200 workmen, and paid for them partly out of her own pocket and partly out of *The Times* fund. The wards were repaired and cleaned in time to receive the wounded.

Not only did she repair the wards; she equipped them. The Purveyor could provide nothing. 'Orderlies were wanting, utensils were wanting, even water was wanting,' she wrote to Sidney Herbert on December 12, 1854. 'I supplied all the utensils, including knives and forks, spoons, cans, towels, etc. ... and was able to send on the instant arrowroot in huge milk pails (two bottles of port wine in each) for 500 men.' The number of sick and wounded finally received was 800. One of the men described his sensations when he at last got off the filthy sick transport and was received by Miss Nightingale and her nurses with clean bedding and warm food – 'we felt we were in heaven', he said.

The affair caused a sensation. Its fame reached the Crimea and was discussed in Colonel Sterling's mess. He was outraged. 'Miss Nightingale coolly draws a cheque. Is this the way to manage the finances of a great nation? *Vox populi?* A divine afflatus. Priestess Miss N. Magnetic impetus drawing cash out of my pocket.' It was the first important demonstration of what men at Scutari called the 'Nightingale power'. Respect for the 'Nightingale power' was increased when it became known that her action had been officially approved by the War Department and the money she had spent refunded to her.

But to Miss Nightingale these victories were only incidental; she never for a moment lost sight of the fact that the object of her mission was to prove the value of women as nurses. But, unhappily, no difficulties with doctors or purveyors were as wearing or as discouraging as her difficulties with her nurses.

'I came out, Ma'am, prepared to submit to everything, to be put on in every way. But there are some things, Ma'am, one can't submit to. There is the Caps, Ma'am, that suits one face and some that suits another. And if I'd known, Ma'am, about the caps, great as was my desire to come out to nurse at Scutari, I wouldn't have come, Ma'am.' This, Miss Nightingale wrote to Dr Bowman on November 14, 1854, was a specimen of the kind of question which had to be adjusted in the midst of appalling horror. 'We are,' she wrote, 'steeped up to our necks in blood.' Mrs Roberts from St Thomas's was worth her weight in gold, Mrs Drake from St John's House was a treasure, but most of the other hospital nurses were not fit to take care of themselves. Nurses had to be forbidden to enter any ward which contained men even moderately well, to be forbidden to be in the wards on any pretext after 8 p.m. To convince any of them, nurses or sisters, of the necessity for discipline was almost impossible. Why should a man who desperately needed stimulating food have to go without because the nurse who had the food could not give it to him until she had been authorized by a doctor? It was felt that Miss Nightingale was callous. It was said that she was determined to increase her own power and cared nothing for the sick.

These difficulties came to a head in December in the case of Sister Elizabeth Wheeler, one of the Sellonites. Sister Elizabeth, who was nursing wards of men suffering from diarrhoea and dysentery, saw the men brought in emaciated and in agony from the fearful pangs of hunger superimposed on diarrhoea, and her heart bled for them. She was a nurse of experience, and in her opinion the amount of food given to the men was inadequate: she was not allowed sufficient milk, eggs, or port wine. A passionate and emotional woman, she had on several occasions forced her way in to Miss Nightingale and made a scene, demanding larger quantities. Miss Nightingale refused to supply anything except what was ordered by the doctor and written and confirmed on the diet roll. Sister Elizabeth was furiously angry. She wrote letters home describing the fearful state of the wards and accusing the doctors of callousness and inhumanity. Unhappily, one of her relatives passed on a letter to *The Times*. It was published on December 8, 1854, as a letter from a heroic Scutari nurse, and was made full use of in the campaign *The Times* was conducting against the Government. Miss Nightingale was horrified; nothing

135

more unfortunate could have occurred. She was trying to convince the doctors of the complete loyalty of her nurses; here was a complete contradiction; she was trying to weld the nurses into a disciplined band by means of her authority, and here her authority was directly attacked. An investigation took place before the Hospitals Commission in December 1854, Miss Nightingale and Sister Elizabeth both giving evidence. Sister Elizabeth's letter was not correct in its facts, for she had represented herself as nursing the wounded, but she had never nursed surgical cases. She had given a very high number of deaths and conveyed the impression that the deaths had occurred in a single ward when in reality they were the deaths for an entire division of the hospital. Her assertions were held to be inaccurate, and she was asked to resign.

Sister Elizabeth's attitude was common to a large number of Miss Nightingale's nurses. Reluctance to accept her authority and obey her instructions was constant from the beginning to the end of her mission, and many of her nurses heartily disliked her.

However, she had managed to establish herself, and now her nurses were fully occupied. She had also acquired two new and loyal workers in Dr and Lady Alicia Blackwood, who had come out at their own expense after the battle of Inkerman. Dr Blackwood obtained an appointment as Military Chaplain; Lady Alicia applied to Miss Nightingale to know where she could be most useful. After a few seconds of silence, and with a peculiar expression of countenance, Miss Nightingale said: 'Do you mean what you say?' Lady Alicia was rather surprised. 'Yes, certainly, why do you ask me that?' 'Oh! because I have had several such applications before, and when I have suggested work, I found it could not be done, or some excuse was made; it was not exactly the sort of thing that was intended, it required special suitability, etc.' 'Well, I *am* in earnest,' said Lady Alicia. On this Miss Nightingale asked her to be responsible for the wretched women who had been allowed to accompany the army and had been sent down from Varna. More than 260 women and infants were living in dark cellars beneath the Barrack Hospital; soldiers' wives, widows, and prostitutes were crowded together, men from the Depot were forced to live with their wives in a room containing fifty or sixty other persons, a soil pipe drained into the corner of one cellar, drinking was incessant and the place was a pandemonium of drunkenness, cursing, and swearing. Lady Alicia removed the more respectable among the women, setting them to work in the laundry, and began a lying-in hospital. The children were separated from the adults, and a system of doling out food through Mr Bracebridge was adopted which, Miss Nightingale said, became the curse of the hospital. Thirty-six women and thirty-six infants under the age of three months howled together daily outside Mr Bracebridge's door – the Turks called them his thirty-six wives.

On December 14 she wrote Sidney Herbert a cheerful letter:

What we may be considered as having effected:

(1) The kitchen for extra diets now in full action.
(2) A great deal more cleaning of wards, mops, scrubbing brushes, brooms and combs given out by ourselves.
(3) 2000 shirts, cotton and flannel, given out and washing organized.
(4) Lying-in hospital begun.
(5) Widows and soldiers' wives relieved and attended to.
(6) A great amount of daily dressing and attention to compound fractures by the most competent of us.
(7) The supervision and stirring-up of the whole machinery generally with the concurrence of the chief medical authority.
(8) The repairing of wards for 800 wounded which would otherwise have been left uninhabitable. (And this I regard as the most important.)

She never wrote quite so cheerfully again. On December 14 she suddenly discovered, through being shown a letter written by Liz Herbert to Mrs Bracebridge, that a party of nurses numbering no fewer than forty had left London under the leadership of Mary Stanley and were actually due to arrive at Scutari the next day.

She had not been consulted or informed, and the despatch of the party was in direct contravention of her agreement with Sidney Herbert. There was an even more serious aspect; at this critical moment, when she was struggling with difficulties caused by Sister Elizabeth and her authority was being questioned, the party was consigned not to her but to Dr Cumming; she had been publicly passed over. The significance of this action was not lost on either her friends or her enemies at Scutari, and Sidney Godolphin Osborne wrote that he feared the Nightingale ministry seemed to be coming to an end.

She was furiously angry, and on December 15 she wrote Sidney Herbert a scathing letter:

DEAR MR HERBERT,

When I came out here as your Supt. it was with the distinct understanding (expressed both in your own hand writing and in the printed announcement which you put in the *Morning Chronicle* which is here in everyone's hands) that nurses were to be sent out at my requisition only, which was to be made only with the approbation of the Medical Officers here. You came to me in great distress and told me you were unable for the moment to find any other person for the office and that, if I failed you, the scheme would fail.

I sacrificed my own judgement and went out with forty females, well knowing that half that number would be more efficient and less trouble, and that the difficulty of inducing forty untrained women, in so extraordinary a position as this (turned loose among 3000 men) to observe any order or even any of the directions of the medical men, would be Herculean.

Experience has justified my foreboding. But I have toiled my way into the confidence of the medical men. I have, by incessant vigilance, day and night, introduced something like order into the disorderly operations of these women. And the plan may be said to have succeeded in some measure, *as it stands*. ...

At this point of affairs arrives at *no one*'s requisition, a fresh batch of women, raising our number to eighty-four.

You have sacrificed the cause, so near my heart. You have sacrificed me – a matter of small importance now – you have sacrificed your own written word to a popular cry. ...

The quartering them *here* is a physical impossibility, the employing them a moral impossibility.

You must feel I ought to resign, where conditions are imposed on me which render the object for which I am employed unattainable – and I only remain at my post until I have provided in some measure for these poor wanderers. You will have to consider where they are to be employed, at Malta, Therapia or elsewhere or whether they are to return to England – and you will appoint a Superintendent in my place until which time I will continue to discharge its duties as well as I can.

Believe me, dear Mr Herbert,

yours very truly,
FLORENCE NIGHTINGALE.

She had not done with him. She added a stinging postscript.

P.S. Had I had the enormous folly to write at the end of eleven days experience to require more women, would it not seem that you, as a Statesman, should have said 'Wait until you can see your way better.' But I made no such request. The proportion of Roman Catholics which is already making an outcry you have raised to 25 in 84. Dr Menzies has declared that he will have two only in the General Hospital – and I cannot place them here in a greater proportion than I have done without exciting the suspicion of the Medical Men and others.

In order that Sidney Herbert should not imagine she had written hastily, she completed the letter by a statement at the foot: 'Written 15 December. Posted 18 December.'

On Sidney Herbert's part there was honest misunderstanding. He was harried, in poor health, and almost worked to death. In the confused, unbalanced mind of Mary Stanley there was a mixture of religious fervour – she had secretly determined to become a Roman Catholic before she started for Scutari – and jealousy. The third person concerned, Liz Herbert, was prone to act emotionally and incalculably, to be easily swept into indiscretions; and she, like Mary Stanley, was blinded by religious fervour; though Liz Herbert did not become a Roman Catholic in her husband's lifetime, she was received into the Roman Church after his death.

Mr Bracebridge spoke angrily of 'Popish plots', and that, Miss Nightingale said, was ridiculous. Yet behind the unreliable fervours of Mary Stanley and the easily persuaded emotionalism of Mrs Herbert was the formidable figure of Manning, who wished to focus on the nuns of his church the fame and the glory which surrounded the Scutari nurses. He had no animus against Miss Nightingale; they remained friends, and she said on several occasions that he had treated her fairly; but the arrival of Mary Stanley's party dealt her mission a blow from which it never completely recovered. Before the arrival of the newcomers on December 15, she was well on the way to complete success. After it, though she achieved personal triumphs, her authority was not established until her mission was almost ended. Her orders were constantly disobeyed, her right to command questioned, and the original purpose of the undertaking became obscured by a fog of sectarian bickering.

The high percentage of Catholics and High Church Anglicans in her original party had already provoked an outcry. Before she arrived at Scutari, a letter published in the *Daily News* on October 28, signed 'Anti-Puseyite', attacked her for recruiting her nurses from the Sellonites and a 'Romanist establishment'. He quoted Sidney Herbert's letter, which he declared to be animated by party spirit. (Parthe and Fanny had indiscreetly passssed the letter round among their friends, and it had been copied.) Liz Herbert had written to the *Daily News* saying that Miss Nightingale was a member of the established Church of England, having been originally brought up a Unitarian. But the storm continued.

'Protestant Churchman' and 'Bible Reader' wrote to the *Standard* to denounce her as an 'Anglican Papist'. Dark references were made to 'Anglo-Catholic ladies at the War Office', 'Jesuit conspiracies', and the activities of 'the pervert Manning'. One parson went so far as to caution his parishioners against sending any help to Scutari through a party composed of female ecclesiastics and Romish nuns instead of common-sense nurses.

If Mary Stanley had publicly announced her intention of joining the Roman Catholic Church, Sidney Herbert would not have allowed her to go to Scutari. But she kept it a secret and took with her Mother Frances Bridgeman of Kinsale, an Irish nun of ardent and rebellious temperament who openly avowed she intended to execute a spiritual as well as a medical mission.

It was Miss Nightingale's fate to be attacked by both sides, to have to endure what she called the 'Protestant Howl' and the 'Roman Catholic Storm'. She belonged to a sect which, as the Dean of Elphin phrased it, is unfortunately a very rare one, the sect of the Good Samaritan.

Mary Stanley had been instrumental in collecting the first party and could not have organized the second in good faith. Its constitution was contrary to Miss Nightingale's rules. The fifteen Irish nuns considered they were under no obligation to obey anyone but their Superior, Mother Bridgeman, and Mother Bridgeman acknowledged only the authority of her Bishop. The fact that the party was consigned to Dr Cumming proved a clear intention to evade Miss Nightingale's authority. The party consisted of 9 ladies, 15 nuns, and 22 nurses, 46 in all. It had been hastily collected. Many of the 'hired nurses' were ludicrously without experience, one old woman, Jane Evans, having spent her life looking after pigs and cows. Out of the whole 46 no fewer than 20 had come out with the intention not of nursing but of being 'assistant ecclesiastics'. Miss Stanley led the party, which was escorted by Dr Meyer, a physician, and the Honourable Jocelyn Percy, M.P., who had conceived a passionate admiration for Miss Nightingale, and left a life of ease and luxury with the object of becoming her fag and her footman. They travelled, like the first party, via Paris and Marseilles, but Mary Stanley had large ideas. A courier went ahead and took rooms for them in the dearest hotel in Paris, and Clarkey was asked to procure a dozen bottles of the best vinegar aromatique in case the nurses and ladies encountered bad smells.

The journey was discouraging. The 'hired nurses' got drunk in the train and horrified the ladies by the vulgar peals of laughter which came from their carriage; one or two were drunk at dinner, several collapsed and revealed they suffered from delicate health, and one appeared in an array of rings and brooches. Mary Stanley's spirits sank. Writing a character of each of her party to Liz Herbert, Mary Stanley admitted that the women chosen were too old; perhaps a closer inquiry should have been made into their antecedents and characters. Among the few 'quiet sensible' women was Miss Polidori, the aunt of Dante Gabriel Rossetti.

On December 15, between three and four in the afternoon, the *Egypt* anchored outside Constantinople. Mr Bracebridge went on board and advised the party not to disembark. There literally was not a vacant corner in Scutari: Miss Nightingale's quarters were already accommodating forty in space adequate for three, and food, water, and fuel were seriously short; nurses required the strictest supervision, and Miss Nightingale could not deal with any more. Dr Meyer and Mr Percy were taken aback, and it was agreed that the nurses and ladies should remain on the boat while the gentlemen went to report the arrival of the party to Dr Cumming. He rebuffed them, declining flatly to employ the nurses and ladies in the hospitals. He could not, even if he had been willing, find them any accommodation at Scutari, which was crammed to overflowing. Lord Stratford had agreed to lend them temporarily a house at Therapia belonging to the Embassy, and there they must go until arrangements could be made to send them home. Next, Dr Meyer and Mr Percy went to Miss Nightingale, who summoned Mr Bracebridge as a witness. The interview was sadly different from anything the romantic Mr Percy had anticipated. She was in a cold fury. She refused to take any responsibility for the party – she had never asked for them, they had come without her consent. Mr Bracebridge took down notes of what passed and later called on the two gentlemen with a memorandum of the conversation which they were requested to sign.

One fact in particular was weighing very heavily on both gentlemen's minds. Owing to the style in which the party had travelled, they had spent the whole of the £1,500 sterling with which they had started and were penniless.

Off they hurried to see the new Military Commandant, Lord William Paulet – Major Sillery had been recalled that week. The Military Commandant of a hospital was all-powerful, and Lord William was implored to assist them. But he could do nothing. It was a physical impossibility to find accommodation for the party in Scutari, and he certainly could not force nurses on doctors who did not want them. There was nothing for it but to retire. Dr Meyer and Mr Percy remained behind in Constantinople trying to find employment and accommodation for the party. Mary Stanley and the other women went to the Embassy house at Therapia, where squabbles immediately broke out between the nurses and ladies.

By the end of December the need for money had become urgent. Dr Meyer and Mr Percy had failed to make any arrangements for the nurses in Constantinople, and Lord and Lady Stratford, though 'kindness itself', did not advance any money. Dr Cumming was applied to, but he refused to make any advance except from his own private funds. On the 21st the two gentlemen brought Mary Stanley to interview Miss Nightingale; Dr Cumming and Mr Bracebridge were present, and notes of the conversation were taken down. Mary Stanley explained her plan: ten of the Protestants were to be appropriated as assistants by the chaplains and ten of the nuns by the priests, 'not as nurses but as female ecclesiastics'. Miss Nightingale absolutely refused to countenance the scheme; it was directly contrary to the instructions she had received from the War Office. She denied any responsibility for the party; it had not been consigned to her, and its direction, maintenance, and employment were not her affair.

She then offered to lend Mary Stanley £90 sterling from her own private income for the immediate necessities of the party, and it was unwillingly accepted. Later she lent another £300. That evening Mary Stanley wrote to Liz Herbert that it needed 'all her love for Flo' not to feel hurt at being treated so *officially* and being made to discuss all arrangements before witnesses. She added that she did not think the Herberts need be anxious about Florence; as far as looks and *power* went she had never seen her in greater force.

*

And before Sebastopol the catastrophe steadily grew, and more and still more sick poured down. Four thousand were received in seventeen days between December 17 and January 3, and the death-rate steadily rose. Mr Bracebridge wrote to Sidney Herbert on December 14: 'Flo has been working herself to death, never sits down to breakfast or dinner without interruption: often never dines ... the attempt to do more will kill her ... to-day 200 sick landed looking worse than any others yet.'

Yet, harassed and distracted as she was, when her first anger was over she saw that it would be disastrous to send the party back. Racial and religious issues were involved, and Lord Napier had gone so far as to say that in his opinion there would be almost a rebellion in Ireland if the Irish nuns under Mother Bridgeman were sent home. A scandal would do the cause she had at heart irreparable harm. She must swallow her grievance.

On December 24 she saw Mary Stanley, Mr Percy, and Dr Meyer again, and suggested a compromise. Some of the Irish nuns should be taken at once into Barrack Hospital, and to make room for them the white sisters from Norwood, who were not experienced in hospital nursing, should be sent home. She would write to Manning and make it clear that no blame attached to the Norwood nuns. This arrangement would not increase the number of Roman Catholic nuns in the hospital, which was something Dr Cumming refused to contemplate. She would also endeavour to get some nurses accepted at the new Convalescent Hospitals which were to open in a few weeks. She refused absolutely to have anything to do with the scheme for religious visiting.

Miss Nightingale was, wrote Mary Stanley to Liz Herbert, 'very low. She feels that to employ the women herself is impossible – to send them back to England is to incur universal odium and perhaps mar for ever her future powers of usefulness.'

On Christmas Day she wrote to Sidney Herbert: 'You have not stood by me but I have stood by you. ... All that I said in my letter to you I say *still* more strongly. Please do read it. ... My heart bleeds for you, that you the centre of the Parliamentary row should have to attend to these miseries, tho' you have betrayed me. ... I believe it may be proved as a logical proposition that it is impossible for me to ride through all these difficulties. My caïque is upset ... but I am sticking on the bottom still. But there will be a storm will brush me off.'

The storm burst immediately. Fine weather was never to

return. The sisters from Norwood, bathed in tears, bitterly resented being sent home, and Father Michael Cuffe, the Roman Catholic chaplain, told Miss Nightingale in an angry interview that she was like Herod driving the Blessed Virgin across the desert. 'Pray confirm Father Michael Cuffe in his position here,' she wrote to Sidney Herbert; 'it is the only agreeable incident that I have had.' Mother Bridgeman refused to allow her nuns to enter the Barrack Hospital without her – it would be 'uncanonical'. She declared they must have their own Jesuit chaplain and refused the ministrations of Father Michael Cuffe, who was the official Roman Catholic chaplain. Miss Nightingale was on her feet for twenty hours at a time and dressing wounds and sores for eight hours at a stretch; but instead of rest she had arguments with Mary Stanley and Mother Bridgeman. Loud-voiced, assertive, voluble, Mother Bridgeman, christened by Miss Nightingale 'Rev. Brickbat', was determined to force an entry into the Barrack Hospital with all her fifteen nuns *vi et armis*. Between them, Miss Nightingale wrote to Sidney Herbert on December 27, they were leading her 'the devil of a life'.

If Roman Catholic anger was aroused, so was Protestant suspicion. 'I grieve to say,' wrote Mrs Bracebridge to Liz Herbert on December 28, 'that Miss Stanley's *false position* is already working fearful mischief, she is acting a very double part and is in league with the Revd. Mother Bridgeman of Kinsale to *force* Flo, if she can, to give way and appoint them together to the General Hospital where they will work their proselytizing unmolested.'

Next a Miss Tebbut at the General Hospital was accused by the Evangelicals of circulating improper books in the wards: she had lent a patient a copy of the *Christian Year*. Miss Nightingale herself was once more denounced by both sides. There had been Father Michael Cuffe's denunciation, and now a Protestant writer, getting information of her close friendship with Rev. Mother Bermondsey, gave the alarm of 'Catholic Nuns transferring their allegiance from the Pope to a Protestant Lady.' When the paper containing this article reached Scutari, Sister Mary Gonzaga, a Bermondsey nun and a most efficient nurse, laughingly called Miss Nightingale 'Your Holiness' and she in turn called Sister Mary 'My Cardinal'. She was heard to say, 'I do so want my Cardinal', and a Protestant scandal ran through the hospital.

Protestants and Catholics not only quarrelled with each other

144

but among themselves. Mother Bridgeman refused to meet the Bermondsey nuns, and her Jesuit Chaplain refused the sacrament to them. A Protestant chaplain wrote to the Secretary of State for War denouncing one of the Protestant nurses as a 'Socinian' – a follower of Socinus, who denied the divinity of Christ – and demanded her instant removal. Dr Blackwood, who had strong Evangelical views, was alleged to have preached against the nuns. One of the Irish nuns converted and rebaptized a soldier on his deathbed, and was promptly sent away by Miss Nightingale. 'I do not intend to let our society become a hot bed of R.C. intriguettes,' she wrote to Sidney Herbert.

Charges and counter-charges from clergymen, priests, private persons, doctors, and nurses flew backward and forward between Whitehall and Scutari. Documents were actually placed before the Secretary of State for War, and *The Times*, in an article published on January 9, 1855, commenting on the progress of Miss Nightingale's mission, stated: 'The success of the experiment as a feature of the medical department of the army cannot be considered as decisively established until certain religious dissensions which have arisen are set at rest. ... There is some danger of the whole undertaking coming to an abrupt conclusion.'

The very practice she had stipulated in her interview with Sidney Herbert as one that must at all costs be avoided – the selection of nurses for sectarian reasons and not for their efficiency as nurses – was thrust on her by the composition of the second party. Presbyterians now wrote demanding that 'some Presbyterian nurses' be sent out, and she felt she must acquiesce. When the nurses arrived, two immediately went out with a pair of orderlies and were brought back hopelessly intoxicated. She had to send them home, but she knew there would be a storm not because she was sending home two nurses but because she was returning two Presbyterians.

'Meanwhile,' she wrote to Lady Canning, ' .. the second party of Nuns who came out now wander over the whole Hospital out of nursing hours, not confining themselves to their own wards, or even to patients but "instructing" (it is their own word) groups of Orderlies and Convalescents in the corridors, doing the work each of ten chaplains.'

However, by January 2 something had been arranged. Dr Cumming had been persuaded to raise the number of nurses to fifty, 'for which', wrote Miss Nightingale, 'we owe him eternal gratitude.' Mother Bridgeman still refused to allow her nuns into the Barrack Hospital, but some of the best 'hired' nurses had been sent for and some of the 'ladies' had gone to the General Hospital, where one of them, Miss Tebbut, finally became Superintendent. Mr Percy had 'sneaked home like a commander who has set so many Robinson Crusoes on a desert island,' and Dr Meyer had obtained a post at a convalescent hospital in Smyrna. 'Enough of this subject, of which among these realities of life and death I am thoroughly sick,' wrote Miss Nightingale to Sidney Herbert.

In the second week of January 1855 she received his answer to her letter of December 18. He accepted full blame, confirmed her authority, implored her not to resign, left everything to her discretion, and, finally, authorized her, if she thought fit, to send the second party home at his personal expense. Liz wrote equally penitent. Miss Nightingale was moved; the letters were 'most Generous and I deeply feel it. At the same time I do not regret what I said.'

She then ceased recriminations and throughout the misfortunes caused by Mary Stanley's party never again reminded Sidney Herbert that he was responsible for their ever having arrived at all.

After two months of hospital life Mary Stanley found herself utterly disillusioned. She no longer wished to go to the Barrack Hospital – it was filthy, vermin-infested; she had found fleas on her dress; and Florence expected far too much in the way of discipline. Her consolations were the frequent visits she paid Lady Stratford at the Embassy and her intimacy with the Irish nuns and their private chaplain, Father Ronan, who was preparing her for reception into the Roman Church; she was actually received some time during the spring. At the end of January it was suggested that the Turkish Cavalry Barracks at Koulali should be turned into a hospital, and Mary Stanley, encouraged by Lady Stratford, determined to take it over with her nuns, nurses, and ladies and run it in her own way. At the same time Lord Raglan suggested that eleven nurses should be sent to the General Hospital at Balaclava. Miss Nightingale did not wish nurses to go, for the Hospitals Commission had reported adversely on the hospital there: it was filthy, inefficient, the orderlies were undisciplined, and Balaclava was even more crammed with troops than Scutari. However, certain nurses, led by an elderly Welsh woman, Elizabeth Davis, from Mary Stanley's party, were determined to escape Miss Nightingale's discipline, and she herself was unwilling to refuse Lord Raglan. She gave way, and eleven volunteers under the control of the Superior of the Sellonite sisters went to Balaclava. Mary Stanley herself, with Mother Bridgeman and ten of her nuns, went off to Koulali, refusing to ask Dr Cumming's permission and declaring she would arrange the purveying of the hospital herself. Five or six 'hired nurses', who were first class, preferred to stay at Scutari, and the remaining five of Mother Bridgeman's nuns were accepted by the General Hospital at Scutari, whence Miss Nightingale received constant complaints of their religious activities. Thus Mary Stanley's party was dissolved.

But Mary Stanley's own reign at Koulali was short. It was to be run on the 'lady' plan. There were to be maids of all work to do the menial tasks. The ladies were not to wear uniform. Liz Herbert sent 'white furred coats' and was asked for straw bonnets. Miss Nightingale was understandably annoyed. 'I have,' she wrote to Sidney Herbert on February 12, '... by strict subordination to the authorities and by avoiding all individual action, introduced a number of arrangements, within the regulations of the service, useful on a large scale but not interesting to individual ladies; e.g. four extra diet kitchens of which two, which I administer, feed about seven hundred of the worst cases, furniture and clothing, washing, bath-house, lock-up cupboards, etc., etc. This is not so amusing as pottering and messing about with little cookeries of individual beef teas for the poor sufferers personally, and my ladies do not like it. I acknowledge it, at the same time it is obvious that what I have done could not have been done had I not worked with the medical authorities and not in rivalry with them. ... Cumming and I work hand in hand, and I have carried through him almost all that was possible under these awful difficulties. And he comes to me every evening. I protest emphatically now, before it is too late against the Koulali plan, i.e. the lady plan. It ends in nothing but spiritual flirtation between the ladies and the soldiers. I saw enough of that here; it pets the particular man, it gets nothing done in the general. ... The ladies all quarrel among themselves. The medical men laugh at their helplessness, but like to have them about for the sake of a little female society, which is natural but not our object.'

Koulali was not ready when Mary Stanley arrived. The second day two steamers suddenly anchored before the hospital, and 300 sick were carried in. There were no beds, no food. Sacks were hastily stuffed with straw; ladies made lemonade. That night Mary Stanley went round the wards and discovered her health would not stand the strain. More sick poured in, and she became hysterical. 'I cannot stay,' she wrote to Liz Herbert. 'I am not strong enough for the work. I have long been forced to give up special ward work.' When she applied to Miss Nightingale for more ladies, the letter was referred to Dr Cumming, who visited Koulali and was not pleased. He found that the ladies did little but stroll about with notebooks in their hands and refused to send any more. Confusion increased. Stores vanished. It was impossible to keep the hospital even decently clean. Mortality steadily rose until Koulali had the highest mortality rate of any hospital, higher even than the Barrack Hospital. Mary Stanley's letters became desperate. 'I feel anxious to come home before the strain has quite worn me out. For my mother's sake I dare not do more. . . .' She even appealed to Miss Nightingale; in the name of their old affection, she must be allowed to explain. In her reply, just, implacable, chilling, Miss Nightingale ended their friendship. 'I have nothing further to say. And for "explanation", I refer you to yourself. I have nothing to forgive. For I have never felt anger. I have never known you. There has been no "difference" between us – except a slight one of opinion as to the distribution of Articles and the manner of doing so to Patients. The pain you have given has not been by differing nor by anything for which forgiveness can be asked, but by not being yourself, or at least what I thought yourself. You say truly how I have loved you. No one will ever love you more. – Florence Nightingale.'

In March Lady Stratford sent for the Head Chaplain, Mr Sabin, and attacked him on the subject of Mary Stanley's grievances. It was monstrous to accuse her of Romanist propaganda; the truth was Miss Nightingale was jealous of Miss Stanley. Mr Sabin lost his temper and told Lady Stratford that she had been 'grossly imposed on'; the rumours of Roman Catholic propaganda at Koulali were true, and Mary Stanley was actively assisting Mother Bridgeman. He himself had received trustworthy information from home that Miss Stanley had in fact been received into the Roman Church, and was only waiting to declare herself. Lady Stratford was horrified – her husband's Protestant views were well known. 'Don't tell Lord Stratford!' she cried. Immediately afterward Mary Stanley went home.

Miss Nightingale had known the truth all along, but her sense of honour would not allow her to make use of it. 'Now observe, dear Mr Herbert,' she wrote on March 5, 1855, 'this bother is none of my making. I have kept strict honour with Lady Stratford and also with Dr Cumming about Mary Stanley's religious opinions. I could so easily have defeated her representations by "telling of her" as the children say, and Mrs Herbert will think that I have. ... Koulali has excited suspicion without me, or in spite of me. Cumming asked the question one day in my room whether Miss Stanley were not an R.C. and I put it off in order that he might not say he heard it from me.' It was a minor consolation that Lady Stratford sent in a bill for £8,200 sterling for purveying Koulali which the authorities had to pay.

At this difficult juncture Miss Nightingale's position was strengthened by Queen Victoria. On December 6 the Queen wrote to Sidney Herbert: 'Would you tell Mrs Herbert that I beg she would let me see frequently the accounts she receives from Miss Nightingale and Mrs Bracebridge, as I hear *no details of the wounded* though I see so many from officers etc., about the battlefield and naturally the former must interest *me* more than anyone. Let Mrs Herbert also know that I wish Miss Nightingale and the ladies would tell these poor noble wounded and sick men that *no-one* takes a warmer interest or feels more for their sufferings or admires their courage and heroism *more* than their Queen. Day and night she thinks of her beloved troops. So does the Prince. Beg Mrs Herbert to communicate these my words to those ladies, as I know that *our* sympathy is valued by these noble fellows.'

'The men were touched,' wrote Miss Nightingale to Sidney Herbert on December 25. ' "It is a very feeling letter" they said. "She thinks of us" (said with tears). "Queen Victoria is a Queen that is very *fond* of her soldiers".' The message was read aloud by the chaplains in the wards, posted up in the hospitals, and published in the newspapers. On December 14 the Queen had sent gifts to the men and a personal message to Miss Nightingale, who was entrusted with the distribution of the gifts. The Queen wished her to 'be made aware that your goodness and self devotion in giving yourself up to the soothing attendance upon these wounded and sick soldiers had been observed by the Queen with sentiments of the highest approval and admiration.' Would she suggest something the Queen could do 'to testify her sense of the courage and endurance so abundantly shown by her sick soldiers?'

Miss Nightingale was already pressing Sidney Herbert to change a regulation affecting the sick soldier's pay; 9*d*. a day was stopped from the pay of the sick soldier in hospital, even though his sickness was the direct result of active service, while the wounded man was stopped only 4½*d*. a day. Now she wrote directly to the Queen asking her to have the stoppage made the same for sickness as for wounds provided the sickness was incurred as the result of duty before the enemy. She also asked that a Firman might be requested from the Sultan making over the military cemeteries at Scutari to the British. The Queen acted immediately on both suggestions. On February 1 it was announced that the men's pay would be rectified as from the battle of the Alma, and in the same month Lord Clarendon, the Foreign Secretary, successfully applied to the Sultan for a Firman transferring the ownership of the cemeteries to the British.

Officials engaged in opposing Miss Nightingale were thus reminded that her influence was very great. If the arrival and conduct of Mary Stanley and her party had shaken Miss Nightingale's prestige, Queen Victoria assisted materially to restore it. 'It did very much having as our friends the great men,' wrote Miss Nightingale to Mrs Herbert in 1855.

In January 1855 the sufferings of the British Army before Sebastopol began to reach a fearful climax. William Howard Russell described the wounded arriving at Balaclava, strapped to the mules lent by the French: 'They formed one of the most ghastly processions that ever poet imagined. ... With closed eyes, open mouths and ghastly attenuated faces, they were borne along two by two, the thin steam of breath visible in the frosty air alone showing that they were alive. One figure was a horror, a corpse, stone dead, strapped upright in its seat ... no doubt the man had died on his way down to the harbour. ... Another man I saw with raw flesh and skin hanging from his fingers, the raw bones of which protruded into the cold, undressed and uncovered.'

Still no stores had reached the army. What had happened to them, the Roebuck Committee demanded later? Huge quantities of warm clothing, of preserved foods, of medical comforts and surgical supplies had been sent out – where did they all go? It was never discovered. The Roebuck Committee found it impossible not to suspect dishonesty, but Miss Nightingale reached a different conclusion. Large quantities unquestionably vanished in the Turkish Customs House, a 'bottomless pit whence nothing ever issues of all that is thrown in,' but she declared notwithstanding that stores were available all the time the men were suffering, never reaching them through the 'regulations of the service'. She cites a number of instances in her *Notes on Matters affecting the Health, Efficiency and Hospital Administration of the British Army*. In January 1855 when the army before Sebastopol was being ravaged by scurvy, a shipload of cabbages was thrown into the harbour at Balaclava on the ground that it was not consigned to anyone. This happened not once but several times. During November, December, and January 1854–5, when green coffee was being issued to the men, there were 173,000 rations of tea in store at Balaclava; 20,000 lb. of lime juice arrived for the troops on December 10, 1854, but none was issued until February. Why? Because no order existed for the inclusion of tea and lime juice in the daily ration.

Again, at the end of December there were blankets enough in store, says Miss Nightingale, to have given a third one to every man. But the men lay on the muddy ground with nothing under them and nothing over them since their blankets had been lost in battle or destroyed in the hurricane, because the regulations did not entitle them to replacement. At Scutari the Hospitals Commission recorded in January 1855: 'Goods have been refused although they were, to our personal knowledge, lying in abundance in the store of the Purveyor. This was done because they had not been examined by the Board of Survey.' Miss Nightingale wrote to Sidney Herbert in March of that year. 'The *Eagle* has now been arrived three weeks, and no use whatever has been made of her stores. Cumming says they have not yet been "sat on".' In February when the men were lying naked in the bitter cold, Mr Wreford, the Purveyor, admitted to the Hospitals Commission that he had received a large quantity of shirts a fortnight ago, but he had done nothing with them, did not even know the quantity as he had not yet had a 'board'. On February 15 Miss Nightingale wrote to Sidney Herbert: 'I received a requisition from the Medical Officers at Balaclava for shirts. ... I went to the Purveyor, as I always do, to give him a chance. The Purveyor answered 1st that he had no shirts. "Yes," I said, "you have received 27,000 landed four days ago." 2nd that he could not unpack them without a board – to which I answered that on every bale I had seen the number within marked, and he could send one or two bales making a memorandum for the Board – 3rd that they were at the General Hospital and he could not get an order in time. It ended by his accepting my offer to send a bale of my shirts which he might replace to me afterwards.'

On January 2, 1,200 sick men arrived in one consignment at Scutari. Eighty-five per cent of these, wrote Miss Nightingale, were cases of acute scurvy. For want of lime juice and vegetables the men's teeth were dropping out; in some cases they were losing toes. On January 4 she wrote to Sidney Herbert enclosing copies of requisitions on the Purveyor, 'properly signed by a 1st class staff surgeon, Dr O'Flaherty,' for supplies required for Barrack Hospital:

Flannel shirts	Answer	None in store
Socks	,,	None in store
Drawers	,,	None in store
N.B. There are some tea-pots and coffee-pots.		
Required for Barrack Hospital		
Plates		None in store
Tin drinking cups		None in store
Earthenware urine cups		Metal plenty
Bedpans		Some
Close stools		Plenty but frames missing
Pails for tea		None at present

In January 1855 there were 12,000 men in hospital and only 11,000 in the camp before Sebastopol; and still the shiploads came pouring down. It was, Miss Nightingale wrote, 'calamity unparalleled in the history of calamity'.

In this emergency she became supreme. She was the rock to which everyone clung, even the Purveyors. She described 'Messrs Wreford, Ward and Reade, veterans of the Spanish War, coming to me for a moment's solace, trembling under responsibility and afraid of informality.' 'Nursing,' she wrote on January 4 to Sidney Herbert, 'is the least of the functions into which I have been forced.'

Her calmness, her resource, her power to take action raised her to the position of a goddess. The men adored her. 'If she were at our head,' they said, 'we should be in Sebastopol next week.' The doctors came to be absolutely dependent on her, and Colonel

Sterling wrote home: 'Miss Nightingale now queens it with absolute power.'

Sidney Herbert had asked her to write to him privately in addition to her official reports, and during her time in Scutari and the Crimea she wrote him a series of over thirty letters of enormous length, crammed with detailed and practical suggestions for the reform of the present system. It is almost incredible that in addition to the unceasing labour she was performing, when she was living in the foul atmosphere of the Barrack Hospital incessantly harried by disputes, callers, complaints, and overwhelmed with official correspondence which had to be written in her own hand, she should have found time and energy to write this long series of vast, carefully thought-out letters, many as long as a pamphlet. She never lost sight of the main issue. At the time of the arrival of the Mary Stanley party she wrote: 'There is a far greater question to be agitated before the country than that of these eighty-four miserable women – eighty-five including me. This is whether the system or no system which is found adequate in time of peace but wholly inadequate to meet the exigencies of a time of war is to be left as it is – *or* patched up temporarily, as you give a beggar half pence – or made equal to the wants not diminishing but increasing of a time of awful pressure.'

On January 8, at the height of the calamity, she wrote: 'I have written a plan for the systematic organization of these Hospitals upon a principle of centralization under which the component parts might be worked in unison. But on consideration deeming so great a change impracticable during the present heavy pressure of calamities here, I refrain from forwarding it, and substitute a sketch of a plan, by which great improvement might be made from within without abandoning the forms under which the service is carried on. ...' Page after page of practical detailed suggestions follow, dealing with the reorganization of the Purveyor's department, the establishment of a corps of medical orderlies, the rearrangement and improvement of the cooking and service of the men's food, the establishment of a medical school at Scutari, where at present there was 'no operating room, no dissecting room; post-mortem examinations are seldom made, and then in the dead house (the ablest Staff surgeon here told me that he considered he had killed hundreds of men owing to the absence of these).' Finally, she made an urgent plea for medical statistics. 'No statistics are kept as to between what ages most deaths occur, as to modes of treatment, appearances of the body after death, etc., etc., etc. ... Our registration is so lamentably defective that often the only record kept is – *a man died* on such and such a day.'

In another immense letter of January 28 she elaborated her scheme for reorganizing the interior administration of the hospitals. The Purveyor was to be abolished, the hospital to have its own storekeeper, the Commissariat to supply all the food under the direction of a 'kind of Hotel-keeper' in the hospital. Each bed in the hospital was to have its own furniture and bedding supplied with it. The hospital was to be an entity in itself, not an appendage produced by the union of several departments.

She asked nothing for herself, nor did she use her influence to make life easier for herself by securing advancement for her friends. The only record of her having solicited promotion was on behalf of Dr McGrigor. She asked Sidney Herbert to promote him a Deputy Inspector-General two years before the proper time and thus do 'a service to humanity at the expense of the Regulations of the Service'. If Dr McGrigor were not promoted, the work he was doing could be stopped by the simple process of the authorities bringing into the hospital someone senior to himself, in which case he would no longer be entitled to give orders. Dr McGrigor was promoted.

Her facts and figures were freely used by Sidney Herbert and other members of the Cabinet, and important changes made in British Army organization during the course of the Crimean War were based on her suggestions. A Medical School was founded during the campaign, and the suggestions respecting the Purveyor, though not carried out immediately, formed the basis of reforms executed at a later date.

*

In spite of the improvements in the Barrack Hospital, something was horribly wrong. The wards were cleaner, the lavatories unstopped, the food adequate, but still the mortality climbed. The disaster was about to enter its second phase. At the end of December an epidemic broke out described variously as 'Asiatic cholera' or 'famine fever', similar to cholera brought over by starving Irish immigrants after the Irish potato famine, and by Miss Nightingale simply as 'gaol fever'. By the middle of January the epidemic was serious – four surgeons died in three weeks, and three nurses and

154

poor old Ward, the Purveyor, and his wife died. The officers on their rounds began to be afraid to go into the wards; they could do nothing for the unfortunates perishing within; they knocked on the door and an orderly shouted 'All right, sir' from inside.

The snow ceased, and faint warmth came to the bleak plateau before Sebastopol on which the British Army was encamped. The number of men sent down by sick transports stopped rising. The percentage of sick was still disastrously, tragically high, but it was stationary.

But in the Barrack Hospital the mortality figures continued to rise. Sister Margaret Goodman saw an *araba*, a rough Turkish tumbril, heaped with what she took to be the carcasses of beasts. They were the naked, emaciated bodies of dead British soldiers. A large square hole of no great depth was dug by Turks, the bodies were tossed into this until they came level with the top; then a layer of earth was shovelled over all, and the Turks stamped it down. They then drove off. The British were unable to bury their dead. A fatigue party could not be mustered whose strength was equal to the task of digging a pit.

In England fury succeeded fury. A great storm of rage, humiliation, and despair had been gathering through the terrible winter of 1854–5. For the first time in history, through reading the despatches of Russell, the public had realized 'with what majesty the British soldier fights'. And these heroes were dead. The men who had stormed the heights at Alma, charged with the Light Brigade at Balaclava, fought the grim battle against overwhelming odds in the fog at Inkerman had perished of hunger and neglect. Even the horses which had taken part in the Charge of the Light Brigade had starved to death.

On January 26 Mr Roebuck, Radical member for Sheffield, brought forward a motion for the appointment of a committee 'to inquire into the condition of the Army before Sebastopol and the conduct of those departments of the Government whose duty it has been to minister to the wants of that Army'. It was a vote of censure on the Government, and it was carried in an uproar by a majority of 157. The Government fell, and Sidney Herbert went out of office, but Miss Nightingale's position was not weakened. The new Prime Minister was her old friend and supporter, Lord Palmerston. The two offices of Secretary for War and Secretary at War were combined and held by Lord Panmure, who was instructed to show consideration for her wishes and opinions. Her

155

reports were regularly forwarded to the Queen and studied by her. Sidney Herbert wrote to assure her that he had no intention of giving up his work for the army because he was out of office. She was still to write to him, and he would see that her reports and suggestions were forwarded to the proper quarters. He would continue to be, she wrote, 'our protector in this terrible great work'.

At the end of February, Lord Panmure sent out a Sanitary Commission to investigate the sanitary state of the buildings used as hospitals and of the camps both at Scutari and in the Crimea. The Commission was formed at the suggestion of Lord Shaftesbury, Lady Palmerston's son-in-law and Miss Nightingale's old friend. Her name did not appear, but the urgency, the clarity, the forcefulness of the instructions are unmistakably hers. 'The utmost expedition must be used in starting your journey. ... On your arrival you will instantly put yourselves into communication with Lord William Paulet. ... It is important that you be deeply impressed with the necessity of not resting content with an order but that you see instantly, by yourselves or your agents, to the commencement of the work and to its superintendence day by day until it is finished.'

This Commission, said Miss Nightingale, 'saved the British Army'. It consisted of Dr John Sutherland, an official of well-known ability and advanced views from the Board of Health, Mr, later Sir, Robert Rawlinson, a civil engineer of eminence, and Dr Gavin. With the ill luck which seemed to dog all Crimean undertakings, Dr Gavin was accidentally killed by his brother letting off a pistol shortly after his arrival, and his place was filled by Dr Milroy. In addition the three Commissioners took with them the Borough Engineer and three sanitary inspectors from Liverpool, where a sanitary act had been in operation longer than anywhere else in the country, and shipped out a large quantity of building material.

They were followed by another Commission, the McNeill and Tulloch Commission of Inquiry into the Supplies for the British Army in the Crimea. This Commission of Inquiry went direct to the Crimea and did not call at Scutari. It consisted of Colonel Alexander Tulloch, R.E., and Sir John McNeill, who had had many years' experience first as a doctor and then as an administrator in India and Persia. He had been Poor Law Commissioner in Scotland, and thanks to his energy and initiative the Highland peasants, though almost as dependent as the Irish on potatoes, had escaped the worst consequences of the failure of the potato crop in 1846–7.

The Sanitary Commission landed at Constantinople at the beginning of March and began work instantly. Their discoveries were hair-raising. They described the sanitary defects of the Barrack Hospital as 'murderous'. Beneath the magnificent structure were sewers of the worst possible construction, mere cess-pools, choked, inefficient, and grossly over-loaded. The whole vast building stood in a sea of decaying filth. The very walls, constructed of porous plaster, were soaked in it. Every breeze, every puff of air, blew poisonous gas through the pipes of numerous open privies into the corridors and wards where the sick were lying. 'It is impossible,' Miss Nightingale told the Royal Commission of 1857, 'to describe the state of the atmosphere of the Barrack Hospital at night. I have been well acquainted with the dwellings of the worst parts of most of the great cities of Europe, but have never been in any atmosphere which I could compare with it.' Nurses had noticed that certain beds were fatal. Every man put in these beds quickly died. They proved to be near the doors of the privies, where the poisonous gases were worst. The water supply was contaminated and totally insufficient. The Commissioners had the channel opened through which the water flowed, and the water supply for the greater part of the hospital was found to be passing through the decaying carcass of a dead horse. The storage of water was in tanks in the courtyard, and these had been built next temporary privies, erected to cope with the needs of men suffering from the prevalent diarrhoea. The privies were open and without any means of flushing or cleaning. The courtyard and precincts of the hospital were filthy. The Commissioners ordered them to be cleared, and during the first fortnight of this work 556 handcarts and large baskets full of rubbish were removed and 24 dead animals and 2 dead horses buried. The Commission began to flush and cleanse the sewers, to limewash the walls and free them from vermin, to tear out the wooden shelves known as Turkish divans which ran round the wards and harboured the rats for which the Barrack Hospital was notorious. The effect was instant. At last the rate of mortality began to fall. In the Crimea spring came with a rush; the bleak plateau before Sebastopol was bathed in sunlight and carpeted with crocuses and hyacinths. The road to Balaclava became pass-

157

able, the men's rations improved, and the survivors of the fearful winter lost their unnatural silence and began once more to curse and swear.

The emergency was passing, and as it passed opposition to Miss Nightingale awoke again.

10

Miss Nightingale's mission falls into two periods. There is first the period of frightful emergency during the winter of 1854-5. In Sidney Godolphin Osborne's opinion, if at that time Miss Nightingale had not been present, the hospitals must have collapsed. Every consideration but that of averting utter catastrophe went by the board, opposition died away, and she became supreme.

But as soon as things had slightly improved, official jealousy re-awoke. In the second period, from the spring of 1855 until her return to England in the summer of 1856, gratitude – except the gratitude of the troops – and admiration disappeared, and she was victimized by petty jealousies, treacheries, and misrepresentations. Throughout this second period she was miserably depressed. At the end of it she was obsessed by a sense of failure.

By the spring of 1855 she was physically exhausted. She was a slight woman who had never been robust, who was accustomed to luxury, and was now living in almost unendurable hardship. When it rained, water poured through the roof of her quarters and dripped through the floor on an officer beneath, who complained that 'Miss Nightingale was pouring water on his head'. The food was uneatable; the allowance of water was still one pint a head a day; the building was vermin-infested, the atmosphere in the hospital so foul that to visit the wards produced diarrhoea. She never went out except to hurry over the quarter of a mile of refuse-strewn mud which separated the Barrack from the General Hospital.

When a flood of sick came in, she was on her feet for twenty-four hours at a stretch. She was known to pass eight hours on her knees dressing wounds. 'She had an utter disregard of contagion,' wrote Sidney Godolphin Osborne. '. . . The more awful to every sense any particular case, especially if it was that of a dying man, the more certainly might her slight form be seen bending over him, administering to his ease by every means in her power and seldom quitting his side until death released him.' It was her rule never to let any man who came under her observation die alone. If he were conscious, she herself stayed beside him; if he were unconscious she sometimes allowed Mrs Bracebridge to take her place. She estimated that during that winter she witnessed 2,000 deathbeds. The worst cases she nursed herself. 'I believe,' wrote Dr Pincoffs, a civilian doctor who worked in the Barrack Hospital, 'that there was never a severe case of any kind that escaped her notice.' One of the nurses described accompanying her on her night rounds. 'It seemed an endless walk. ... As we slowly passed along the silence was profound; very seldom did a moan or cry from those deeply suffering fall on our ears. A dim light burned here and there. Miss Nightingale carried her lantern which she would set down before she bent over any of the patients. I much admired her manner to the men – it was so tender and kind.'

Her influence was extraordinary. She could make the men stop drinking, write home to their wives, submit to pain. 'She was wonderful,' said a veteran, 'at cheering up anyone who was a bit low.' The surgeons were amazed at her ability to strengthen men doomed to an operation. 'The magic of her power over men was felt,' writes Kinglake, 'in the room – the dreaded, the blood-stained room – where operations took place. There perhaps the maimed soldier if not yet resigned to his fate, might be craving death rather than meet the knife of the surgeon, but when such a one looked and saw that the honoured Lady in Chief was patiently standing beside him – and with lips closely set and hands folded – decreeing herself to go through the pain of witnessing pain, he used to fall into the mood of obeying her silent command and – finding strange support in her presence – bring himself to submit and endure.'

The troops worshipped her. 'What a comfort it was to see her pass even,' wrote a soldier. 'She would speak to one, and nod and smile to as many more; but she could not do it all you know. We lay there by hundreds; but we could kiss her shadow as it fell and lay our heads on the pillow again content.'

For her sake the troops gave up the bad language which has always been the privilege of the British private soldier. 'Before she came,' ran another letter, 'there was cussing and swearing but after that it was as holy as a church.'

When the war was over Miss Nightingale wrote: '... The tears come into my eyes as I think how, amidst scenes of loathsome disease and death, there rose above it all the innate dignity, gentleness and chivalry of the men (for never surely was chivalry so strikingly exemplified) shining in the midst of what must be considered the lowest sinks of human misery, and preventing instinctively the use of one expression which could distress a gentlewoman.'

It was work hard enough to have crushed any ordinary woman; yet, she wrote, it was the least of her functions. The crushing burden was the administrative work. Her quarters were called the Tower of Babel. All day long a stream of callers thronged her stairs, captains of sick transports, officers of Royal Engineers, nurses, merchants, doctors, chaplains, asking for everything from writing-paper to advice on a sick man's diet, demanding shirts, splints, bandages, port wine, stoves, and butter.

She slept in the storeroom in a bed behind a screen; in the daytime she saw callers sitting and writing at a little unpainted deal table in front of it. She wore a black woollen dress, white linen collar and cuffs and apron, and a white cap under a black silk handkerchief. Every time there was a pause she snatched her pen and went on writing.

No one in the party was capable of acting as her secretary. The requisitions, the orders, the records, the immense correspondence entailed by the acknowledgement and recording of the 'Free Gifts' (the voluntary contributions sent out from home), the reports, the letters, must all be written by herself. Mrs Bracebridge had superintendence of the 'Free Gift Store'; otherwise she had no assistance of any kind.

It was terribly cold, and she hated cold. There was no satisfactory stove in her quarters – one had been sent out from England, but it would not draw and she used it as a table and it was piled with papers. Her breath congealed on the air; the ink froze in the well; rats scampered in the walls and peered out from the wainscoting. Hour after hour she wrote on; the staff of the hospital declared that the light in her room was never put out. She wrote for the men, described their last hours and sent home their dying messages; she told wives of their husband's continued affection, and mothers that their sons had died holding her hand. She wrote for the nurses, many of whom had left children behind. She wrote her enormous letters to Sidney Herbert; she wrote official reports, official letters; she kept lists, filled in innumerable requisitions. Papers were piled round her in heaps; they lay on the floor, on her bed, on the chairs. Often in the morning Mrs Bracebridge found her still in her clothes on her bed, where she had flung herself down in a stupor of fatigue.

She spared herself nothing – but the joy had gone out of the work. The high spirit, the faith which had sustained her through the first months faded as she learned the power of official intrigue.

'Alas among all the men here,' she wrote to Sidney Herbert in February 1855, 'is there one really anxious for the good of these hospitals? One who is not an insincere animal at the bottom, who is not thinking of going in with the winning side whichever that is? I do believe that of all those who have been concerned in the fate of these miserable sick you and I are the only ones who really cared for them.' A month later she wrote: 'A great deal has been said of our self sacrifice, heroism and so forth. The real humiliation, the real hardship of this place, dear Mr Herbert, is that we have to do with men who are neither gentlemen nor men of education nor even men of business, nor men of feeling, whose only object is to keep themselves out of blame.'

She had crossed the path of such a man, and the great conflict of her mission was about to begin.

Dr John Hall, Chief of Medical Staff of the British Expeditionary Army, had been kept occupied in the Crimea, but the hospitals of Scutari were under his control and he had no intention of allowing them to get out of hand. His name had been associated with an unsavoury case in which a private stationed at Hounslow Barracks had died after receiving a flogging of 150 lashes, and he was known throughout the army as a strict disciplinarian averse to pampering the troops. He did not believe in chloroform, and in his letter of instructions to his officers at the opening of the campaign on August 3 he warned them against its use. 'The smart use of the knife is a powerful stimulant and it is much better to hear a man bawl lustily than to see him sink silently into the grave.' He was revengeful, powerful, a master of the confidential report. Miss Nightingale wrote to Lady Cranworth that a doctor's promotion depended 'upon a trick, a caprice of the Inspector General (i.e. Dr Hall) ... and may be lost for an offensive word reported perhaps by an orderly and of which he never hears and which he may never have said.' In May 1856 she wrote: 'In the last two months at this hospital alone, two medical officers have been superseded upon evidence collected in the above manner.'

Dr Hall entered upon his duties in the Crimea with a sense of injustice. He had been in Bombay, he had been due for promotion, and he thought he deserved a post at home. He had solicited such a post and heard with disgust that he had been appointed Chief of Medical Staff of the British Expeditionary Army. In October 1854 he was sent by Lord Raglan to inspect the hospitals at Scutari. The hospitals were then filthy and destitute. However, Dr Hall wrote on October 20 to Dr Andrew Smith stating he had 'much satisfaction in being able to inform him that the whole hospital establishment here (i.e. at Scutari), has now been put on a very creditable footing and that nothing is lacking'.

It was a fatal statement. He had committed himself. Henceforward he had to stand by what he had said, and his subordinates had to back him up. Dr Menzies dared not contradict Dr Hall's specific statement. He repeated it parrot-like to Lord Stratford, to Dr Andrew Smith. It was not until Sidney Herbert received Miss Nightingale's first report that the truth was known. In December 1854 he told Lord Raglan: 'I cannot help feeling that Dr Hall resents offers of assistance as being slurs on his preparations.'

In the spring of 1855 Dr Hall was boiling with rage. The Hospitals Commission had reported unfavourably on his hospitals and, worse, he had been censured by Lord Raglan.

The most notorious of the sick transport scandals was the case of the *Avon*. The first man had been put on board the *Avon* at Balaclava on November 19, 1854, the last man on December 3. The men were laid on the bare deck without any covering but greatcoat or blanket. One young assistant surgeon was instructed to attend to several hundred men, and so they were left for a fortnight. The state of the ship and the condition of the men was then indescribable. A regimental officer was induced to visit the ship and, horrified by what he saw, galloped at once to Lord Raglan. Though it was midnight Lord Raglan sent at once to Dr Hall demanding immediate action. An inquiry was held, Dr Lawson, the Principal Medical Officer at Balaclava, was held responsible and severely censured for 'apathy and lack of interest in the welfare of the sick', and Dr Hall was recommended to relieve him of his duties. Further, in a General Order of December 13, 1854, Lord Raglan stated he could not acquit Dr Hall himself of blame in this matter. Dr Hall judged the time had come to assert himself. He was by no means beaten. He knew his powers, he had his friends, and within his own department he was invincible. Dr Menzies, the Senior Medical Officer at the Barrack Hospital, had been succeeded by Dr Forrest. After a few weeks Dr Forrest resigned and went home in despair, and Dr Hall then appointed Dr Lawson to take his place. The man responsible for the *Avon* was to be Senior Medical Officer at the Barrack Hospital.

Miss Nightingale received the news with horror. 'Before destroying our work Dr Hall begins to caress us with his paws,' she wrote, and she warned Sidney Herbert: 'The people here will try the strength of the old system against Government reforms with a strength of purpose and a cohesion of individuals which you are not likely to give them credit for.'

Dr Lawson was a walking reminder of what the medical department could do. He had been censured and was to be relieved of his duties; he had been relieved of his duties – to assume them in a different place. Dr Hall knew how to protect his own, and he knew how to punish the disloyal. Dr Smith and Dr Hall were absolute masters of the Army Medical Department, and no Nightingale power, no Sidney Herbert could save those unhappy slaves who offended their masters.

A wave of terror swept over the medical staff at Scutari. Dr Cumming continued to call on Miss Nightingale every day, but he became nervous. He had been a member of the Hospitals Commission, but presently he was refusing to carry out his own recommendations. For example, the Hospitals Commission had stressed the urgent necessity of equipping the wards with bedding and utensils, and in March 1855 a large quantity of hospital stores arrived with which Dr McGrigor had the wards equipped. Dr Cumming ordered the new equipment to be removed.

Another broken reed was Lord William Paulet, who frankly detested his job. He had been sent out because he had wealth, position, and prestige, and Major Sillery had failed because he had none of these. 'Lord Wm Paulet is appalled at the view of evils he has no idea what to do with,' wrote Miss Nightingale; '... and then he shuts his eyes and hopes when he opens them he shall see something else.' As things became more difficult, he withdrew – he put his head, she said, under his wing, spending his time with Lady Stratford picnicking along the picturesque shores of the Bosphorus, accompanied by hampers of the delicacies for which the Embassy chef was famous, ostensibly for the purpose of inspecting possible sites for convalescent hospitals. Nothing was to be expected from Lord William Paulet.

Dr McGrigor began to succumb to Lawson's influence. He avoided Miss Nightingale; he ceased to be urgent in pressing the fulfilment of the recommendations of the Hospitals Commission. He was, she wrote, 'the one of all others who really wished to help – but he was weak.' She felt betrayed, though she still had her triumphs. A whole corridor which the Purveyor had declared himself before witnesses unable to equip was fitted out by her and Mr Macdonald from Constantinople by nightfall. 'What I have done I shall continue doing,' she wrote, '... but I am weary of this hopeless work.'

Within the hospital the work of the Sanitary Commission was having rapid effect. The fearful mortality rate of February had fallen in the three weeks ending April 7 to 14½ per cent, by April 28 to 10·7 per cent, and by May 19 to 5·2 per cent.

Thanks to Miss Nightingale's purveying – the Purveyor's stores were still empty, and the authorities were slipping back into a state of mind when equipment was thought an unnecessary extravagance for a hospital – there were plenty of drugs, surgical instruments, baths, hot-water bottles, and medical comforts. Dr Pincoffs noted that these were present in satisfactory quantities when he joined the hospital in the spring. There were also operating tables, supplied by her for the second time: the first set had been burned as firewood in the great cold of January 1854.

Food had been miraculously improved by Alexis Soyer, the famous chef of the Reform Club, who arrived in March 1855 with full authority from Lord Panmure. Soyer came out at his own expense attended by a 'gentleman of colour' as his secretary. In manner and appearance he was a comic opera Frenchman, but Miss Nightingale recognized his genius and became his friend. 'Others,' she wrote, 'have studied cookery for the purpose of gormandizing, some for show. But none but he for the purpose of cooking large quantities of food in the most nutritive and economical manner for great quantities of people.' Though the authorities received him 'very coolly', Soyer was armed with authority and he proceeded to attack the kitchens of the Barrack Hospital. He composed recipes for using the army rations to make excellent soup and stews. He put an end to the frightful system of boiling. He insisted on having permanently allocated to the kitchens soldiers who could be trained as cooks. He invented ovens to bake bread and biscuits and a Scutari teapot which made and kept tea hot for fifty men. As he walked the wards with his tureens of soup, the men cheered him with three times three. Finally, he gave a luncheon attended by Lord and Lady Stratford and their suite, at which he served delicious dishes made from army rations.

In one thing Soyer failed. Like Miss Nightingale, he strongly objected to the way the meat was divided; since weight was the only criterion one man might get all bone; why should not the meat be boned, and each man receive a boneless portion, with the bones being used for broth? The answer from Dr Cumming was that it would need a new Regulation of the Service to bone the meat.

In May 1855 Miss Nightingale wrote to Sidney Herbert to describe 'the first really satisfactory reception of sick'. Two hundred men from the *Severn* transport were received, bathed, and their hair cut and cleansed. Their filthy clothes and blankets were taken from them, they were given clean hospital gowns, put into decent beds and given well-cooked nourishing food. In spite of obstacles, disappointments, opposition, she had, to this degree, succeeded.

And now that the Barrack Hospital was reasonably satisfactory, she determined to go to the Crimea. There were two large hospitals at Balaclava. One, the General Hospital, had been established at the time of the British occupation in September 1854 and, like the General Hospital at Scutari, had been intended to be the only hospital. This was the hospital in Dr John Hall's personal charge on which the Hospitals Commission had reported adversely. The enormous numbers of sick had necessitated further accommodation, and a hospital of huts called the Castle Hospital had been erected on the heights above Balaclava harbour. Both had a staff of female nurses, and disquieting news had reached Miss Nightingale of the nurses' conduct, particularly at the General Hospital.

And now the fatal flaw in her instructions appeared, and her authority in the Crimea proved to be by no means established. Precise information as to her standing, her instructions, and the assistance to be afforded to her had been sent to Lord Raglan, Lord Stratford, and Dr John Hall. But Lord Raglan was occupied with the problems of a disastrous campaign; Lord Stratford was indifferent; Dr John Hall was malicious. He asserted that, as her instructions named her 'Superintendent of the Female Nursing Establishment in the English Military General Hospitals in *Turkey*', she had no jurisdiction over the Crimea.

The seriousness of the situation was not appreciated at home. Mr Augustus Stafford wrote: 'The nature of her difficulties is NOT understood and perhaps never will be.' Supported by Dr Hall, nurses in the Crimea were defying her authority. One of them, Miss Clough, a 'lady' of Miss Stanley's party, had broken away and gone to join Sir Colin Campbell's Hospital above Balaclava, inspired by romantic enthusiasm for the Highland Brigade. 'She must be a funny fellow, she of the Highland Heights,' commented Miss Nightingale. A constant rebel was Mrs Elizabeth Davis, the Welshwoman brought out by Mary Stanley. She had begun to dislike Miss Nightingale before she saw her. 'I did not like the name of Nightingale. When I first hear a name I am very apt to know by my feelings whether I shall like the person who bears it,' she wrote. She had had experience in nursing and was selected for the Barrack Hospital. Once there she proved a storm centre. She refused to obey orders or to conform to the system for the distribution of the 'Free Gifts'. She accused Miss Nightingale of using these for her own comfort and alleged that, while the nurses were fed on filaments of the meat which had been stewed down for the patients' soup, Miss Nightingale had a French cook and three courses served up every day. Finally, she joined the party of eleven volunteers who went, against Miss Nightingale's wishes, to Balaclava in January 1855.

Once there she made an alliance with Dr John Hall, and another important personage in the Crimea, Mr David Fitz-Gerald, the Purveyor-in-Chief. Mr Fitz-Gerald was as angrily opposed to Miss Nightingale as was Dr Hall, and as equally determined to keep her out of the Crimea.

Elizabeth Davis, an excellent cook, had assumed command of the kitchen in Balaclava General Hospital, which she conducted with rollicking extravagance, rejoicing in feeding up the handsome young officers who were her special pets. It was Miss Nightingale's rule that none of her nurses should attend on or cook for officers except by special arrangement. At one issue Mrs Davis received '6 dozen port wine, 6 dozen sherry, 6 dozen brandy, a cask of rice, a cask of arrowroot, a cask of sago and a box of sugar'; and her requisitions for the General Hospital were filled at once by Mr Fitz-Gerald without being countersigned by

Dr Hall. The situation became too much for the Superintendent, the Superior of the Sellonites, who, Miss Nightingale said, 'lost her head and her health', collapsed, and went home. In her place another of Mary Stanley's party was appointed, Miss Weare, a fussy, gentle old spinster who swiftly became dominated by Mrs Davis and Dr Hall. Miss Weare confided to Dr Hall how much more *natural* she found it to obey a gentleman. Miss Nightingale was very wonderful, of course, but she could not get used to taking orders from a lady.

With the 'Free Gifts' Mrs Davis and her allies were even more open-handed. In an orgy of distribution ninety bales and boxes were given away without any record of who had received them.

The 'Free Gifts' – 'these frightful contributions', Miss Nightingale called them, together with the labour of acknowledging them, storing them in safety, and distributing them satisfactorily, were becoming the bane of her life. Ever since November 1854 parcels had been sent from England for the troops. 'There is not a small town, not a parish in England from which we have not received contributions,' she wrote in May 1855, 'not one of these is worth its freight, but the smaller the value, of course, the greater the importance the contributors attach to it. If you knew the trouble of landing, of unpacking, of acknowledging! The good that has been done here has been done by money, money purchasing articles in Constantinople.'

Among the 'Free Gifts' were articles of value. Queen Victoria had sent a number of water-beds; there were also provisions, groceries, wine, brandy, soup, and clothing. To keep a check was difficult; the store, like every other place in Scutari, was overrun by rats, and the Maltese, Greek, and Turkish labourers who worked round the hospital were dishonest almost without exception. After her arrival on November 5, 1854, Miss Nightingale kept an exact record of every article received and issued by her. After February 15, 1855, Mrs Bracebridge was left in sole charge.

On May 2, 1855, she sailed from Scutari for Balaclava in the *Robert Lowe*. 'Poor old Flo,' she wrote to her mother, 'steaming up the Bosphorus and across the Black Sea with four nurses, two cooks, and a boy to Crim Tartary ... in the Robert Lowe or Robert Slow (for an exceedingly slow boat she is) ... taking back 420 of her patients, a draught of convalescents returning to their regiments to be shot at again. "A Mother in Israel," Pastor

Fliedner called me; a Mother in the Coldstreams, is the more appropriate appellation.'

Besides Soyer and a French chef, the party included Soyer's secretary, the 'gentleman of colour', Mr Bracebridge and a boy named Robert Robinson, an invalided drummer from the 68th Light Infantry. He described himself as Miss Nightingale's 'man' – Soyer could not resist asking him whether he was twelve years old yet – and was accustomed to explain that he had 'forsaken his instruments in order to devote his civil and military career to Miss Nightingale'. He carried her letters and messages, escorted her when she went from the Barrack Hospital to the General Hospital, and had charge of the lamp which she carried at night. Among the Nightingale papers is a manuscript account of his experiences during the campaign, entitled 'Robert Robinson's Memoir'. He was, said Soyer, 'a regular *enfant de troupe*, full of wit and glee'.

On May 5, six months after her arrival at Constantinople – 'and what the disappointments of those six months have been no one could tell,' she wrote, 'but still I am not dead but alive' – the *Robert Lowe* anchored in Balaclava harbour. Balaclava was crammed to overflowing, and she was invited by the Captain to make her quarters on board the ship, which soon, wrote Soyer, resembled a floating drawing-room, as doctors, senior officers, and officials, including Sir John McNeill of the Tulloch and McNeill Commission and Dr Sutherland of the Sanitary Commission, came to pay their respects. In the afternoon, escorted by a number of gentlemen, she went ashore to report herself to Lord Raglan. She appeared, says Soyer, in a 'genteel Amazone', and rode a 'very pretty mare which by its gambols and caracoling seemed proud to carry its noble charge'. Lord Raglan being away for the day, she decided to visit the mortar battery outside Sebastopol. The astonishing sight of a lady in Balaclava accompanied by a crowd of gentlemen, many of them in glittering uniforms, produced 'an extraordinary effect'. The news spread like wildfire that the lady was Miss Nightingale, and the soldiers rushed from their tents and 'cheered her to the echo with three times three'. At the Mortar Battery Soyer requested her to ascend the rampart and seat herself on the centre mortar, 'to which she very gracefully acceded'. He then 'boldly exclaimed, "Gentlemen, behold this amiable lady sitting fearlessly upon the terrible instrument of war! Behold the heroic daughter of England, the soldiers' friend!"' Three cheers were given by all. Meanwhile five or six of her escort had picked bouquets of the wild lilies and orchids which carpeted the plateau. She was requested to choose the one she liked best and responded by gathering them all in her arms.

The party then cantered home, Miss Nightingale looking strangely exhausted. It was, she said, the unaccustomed fresh air.

The next morning, accompanied by Soyer, she began her inspection. It was a depressing task. The hospitals were dirty and extravagantly run, the nurses inefficient and undisciplined. She was received with hostility and, at the General Hospital, with insolence. 'I should have as soon expected to see the Queen here as you,' said Mrs Davis.

She ignored hostility and rudeness. She got out plans with Soyer's assistance for new extra diet kitchens at the General Hospital. She decided Miss Weare must be replaced – the General Hospital was evidently out of hand. She then went up to the Castle Hospital, the new hospital of huts where Mrs Shaw Stewart (the 'Mrs' was a courtesy title), a difficult woman herself, was having a difficult time. Mrs Shaw Stewart, one of Mary Stanley's party, was one of the few women of social position who had any real experience in nursing. She was the sister of Sir Michael Shaw Stewart, M.P., and had undergone training in Germany and nursed in a London hospital. She was skilful, kind, a magnificent worker, but she would be a martyr. Do what her friends would, conciliate her, defer to her, coax her, she maintained she was being ill-treated. At the Castle Hospital she had no need to imagine persecution, for Dr Hall was making her work as difficult as possible. He caused immense inconvenience by insisting that all her requisitions must be sent to him personally. Work which the Sanitary Commission had directed was not even started, her kitchens were inadequate, the Purveyor habitually held up her supplies, and, finally, Dr Hall made a practice of sending her messages of criticism through her staff.

Miss Nightingale gathered herself together to do battle, but before anything could be accomplished she collapsed. After seeing Mrs Shaw Stewart, she had admitted great weakness and fatigue, and the next day, while interviewing Miss Weare, she fainted. The Senior Medical Officer from the Balaclava General Hospital was hastily summoned; after he had called two other doctors into consultation, a statement was issued that Miss Nightingale was suffering from Crimean fever.

All Balaclava, says Soyer, was in an uproar. It was decided that she must be removed from the ship. The harbour was being cleansed by the Sanitary Commission, and the men working to remove the ghastly debris found the stench so horrible that they constantly fainted and had to receive an official issue of brandy. She must be taken to the pure air of the Castle Hospital on the heights. A solemn cortège transported her from the ship, four soldiers carrying her on a stretcher and Dr Anderson and Mrs Roberts walking by her side; Soyer's secretary – Soyer himself was away – held an umbrella over her head, and Robert Robinson walked behind in tears, being, in his own words, 'not strong enough to help carry or tall enough to hold the umbrella'. By this time she was delirious and very ill. At Balaclava the troops seemed in mourning, and at Scutari the men when they heard the news, 'turned their faces to the wall and cried. All their trust was in her,' a Sergeant wrote home.

For more than two weeks, nursed by Mrs Roberts, she hovered between life and death. In her delirium she was constantly writing. It was found impossible to keep her quiet unless she wrote, so she was given pen and paper; among the Nightingale papers are sheets covered with feverish notes. She thought her room was full of people demanding supplies, that an engine was inside her head, that a Persian adventurer came and stood beside her bed and told her that Mr Bracebridge had given him a draft for £300,000 sterling, and she wrote to Sir John McNeill asking him to deal with the man because he had been in Persia. In the height of the fever all her hair was cut off. The news went round the camp, and Colonel Sterling wrote that he heard the Bird had had to have her head shaved – would she wear a wig or a helmet!

At home the tidings were received with consternation, and when it was known that she was recovering strangers passed on the good news to each other in the streets.

On May 24 a horseman wrapped in a cloak rode up to her hut and knocked. Mrs Roberts sprang out – 'Hist, hist, don't make such a horrible noise as that, my man.' He asked if this were Miss Nightingale's hut. Mrs Roberts said it was, and he tried to walk in. Mrs Roberts pushed him back. 'And pray who are you?' she asked. 'Oh, only a soldier, but I must see her, I have come a long way, my name is Raglan, she knows me very well.' 'Oh, Mrs Roberts, it is Lord Raglan,' called Miss Nightingale. He came in and, drawing up a stool to her bedside, talked to her at length. That night he telegraphed home that Miss Nightingale was out of danger, and on May 28 Queen Victoria was 'truly thankful to learn that that excellent and valuable person Miss Nightingale is safe.'

She was frantic to settle the urgent problems at Balaclava, but her weakness was so extreme that she could not feed herself or raise her voice above a whisper. The doctors advised her to go to England, or failing that, to Switzerland. She refused, and Mrs Bracebridge, who had hastened from Scutari to look after her, pointed out that she was such an execrable sailor that a long sea voyage in her present state might well kill her. It was arranged that she should be taken to Scutari on a transport and occupy a house belonging to Mr Sabin, who had gone home on sick leave.

A curious incident followed. Dr Hadley, the Senior Medical Officer at the Castle Hospital, had attended her. Dr Hadley was a friend of Dr John Hall, and the two doctors selected the transport, the *Jura*, on which she was to go to Scutari. She was actually on board when Mr Bracebridge discovered that the *Jura* was not calling at Scutari but going direct to England. Miss Nightingale was hurried off the transport in a fainting condition by Mr Bracebridge and Lord Ward, and crossed to Scutari on Lord Ward's steam yacht. On October 19, 1855, she wrote to Sidney Herbert: 'It was quite true that Doctors Hall and Hadley sent for a list of vessels going home, and chose one, the Jura, which was not going to stop at Scutari *because* it was not going to stop at Scutari, and put me on board her for England.'

The voyage was rough, the yacht was kept at sea an extra day, and Miss Nightingale was dreadfully ill. At Scutari her weakness and exhaustion were such that she was unable to speak. She was terribly changed, emaciated, white-faced under the handkerchief tied closely round her head to conceal her shorn hair. Two relays of guardsmen carried her to Mr Sabin's house on a stretcher. Twelve private soldiers divided the honour of carrying her baggage. The stretcher was followed by a large number of men, absolutely silent and many openly in tears. 'I do not remember anything so gratifying to the feelings,' wrote Soyer, 'as that simple though grand procession.'

Mr Sabin's house had windows opening on to the Bosphorus – the most famous view in the world which, she said, she had never had time to look at – and a green tree in a garden behind. Here she began slowly to recover.

For the next few weeks she lived in the world of the convalescent, a world filled with small things. Sidney Herbert had sent her a terrier from England, and she had an owl, given her by the troops to take the place of Athena, and a baby. The baby belonged to a Sergeant Brownlow, and while its mother was washing for the hospital used to spend its day in a sort of Turkish wooden pen which she could see from her bed. Its merits, she wrote afterwards, were commemorated in the chapter on 'Minding Baby' in *Notes on Nursing*. Parthe composed and illustrated and sent her 'The Life and Death of Athena, an Owlet'. Mrs Bracebridge read it aloud while Miss Nightingale alternately laughed and cried and noticed how the terrier kept fidgeting and drawing attention to himself, 'knowing by instinct we were reading about something we loved very much and being jealous'. By July she was better and had decided she was not going away anywhere. 'If I go, all this will go to pieces,' she wrote to Parthe on July 9. Dr Sutherland told her the fever had saved her life by forcing her to rest and implored her to spare herself. She dared not. She had been compelled to leave the Crimea before she had settled anything, and she was receiving reports that the situation was going from bad to worse. Every day her authority was being more flagrantly disregarded. As soon as was humanly possible, she must go back to Balaclava and fight it out.

She spent a few days at Therapia with Mrs Bracebridge, then returned to Mr Sabin's house and resumed ordinary life. She contrived to give an impression of complete recovery. Lothian Nicholson visited her on his way up to the Crimea and was 'quite enthusiastic about her good looks'. Her cropped hair was growing in little curls which gave her a curiously touching and childish appearance.

But as she recovered the stormclouds gathered. She was about to enter the most difficult and exhausting phase of her mission. During her illness Lord Raglan died and was succeeded by General Simpson, a soldier of many years' seniority who had barely seen active service.

General Simpson's intelligence was not great, his social position inferior; he was against new-fangled notions of pampering the troops, and Miss Nightingale never succeeded in establishing the personal contact she had enjoyed with Lord Raglan. 'The man who was Lord FitzRoy Somerset' (Lord Raglan as youngest son of the Duke of Beaufort bore the title of Lord FitzRoy Somerset before being created Baron Raglan in 1852) 'would naturally not be above interesting himself in hospital matters and a parcel of women – while the man who was James Simpson would essentially think it infra dig,' she wrote in November 1855. Moreover, for some reason the official instructions as to her position and authority which had been sent by Sidney Herbert when Secretary at War to Lord Raglan were not passed on to General Simpson.

She learned that in the Crimea the kitchens which she had planned with Soyer had not been built, supplies were still being withheld from Mrs Shaw Stewart, the conduct of the nurses was still unsatisfactory. In July she sent up a French man-cook, to whom she paid £100 sterling a year out of her own private income, but the authorities refused to employ him. She requested that the ineffectual Miss Weare should be relieved as Superintendent of the General Hospital. Dr Hall's reply was to appoint Miss Weare Superintendent of the Monastery Hospital, a new hospital for ophthalmic cases and convalescents, and ignore her request.

As she was bracing herself to gather strength and return to the Crimea, a fresh blow fell. The Bracebridges wished to go home. For nine months they had shared the fearful sights, the horrible smells, the uneatable food, the insolence, the petty slights, and the perpetual rudeness. They had endured, toiled, sacrificed themselves, and yet – they had not been a complete success. Their devotion was as strong as ever, Miss Nightingale's affection as grateful. 'No one can tell what she has been to me,' she wrote of Selina, but Selina had muddled the 'Free Gift' store, and Mr Bracebridge's relations with the officials were increasingly unhappy.

Though she was barely convalescent, she would not hear of delay in the Bracebridges' departure. Everything was made easy. It was given out that they were going home for a few months and would come back in the autumn, but she knew they would never return. As soon as they sailed on July 28 she went back to her quarters at the Barrack Hospital, retaining Mr Sabin's house and sending her nurses there by turns to have a rest.

The medical authorities did not welcome her. They felt that the state of the hospital was now satisfactory and her help was not needed; there was an unwillingness to consult her and an outbreak of complaints. Orderlies caught in wrongdoing had only to say Miss Nightingale had given the order to be exonerated. Some of the admirable work of the Sanitary Commission was being undone. The engineering works were not completed, and the men began once more to drink water that looked like barley water. Trouble in controlling the nurses was continuous. Two nurses broke out one Saturday night and were brought back dead drunk. 'A great disappointment to me,' wrote Miss Nightingale, 'as they were both good-natured hard-working women.'

Nurses who did not drink got married. Lady Alicia Blackwood related that one morning six of Miss Nightingale's best nurses came into her room followed by six corporals or sergeants to announce their impending weddings. On one occasion an emissary from a Turkish official called to Miss Nightingale with an offer to purchase a particularly plump nurse for his master's harem.

She lost one of her best nurses on August 9 when Mrs Drake, from St John's House, died of cholera at Balaclava. Next she was involved in unpleasantness through the death of Miss Clough, who had got into difficulties on the 'Highland Heights'. She disliked living in a hut, could not control the orderlies, was accused of financial irregularities, quarrelled with everyone, fell ill, and asked to be sent home. On the boat she became worse and was put ashore at Scutari, where she died. Miss Nightingale had to receive her body, arrange her funeral, communicate news of her death to her relations at home, and straighten her affairs.

Much more serious trouble followed. After Mrs Bracebridge went home, Miss Nightingale appointed a Miss Salisbury to take charge of the 'Free Gift' store at a salary. From the moment she took up her post, she began writing letters home accusing Miss Nightingale of neglecting the patients, of wasting the 'Free Gifts', and of having been concerned in Miss Clough's sudden death. These letters found their way to Mary Stanley, who was now in London. Miss Salisbury next began thieving from the store on a considerable scale. A search was ordered not only of Miss Salisbury's room but of the room of two Maltese kitchen-workers whom she had introduced. The results were staggering. The beds of the Maltese were found to be entirely constructed of piles of stolen goods, while in Miss Salisbury's room every box, every package, every crevice and cranny was crammed.

Miss Nightingale summoned the Military Commandant. Lord William Paulet had just gone home and had been replaced by General Storks, a man of first-rate ability and one of her staunch admirers. The wretched Miss Salisbury was now grovelling on the floor, sobbing, screaming, and clutching at Miss Nightingale's feet, imploring her not to prosecute, but to send her home, now, at once, immediately. A grave mistake was made. Miss Nightingale wished above all things to avoid a scandal, and she and General Storks agreed that the wisest course was to send Miss Salisbury home with as little fuss as possible. She sailed immediately. But after she had gone, it was discovered that she had been stealing not only Free Gifts but government stores as well. General Storks suggested that in order to trace the stores and discover her accomplices her desk, which in the flurry of departure she had left behind, should be opened and searched and letters that came for her should be opened and read.

When Miss Salisbury arrived in England, she declared she had been ill-treated. The gifts were decaying in the store because Miss Nightingale refused to let them be used, or used them herself, and Miss Salisbury had abstracted them in order to give them to the poor fellows for whom they were intended. Why, she demanded, had not the police been called in if what Miss Nightingale asserted was true? Miss Salisbury was soon in conference with Mary Stanley, and a formal complaint against Miss Nightingale was drawn up and submitted to the War Office.

Within the War Office there were two parties, a reform party and an anti-reform party. Sending out four Commissions of Inquiry, sending out even Miss Nightingale herself, had not been accomplished without battles. The anti-reform party had been defeated and were ready to use any weapon that came to hand. At their head was Mr Benjamin Hawes, Permanent Under-Secretary at the War Office.

Miss Salisbury's complaint was submitted to Mr Hawes, and he chose to take it very seriously. An official letter was written to Miss Nightingale and General Storks – who had schemes for the reform of army administration which Mr Hawes did not find sympathetic – not inviting a report but requesting them to justify their conduct.

Miss Nightingale had now to add to her labours the fearful task of straightening out the 'Free Gift' store. Miss Salisbury's accusations and the action of the War Office became known in London, and her family blamed the Bracebridges; W. E. N., wrote Uncle Sam, '*would* give out against good B.' Someone must go out to be with Florence. Aunt Mai tactfully suggested she should go out for a short time until the Bracebridges returned –

ostensibly in the autumn – and Uncle Sam rather unwillingly consented.

On September 16 Aunt Mai arrived at Scutari. She burst into tears at her first sight of Florence, altered by her illness, thin and worn, and with her hair cut short looking curiously like the child of thirty years ago. The web of partisan intrigue, the petty thwartings, irritations, and discourtesies in which she was forced to live horrified Aunt Mai. 'The public generally imagine her by the soldier's bedside,' she wrote on September 18, 1855; '... how easy, how satisfactory if that were all. The quantity of writing, the quantity of talking is the weary work, the dealing with the mean, the selfish, the incompetent.'

The pressure of work was enormous. During her first week Aunt Mai recorded getting up at 6 a.m. and copying until 11 p.m., and next day getting up at 5 a.m. and copying again until 11 p.m.

At the beginning of October Miss Nightingale went back to the Crimea, where a new tempest had blown up, with, in its centre, Rev. Mother Bridgeman – 'Mother Brickbat'. Miss Nightingale had never succeeded in persuading Mother Bridgeman to acknowledge her authority. Mother Bridgeman had gone with her nuns to Koulali, where they issued 'extras', wine, invalid food, and clothing at their own discretion and without a requisition from the doctor in charge. Lord Panmure, on his appointment as Secretary of State for War, had asked Miss Nightingale to relinquish Koulali, and she had consented. But the lavishness there became such a scandal that the Principal Medical Officer insisted that the Scutari system must be adopted. The nuns then resigned, saying their usefulness was destroyed.

At the end of September 1855 Miss Nightingale had learned that Mother Bridgeman and her nuns, without either informing her or asking her permission, had gone to the General Hospital, Balaclava, where Mother Bridgeman was to be Superintendent. She asked Dr Hall for an explanation and he alleged that he had written her a letter asking for more nurses, but had had no reply and had been forced to take action. No such letter had been received.

Mother Bridgeman then wrote announcing that four of her nuns who were still working at the General Hospital, Scutari, were to proceed to Balaclava. Miss Nightingale pointed out that to remove nurses who were engaged in her service was against all

177

rules. Mother Bridgeman refused to give way, and Miss Nightingale appealed to Lord Stratford; it was, she wrote, impossible for her to carry on her work if interference with the control of her nurses was permitted. At the same time, in a private letter, she told him that she was quite ready to arrange for the nuns to go to Balaclava; if any women were to be at the General Hospital, Balaclava, she thought nuns the least undesirable, but arrangements must be made through her and not over her head. Lord Stratford hastened in complimentary terms to assure her of his entire agreement, but informed her that she should approach not himself but General Storks.

When Miss Nightingale considered the situation, she came to the conclusion that her personal resentment must be swallowed. Wide implications were involved. The new recruits brought out to replace the army which had perished in the winter of 1854–5 were largely Irish and Catholics, and it was already being said that they were being deprived of spiritual ministrations. 'Had we more nuns,' she wrote to Mrs Herbert in November 1855, 'it would be very desirable, to diminish disaffection. But *just not* the Irish ones. The wisest thing the War Office could do now would be to send out a few more of the Bermondsey nuns to join those already at Scutari and counter balance the influence of the *Irish* ones, who hate their soberer sisters with the mortal hatred, which, I believe, only Nuns and Household Servants *can* feel towards each other.'

She returned to the Crimea determined, in her favourite phrase, to 'arrange things'. On September 8 Sebastopol had quietly and ingloriously fallen, evacuated by the enemy, and the end of the war was only a question of time. General Simpson had resigned his command and gone home suffering from Crimean diarrhoea and been succeeded by Sir William Codrington. She was desperately anxious to keep things together, not to come to shipwreck at the eleventh hour. She was ready to conciliate – to conciliate Dr Hall, conciliate Mother Bridgeman, conciliate Mr FitzGerald, the Purveyor.

The weather was bad, sailing delayed and the passage finally made in a gale. She was prostrated. Outside Balaclava it proved impossible to make the narrow opening to the harbour or even to bring out a tug. While the transport rose and fell on huge swells, a small boat was brought alongside. A sailor held her over the side of the ship, and as the boat rose dropped her into it.

At first it seemed that she might succeed in 'arranging things'. It was an advantage to be without Mr Bracebridge: 'I find much less difficulty in getting on here without him than with him,' she wrote in November 1855. 'A woman obtains that from military courtesy (if she does not shock either their habits of business or their caste prejudice), which a man who pitted the civilian against the military element and the female against the doctors, partly from temper, partly from policy, effectually hindered.' On the surface she was on friendly terms with Dr Hall and Mr Fitz-Gerald. In fact, Mr FitzGerald went so far as to confess to her he hoped that Mother Bridgeman's nuns would not import extravagant Koulali habits into Balaclava.

And then a copy of *The Times* for October 16, 1855, arrived at Balaclava, and all her work was undone. It contained a report of a lecture given by Mr Bracebridge at the Town Hall, Coventry. Everything Mr Bracebridge had previously said, which she had implored him to refrain from saying, he had now repeated publicly. The lecture was a furious and inaccurate attack on the British Army authorities and the British Army doctors. The harm done was incalculable. Other papers reprinted Mr Bracebridge's allegations, and it was believed that Miss Nightingale had instigated a Press attack on the Army Medical Department. Everything asserted of her by Dr Hall was felt to be justified.

'When one reads such twaddling nonsense,' wrote Dr Hall to Dr Andrew Smith, 'as that uttered by Mr Bracebridge and which was so much lauded in the "Times" because the garrulous old gentleman talked about Miss Nightingale putting hospitals containing three or four thousand patients in order in a couple of days by means of the "Times" fund, one cannot suppress a feeling of contempt for the man who indulges in such exaggerations and pity for the ignorant multitude who are deluded by these fairy tales.'

Angry as Dr Hall was, he was no more furious than Miss Nightingale herself. On November 5 she told Mr Bracebridge she wished for no 'mere irresponsibility of opposition'. She objected in the strongest possible manner to his lecture, '*First*, because it is not our business and I have expressly denied being a medical officer ... *secondly*, because it justifies all the attacks made against us for unwarrantable interference and criticism, and *thirdly*, because I believe it to be utterly unfair.' Alas, the damage

179

had been done, and it was irremediable. She contemplated the wreckage of her endeavours with despair.

'I have been appointed a twelvemonth today,' she wrote to Aunt Mai, 'and what a twelvemonth of dirt it has been, of experience which would sadden not a life but eternity. Who has ever had a sadder experience. Christ was betrayed by one, but my cause has been betrayed by everyone – ruined, destroyed, betrayed by everyone alas one may truly say excepting Mrs Roberts, Rev. Mother and Mrs Stewart. All the rest, Weare, Clough, Salisbury, Stanley et id genus omne where are they? And Mrs Stewart is more than half mad. A cause which is supported by a mad woman and twenty fools must be a falling house. ... Dr Hall is dead against me, justly provoked but not by me. He descends to every meanness to make my position more difficult.'

As if she had not enough to endure, she was taken ill again and forced to enter the Castle Hospital with severe sciatica. Minus the pain, which was great, she wrote to Mrs Bracebridge that the attack did not seem to have damaged her much. 'I have now had all that this climate can give, Crimean fever, Dysentery, Rheumatism and believe myself thoroughly acclimatized and ready to stand out the war with any man.'

In a week she was up and working again, ignoring personal humiliations as long as female nursing in military hospitals might emerge as a unified undertaking at the end of the War. No official statement came to establish her authority, and Dr Hall gave out that she was an adventuress and to be treated as such. Minor officials treated her with vulgar impertinence. The Purveyor refused to honour her drafts. When she went to the General Hospital, she was kept waiting.

But she would not be provoked. She persisted in visiting Mother Bridgeman, and when Sister Winifred, a lay sister from Mother Bridgeman's party, died of cholera, she went to the funeral and joined in the prayers. 'Mother Brickbat's conduct has been neither that of a Christian, a gentlewoman, or even a woman,' Miss Nightingale wrote to Mrs Herbert. 'At the same time I am the best personal friends with the Revd Brickbat and I have even offered to put up a cross to poor Winifred to which she has deigned no reply. But anything to avoid a woman's quarrel which *can* be done or submitted to on my part *shall* be done – and submitted to.'

All she had accomplished by coming to the Crimea, she wrote, was that the extra diet kitchens which should have been erected in May were erected in November. At the end of November she was hastily summoned back to Scutari, where a new cholera epidemic had broken out. Before she left, she wrote to Sidney Herbert: 'There is not an official who would not burn me like Joan of Arc if he could, but they know the War Office cannot turn me out because the country is with me – that is my position.' The admiration and affection with which the people of England regarded her roused in the Crimean authorities dislike and distrust. But their masters at home, Ministers to whom public opinion was of importance, had a different outlook, and in November, when her prestige in the Crimea had never been so low or her difficulties so great, an astonishing demonstration of public feeling and affection in England placed her in the position of a national heroine whom no one could afford to ignore.

II

A LEGEND had been growing up in England, the result of the survivors of the British Army coming home and telling up and down the country the story of Miss Nightingale and the Barrack Hospital. The legend was born and gained strength in cottages, tenements, and courts, in beer houses and gin shops. The rich might grow romantic, and dukes, in the slang of the day, declare themselves 'fanatico for the new Joan of Arc', but the legend of Florence Nightingale belonged to the poor, the illiterate, the helpless, whose sons and lovers she refused to treat as the scum of the earth. 'The people love you,' wrote Parthe, 'with a kind of passionate tenderness which goes to my heart.'

The hacks of Seven Dials, where topical doggerel was produced for the mob, hymned her in innumerable songs. 'The Nightingale in the East',* decorated with a wood-cut of, apparently, a lady reposing in a tent bed, and to be sung to the tune of the 'Cottage and the Wind Mill', was still popular at regimental reunions fifty years later. One of its eight verses runs:

> Her heart it means good for no bounty she'll take,
> She'd lay down her life for the poor soldier's sake;
> She prays for the dying, she gives peace to the brave,
> She feels that the soldier has a soul to be saved,
> The wounded they love her as it has been seen,
> She's the soldier's preserver, they call her their Queen.
> May God give her strength, and her heart never fail,
> One of Heaven's best gifts is Miss Nightingale.

Another adapted the popular song 'The Pilot that weathered the storm' to 'Fair Florence who weathered the storm'. Another en-

* Contrary to present custom, Scutari, the Crimea, and even Constantinople, are described by Miss Nightingale and her contemporaries as being 'in the East'. She writes of 'my time in the East', the Nightingale Fund Committee speaks of her 'services in the Hospitals of the East', the War Office addresses orders 'to the Army in the East'.

titled 'God Bless Miss Nightingale' contained sentiments which, in the circumstances, were ironical:

> God bless Miss Nightingale,
> May she be free from strife;
> These are the prayers
> Of the poor soldier's wife.

Others were 'Angels with Sweet Approving Smiles', 'The Star in the East', 'The Shadow on the Pillow', 'The Soldier's Cheer'.

Quantities of a biography, printed in Seven Dials, were sold, price one penny, *'The only and unabridged edition of the Life of Miss Nightingale. Detailing her Christian and Heroic Deeds in the Land of Tumult and Death which has made her Name Deservedly Immortal, not only in England but in all Civilised Parts of the World, winning the Prayers of the Soldier, the Widow and the Orphan.'*

A Staffordshire figure labelled 'Miss Nightingale' depicts her not with the famous lamp, but carrying two cups on a small tray and romantically dressed in a long, white flowered skirt, a blue bodice with a pink bow, and wearing red slippers. Her portrait was eagerly demanded, but the family did not dare supply it because she had an objection to having her portrait circulated. The likenesses of her were imaginary; one print shows her as a lady with a Spanish comb in her hair, dark and passionate; another depicts a golden-haired Miss in a bower of roses. Strangers called at Embley and asked to be allowed to see her desk. Shipowners named their ships after her. A life-boat was called the *Florence Nightingale*, one of the crew writing first to make sure the name was 'got all correct'. Sir Edward Cook quotes a newspaper cutting which records that 'The Forest Plate Handicap was won by Miss Nightingale beating Barbarity and nine others'. A popular tableau at Madame Tussaud's presented 'A Grand Exhibition of Miss Florence Nightingale administering to the Sick and Wounded'.

The successive tidings of her illness, her recovery, and her determination to stay at her post until the end of the war raised public feeling to boiling-point, and Sidney Herbert felt the authorities might usefully be reminded that she had the country at her feet. A committee was formed of which Richard Monckton Milnes was a member and Sidney Herbert honorary secretary, and on November 29, 1855, a public meeting was called at Willis's Rooms, in St James's Street, 'to give expression to a general feeling that the services of Miss Nightingale in the Hospitals of the East demand the grateful recognition of the British people'. The Duke of Cambridge was chairman, the Duke of Argyll, Lord Stanley, Sidney Herbert, and Richard Monckton Milnes made speeches, and Sidney Herbert read the letter from Scutari in which a soldier described the men kissing Miss Nightingale's shadow as she passed, which suggested the poem 'Santa Filomena' to Longfellow. The meeting was crowded to suffocation and wildly enthusiastic, and similar meetings were held throughout the country. The first intention was to present an article of gold or silver suitably inscribed, 'something of the teapot and bracelet variety', wrote Parthe, but so much money came in that it was decided to establish a Nightingale Fund, to enable Miss Nightingale to 'establish and control an institute for the training, sustenance and protection of nurses paid and unpaid'. The Nightingales did not attend the meeting – Parthe and Fanny were afraid they would be overcome by emotion, W. E. N. that he might be asked to speak. After the meeting Fanny held a reception of 'notabilities' in her sitting-room at the Burlington Hotel, and wrote: 'The 29th of November. The most interesting day of thy mother's life. It is very late, my child, but I cannot go to bed without telling you that your meeting has been a glorious one ... the like has never happened before, but will, I trust, from your example, gladden the hearts of many future mothers.' Miss Nightingale wrote back quietly: 'My reputation has not been a boon in my work; but if you have been pleased that is enough.'

The formation of the Nightingale Fund was mentioned in General Orders to the Army in the East, and it was suggested that subscriptions should take the form of contributing a day's pay. Dr Hall refused, but otherwise the response was good and nearly £9,000 sterling was subscribed by the troops.

After the formation of the fund Queen Victoria, to 'mark her warm feelings of admiration in a way which should be agreeable', presented a brooch designed by the Prince Consort, a St George's Cross in red enamel surmounted by a diamond crown; the cross bears the word 'Crimea', and is encircled with the words 'Blessed are the merciful'. On the reverse is the inscription, 'To Miss Florence Nightingale, as a mark of esteem and gratitude for her devotion towards the Queen's brave soldiers from Victoria R. 1855.'

Miss Nightingale was not gratified: praise, popularity, prints, jewels left her unmoved. She wrote to Parthe in July: 'My own effigies and praises were less welcome. I do not affect indifference towards *real* sympathy, but I have felt painfully, the more painfully since I have had time to hear of it, the éclat which has been given to this adventure of mine. ... Our own old party which began its work in hardship, toil, struggle and obscurity has done better than any other. ... The small still beginning, the simple hardship, the silent and *gradual* struggle upwards; these are the climate in which an enterprise really thrives and grows. ...'

In reply to the resolution sent by the committee of the Nightingale Fund she wrote that she could not contemplate undertaking anything in addition to her present work, and would only accept the fund on condition that it was understood that there was great uncertainty as to when she would be able to employ it. The fact was that the organization and reform of nursing no longer filled her whole horizon. Nursing had become subsidiary to the welfare of the British Army.

She had set herself a new and gigantic task – she had determined to reform the treatment of the British private soldier. A mystical devotion to the British Army had grown up within her. In the troops she found the qualities which moved her most. They were victims; her deepest instinct was to be the defender of victims. They were courageous, and she instantly responded to courage. Their world was not ruled by money, and she detested materialism. The supreme loyalty which made a man give his life for his comrade, the courage which enabled him to advance steadily under fire, were displayed by men who were paid a shilling a day.

She did not sentimentalize the British private soldier. 'What has he done with the £1 – drank it up I suppose,' she scribbled at Scutari. 'He asks us to find a post for his wife,' runs another note; 'he had better say which wife.' Queen Victoria offered to send eau de cologne for the troops, but she said someone had better tell her a little gin would be more popular. She was told one of the wounded wanted company and observed she knew the company he pined for, that of a brandy bottle under his pillow. She was content to accept and love the troops as she accepted and loved children and animals. She called herself the mother of 50,000 children.

'I have never,' she wrote to Parthe in March 1856, 'been able to join in the popular cry about the recklessness, sensuality, and helplessness of the soldiers. On the contrary I should say (and perhaps few women have ever seen more of the manufacturing and agricultural classes of England than I have before I came out here) that I have never seen so teachable and helpful a class as the Army generally. Give them opportunity promptly and securely to send money home and they will use it. Give them schools and lectures and they will come to them. Give them books and games and amusements and they will leave off drinking. Give them suffering and they will bear it. Give them work and they will do it. I would rather have to do with the Army generally than with any other class I have attempted to serve.'

At Scutari – and Scutari was a typical army depot – the troops were given no opportunities but to drink. When a man became convalescent, he was discharged to the Depot, and inevitably he drank in the spirit shops and drank liquor so poisonous that men frequently fell down after swallowing only a small quantity. A large proportion of every batch of convalescents was carried back drunk within twenty-four hours. 'Dead drunk,' said Miss Nightingale, 'for they die of it and the officers look on with composure.'

It became clear to her that she must look after the troops not only when they were ill but when they were well. What she did for them outside the hospital was as important as what she did inside the hospital.

In May 1855 after strenuous opposition, she opened a small reading-room for men able to walk but not to leave the hospital. The authorities feared that the men would get above themselves if they read instead of drinking, and she was accused of 'destroying discipline'. However, their conduct was excellent. She found that a great many of the men could neither read nor write, and she asked if she might engage a schoolmaster. This was absolutely refused. 'You are spoiling the brutes,' Lord William Paulet told her.

She discovered the men drank their pay away because they were dissatisfied with the official method of sending money home through the Paymaster. Rightly or wrongly they believed they were defrauded and their ignorance exploited. She made it a practice to sit in her room for one afternoon a week and receive the money of any soldier in the hospital who desired to send it home to his family. The money went to Uncle Sam, who bought postal orders which he dispatched to the various addresses. About £1,000 sterling a month was brought in. When the men were discharged from hospital and rejoined their regiments in the Crimea, they wished to continue sending money home through the post. She submitted a scheme to the authorities, but it was refused.

In November 1855 in her letter of thanks for Queen Victoria's brooch, she laid before the Queen the causes and remedies of the prevalent drunkenness in the army and the men's difficulties in remitting money home. On December 21 the Queen sent the letter down to a Cabinet meeting. Palmerston, the Prime Minister, thought it excellent, and Clarendon, the Foreign Secretary, said it was full of real stuff; but Panmure said it only showed that she knew nothing of the British soldier. The same day he wrote to Sir William Codrington, the Commander-in-Chief in the Crimea: 'the great cry now, and Miss Nightingale inflames it, is that the men are too rich; granted, but it is added that they have no means to remit their money home. In vain I point out that this is not true. . . . We have now offered the Post Office to them, but I am sure it will do no good. The soldier is not a remitting animal.'

Lord Panmure proved wrong. Offices where money orders could be obtained were opened at Constantinople, Scutari, Balaclava, and the headquarters of the camp outside Sebastopol, and £71,000 sterling were sent home in less than six months. It was, said Miss Nightingale, all money saved from the drink shops.

When Lord William Paulet was replaced by General Storks she found an enthusiastic collaborator; working hand in hand, they brought discipline and order to the Barrack Hospital and its neighbourhood. First, the drink shops were closed, and the streets of the village and the surrounding neighbourhood were patrolled after dark. Next, in September 1855, a large recreation room for the army called the Inkerman Coffee House was opened, with the aid of private funds, in a wooden hut between the two hospitals. A second recreation room for patients in the Barrack Hospital was opened in a wooden hut in the courtyard. There were no rules except that women were not allowed. The walls were hung with maps and prints; the room was furnished with armchairs and writing-tables; newspapers and writing materials were paid for by Miss Nightingale. 'The men,' she wrote, 'sat there reading and writing their letters, and the Library of the British Museum could not have presented a more silent or orderly scene.' The officers had told her that the men would steal the notepaper and sell it for drink, but this never occurred.

By the spring of 1856, four schools, conducted by professional schoolmasters, had been opened. 'The lectures,' she wrote, 'were crowded to excess so that the men would take the door off the hut to hear. Singing classes were formed and the members allowed to sing in the Garrison Chapel. The men got up a little theatre for themselves, for which dresses and materials were lent by a private hand, and this theatre was, I believe, always perfectly orderly. Football and other games for the healthy, dominoes and chess for the sick, were in great request. ... A more orderly population than that of the whole Command of Scutari in 1855–1856, though increased by the whole of the Cavalry being sent down there for winter quarters, it is impossible to conceive.'

It was an astonishing achievement, and during the winter of 1855–6 the picture of the British soldier as a drunken intractable brute faded away never to return. 'She taught,' said an eyewitness, 'officers and officials to treat the soldiers as Christian men.'

But the achievement had been accomplished only at the price of unremitting toil. Throughout the summer of 1855, when she was desperately ill, through the autumn when she was alone, weak, and crushed by the enormous demands of her official work, she had somehow to accomplish the additional heavy correspondence, the persuading, the interviewing, the accounting and listing involved in her welfare work. She had never completely recovered from her illness in May. She still had sciatica, she was losing weight rapidly, and her ears gave trouble.

Lady Hornby, wife of the British Commissioner to Turkey, met her at a Christmas party given by Lady Stratford. 'I felt quite dumb,' Lady Hornby wrote, 'as I looked at her wasted figure and the short brown hair combed over her forehead like a child's, cut so when her life was despaired of from a fever but a short time ago. Her dress ... was black, its only ornament being a large enamelled brooch, which looked to me like the colours of a regiment. ... To hide the close white cap a little, she had tied a white crepe handkerchief over the back of it, only allowing the border of lace to be seen. . . . She was still very weak, and could not join in the games, but she sat on a sofa, and looked on, laughing until the tears came into her eyes.'

At home there was no conception of the situation confronting Miss Nightingale. The welfare work had succeeded, but in every other direction she was failing. The good she had done was being undone, the decisions she had formed were being reversed, and she was not only helpless but perpetually tormented by official spite.

The Depot within the Barrack Hospital building had been condemned by the Sanitary Commission and the troops evacuated. Faced with difficulty in procuring recruits to replace the army lost before Sebastopol, the Government raised a German Legion of mercenaries, and, ignoring all protests, the military authorities put the Legion into the Depot. Cholera broke out and spread to the hospital, and one of the first to die was Dr McGrigor. His successor refused to allow the nurses to administer medicine and restricted their duties to feeding the patients and changing their beds. The hospital at Koulali finally collapsed, and Miss Nightingale found herself saddled with the unpleasant task of winding up its affairs. The Superintendent of a hospital for officers at one of the Sultan's Palaces died, and she was requested to take the patients into the Barrack Hospital. She agreed, and two scandals ensued. One nurse was accused of ill-treating her patients, another of being too kind to them and receiving visits at midnight. Most sordid, most heart-breaking of all, was the case of Miss Salisbury.

Night after night, when the enormous mass of her daily work had been done, she must sit up wrestling with her statement for the War Office. It was bitterly cold; the stove sent out from England would not draw, and the charcoal brazier made her head ache. 'They are killing me,' she told Aunt Mai. When the New Year of 1856 dawned, her health had still further deteriorated; she had earache, continual laryngitis, and found it difficult to sleep. In the dark icy cold she paced her room obsessed by failure. 'The victory is lost already,' she told Aunt Mai. 'But it is won on some points.' Aunt Mai reminded her, and added, when reporting the conversation to Uncle Sam, 'if you could hear what the Hospital was and what it is through her struggles you would say so.' It was unbelievable, wrote Aunt Mai again, how she worked; Aunt Mai could never have imagined any labour so unceasing, so unending. 'Food, rest, temperature never interfere with her doing her work. You would be surprised at the temperature in which she lives ... she who suffers so much from cold. ... She has attained a most wonderful calm. No irritation of temper, no hurry or confusion of manner ever appears for a moment.'

But the calm was only on the surface. Aunt Mai wrote confidentially to Mrs Herbert that after a long difficult interview Florence often seemed about to faint with exhaustion. After the interviews in connexion with the Salisbury case she had collapsed on several occasions. She lay on the sofa unable to speak or eat, and yet if anyone came to see her on business she pulled herself together and appeared normal.

In January 1856 the McNeill and Tulloch Commission into the Supplies for the British Army in the Crimea laid its final report before Parliament and confirmed what Miss Nightingale had already told Sidney Herbert. The disaster of the winter of 1854–5 had been unnecessary, a compound of indifference, stupidity, inefficiency, and bureaucracy. The report, though restrained and dispassionate, named a number of senior officers as being negligent, indifferent, and inefficient. Among them were Lord Cardigan, Lord Lucan, and Sir Richard Airey, the Quartermaster-General. The facts in the report had been communicated to the Government six months ago; yet almost all these officers had been promoted or decorated, and none had been removed from his position. The publication of the report created a storm, and Lord Panmure directed a Board of General Officers to assemble at Chelsea to 'allow the officers adverted to in the report to have an opportunity of defending themselves'. Extensive whitewashing was to be done. Immediately following the establishment of the Board, a fresh list of decorations and promotions was published. Benjamin Hawes got a K.C.B. and so did Dr John Hall. 'Knight of the Crimean Burial grounds I suppose,' wrote Miss Nightingale.

It seemed the triumph of all she had been fighting against, the final defeat of justice by power. 'I am in a state of chronic rage,' she wrote on March 3, 1856, 'I who saw the men come down through all that long dreadful winter, without other covering than a dirty blanket and a pair of old regimental trousers, when we knew the stores were bursting with warm clothing, living skeletons devoured by vermin, ulcerated, hopeless, speechless, dying like the Greeks as they wrapped their heads in their blankets and spoke never a word. ... Can we hear of the promotion of the men who caused this colossal calamity, we who saw it? Would that the men could speak who died in the puddles of Calamita Bay!'

There seemed no end to weariness, disillusion, falseness. Lady Stratford, fêting her on Christmas Day, turned aside to assure a visitor she fully believed Miss Salisbury's story. Mary Stanley, busily spreading rumours at home, yet wrote letters breathing devoted affection. Mother Bridgeman was still unsubdued and rebellious in the Crimea, and Miss Nightingale's own position was still officially unsupported. Now a new misery was added. At the beginning of December Mr FitzGerald, the Chief Purveyor in the Crimea, encouraged by Miss Salisbury's success, wrote a 'Confidential Report' on Miss Nightingale and her nurses which were forwarded by Sir John Hall to sympathetic quarters at the War Office. It was not a report but a series of accusations. She herself was accused of insubordination; her nurses were described as dishonest, extravagant, disobedient, inefficient, and immoral. Mother Bridgeman and her nuns were warmly commended for zeal, skill, economy, and obedience.

It was, said Miss Nightingale, 'a tissue of unfounded assertions, wilful perversions, malicious and scandalous libels,' but the readiness with which its malice was received and exploited in official quarters added enormously to her difficulties.

Fact was disregarded. She was accused of unjustly removing Mrs Shaw Stewart from being Superintendent of the General Hospital, Balaclava, when in fact Mrs Shaw Stewart, Miss Nightingale's personal friend and staunch supporter, had been urgently requisitioned by the medical authorities to put the newly opened Castle Hospital on its feet. The Castle Hospital contained twice as many patients as the General Hospital and was a promotion.

In May 1855 she had wished to replace Miss Weare as Superintendent of the General Hospital. Dr Hall refused, advancing as a reason Miss Weare's successful management of the sick officers. In the Confidential Report it was stated that Miss Nightingale persisted in maintaining Miss Weare at her post in spite of the fact that Dr Hall had demanded her removal on account of her unsuccessful treatment of sick officers.

So immoral were her nurses, alleged the Confidential Report, that five had been sent home for promiscuous conduct on a single ship. Of the five nurses named, four were her 'very best nurses', honourably invalided home 'broken by their exertions'. One had actually been officially commended by Mr FitzGerald himself. The fifth had not gone home at all but was working at Scutari.

Mother Bridgeman's nuns were commended for their economy and obedience to the medical authorities, though in fact the nuns had left Koulali on account of their enormous expenditure, and Mr FitzGerald himself had said he hoped they would not bring their system with them to Balaclava.

Miss Nightingale was made aware of the existence of the report through Lady Cranworth, who was a friend of Sir Benjamin Hawes. The report was shown to Lady Cranworth, and it was intimated that Miss Nightingale would be wise to make a reply. She was not allowed to see the report, the substance only of the allegations was conveyed. However, she wrote a full refutation. In reply she was told that her statement 'in some cases did not meet the exact point'. There was no other comment. Since she had never seen the original document, the result was not surprising.

In February 1856 her difficulties reached a climax. She was asked to send nurses to the Crimea by the Chief Medical Officer of the Land Transport Corps. But the situation was such that she doubted if she dared send nurses. Mr FitzGerald, elated by the success of his Confidential Report, was refusing to honour her drafts; she was owed £1,500 sterling which she could not get; Mother Bridgeman reigned at Balaclava; the Hall and FitzGerald party was openly declaring they intended to root her out of the Crimea.

She had already written, on January 7, an official letter to the War Office complaining of Sir John Hall's action in sending Mother Bridgeman to Balaclava over her head, but it had brought no result. Before she went further, she wrote to Dr Sutherland, who was at Balaclava, asking him if he thought it wise for her to bring nurses to the Crimea. On February 4 Dr Sutherland told her she should make no such attempt. He advised her to 'state a case fully to the War Department and ask them to place you on a proper footing with the authorities here'. The position, he said, turned on the employment of the words 'in Turkey', which the Crimean authorities contended did not include the Crimea.

However, there were indications that Sir John Hall was uneasy. He withdrew his support from Mr FitzGerald: the statements in the Confidential Report, he said, were made on Mr FitzGerald's personal responsibility. 'Mr FitzGerald is in fact thrown overboard,' wrote Dr Sutherland. Miss Nightingale scribbled on the margin of the letter, 'I am not at all surprised. ... I always expected it.'

On February 20, 1856, she wrote formally to Sidney Herbert enclosing Dr Sutherland's letter and asking him to urge the War Department to telegraph a statement of her powers to the military and medical authorities in the Crimea. 'It is obvious that my usefulness is destroyed, my work prevented or hindered, and precious time wasted, by the uncertainty of the relations in which I am left with the Crimean authorities.'

On the same day she wrote him a private letter. She was very angry. 'The War Office gives me tinsel and plenty of praise, which I do not want, and does not give me the real business like efficient standing which I do want. ... The War Office sent me here. And surely it should not leave me to fight my own battle. ... If they think I have not done my work well, let them re-call me. But, if otherwise, let them not leave me to shift for myself, in an ever recurring and exhausting struggle for every inch of the ground secured to me by the original agreement.' She demanded that he should read the correspondence relating to the Confidential Report which was at Lady Cranworth's house and move for the production of papers in the House of Commons. 'This is bad treatment. ... I am assured that the people of England would not suffer this for me.'

In reply he assured her that her position was to be cleared up. By some mischance it appeared that the new Commander-in-Chief, Sir William Codrington, like General Simpson, was unaware of her official status. As for moving for papers in the House of Commons, he refused. 'I am going to criticize you and scold you,' he wrote. 'You have been overdone with your long, anxious, harassing work. ... The Salisbury party, the Stanley party would, of course, take up the Hall and FitzGerald view and press their particular cases, and the public, distracted, indolent, weary would settle that it was a pack of women quarrelling among themselves, that it is six of one and half a dozen of the other, and everyone is equally to blame all round. ... These are misrepresentations and annoyances to which all persons in office, and you are in office, are exposed – a single flower of the sort from which the bed of roses on which Secretaries of State repose is made.' She answered on March 6 that she was being asked to do the work of a Secretary of State without the status of a Secretary of State. The War Office was feeble and treacherous, she wrote, and she pictured them saying: 'Could we not shelve Miss N? I daresay she does a great deal of good but she quarrels with the authorities and we can't have that.'

On March 10 Sir John Hall wrote her a suave and courteous letter inviting her to bring ten nurses to the hospital of the Land Transport Corps. She accepted the invitation, but attached so little importance to his good-will that she took with her everything she and her nurses could need, not only food but stoves.

On the day she left, March 16, 1856, a dispatch, establishing her position in terms far beyond anything of which she had ever dreamed, reached the Crimea.

The dispatch had a curious history. In October 1855 a certain Colonel Lefroy, with the title of 'Confidential Adviser to the Secretary of War on Scientific Matters', appeared, first in Scutari then in the Crimea. He was, in fact, engaged on a secret mission. He was to observe and report to Lord Panmure the truth about the state of the hospitals. He conceived the greatest admiration for Miss Nightingale; they became intimate friends, and he assisted in her welfare work for the troops. Colonel Lefroy reached home at the beginning of February. He possessed, and was said to be the only man who did possess, very great influence over Lord Panmure. He pressed her case with warmth. In an official minute he wrote that the medical men were jealous of her mission – Sir John Hall would gladly upset it tomorrow if he could. She had asked for a telegram defining her position, but Colonel Lefroy went further; he wished her to have the unique distinction of her name in General Orders: the bulletin issued daily by the Commander-in-Chief and posted in every barrack and mess. 'A General Order,' he wrote, '... is due to all she has done and sacrificed. Among other reasons for it, it will put a stop to any spirit of growing independence among those ladies and nurses who are still under her, a spirit encouraged with no friendly intention in more than one quarter.'

A battle ensued. Minutes flew backward and forward. It was pointed out to Lord Panmure that the dispatch as worded amounted to a censure on Sir John Hall, but Panmure refused to alter it. Curious information was reaching him as to the state of the General Hospital, Balaclava, from sources other than Colonel Lefroy. At the end of December Sir William Codrington had written complaining of the amount of liquor consumed by the sick. Lord Panmure decided that the dispatch was to be issued as it stood. On February 25 the dispatch left the War Office and was published by Sir William Codrington in General Orders on March 16. 'The Secretary of State for War has addressed the following despatch to the Commander of the Forces, with a desire that it should be promulgated in General Orders: "It appears to me that the Medical Authorities of the Army do not correctly comprehend Miss Nightingale's position as it has been officially recognized by me. I therefore think it right to state to you briefly for their guidance, as well as for the information of the Army, what the position of that excellent lady is. Miss Nightingale is recognized by Her Majesty's Government as the General Superintendent of the Female Nursing Establishment of the military hospitals of the Army. No lady, or sister, or nurse, is to be transferred from one hospital to another, or introduced into any hospital without consultation with her. Her instructions, however, require to have the approval of the Principal Medical Officer in the exercise of the responsibility thus vested in her. The Principal Medical Officer will communicate with Miss Nightingale upon all subjects connected with the Female Nursing Establishment, and will give his directions through that lady."'

It was triumph. It was more than she had ever asked. It was complete defeat for the Hall party, the Stanley party, the Salisbury party. She reached Balaclava on March 24, in a blinding snowstorm, and was formally welcomed by Sir William Codrington. Next day she took her ten nurses to the Hospital of the Land Transport Corps about a mile and a half from Balaclava.

Her struggle was over, and the war was all but over, too. A Peace Conference was meeting in Paris, hostilities had ceased, and a formal declaration of peace was expected at any moment. Once more she strove to compose her differences with Mother Bridgeman. Anything Mother Bridgeman wished Miss Nightingale would do. Once more she failed. Mother Bridgeman refused to submit to Miss Nightingale's control, she refused to be 'humbled' or 'mortified', and she insisted on going home at once. 'I have piped to her and done the Circe in vain,' Miss Nightingale wrote to Sidney Herbert. On March 28, with a passage arranged by Sir John Hall and glowing testimonials from him and Mr FitzGerald in her pocket, Mother Bridgeman sailed for home.

Yet, though defeated, Sir John Hall and Mr FitzGerald were still able to make themselves unpleasant. Sir John Hall questioned and delayed Miss Nightingale's requisitions, made difficulties about the nurses' duties, raised points that had been settled months ago, even objected to using the extra diet kitchens. Mr FitzGerald in turn contrived to deprive her party of rations. She applied to Sir John Hall and was informed that he could entertain no complaints relative to the Purveyor. On April 4 she wrote to Sidney Herbert: 'We have now been ten days without rations. ... I thank God my charge has felt neither cold nor hunger. ... I have, however, felt both. ... During these ten days I have fed and warmed these women at my own private expense by my own private exertions. I have never been off my horse until 9 or 10 at night, except when it was too dark to ride home over these crags even with a lantern, when I have gone on foot. During the greater part of the day I have been without food necessarily, except a little brandy and water (you see I am taking to drinking like my comrades of the Army). But the object of my coming has been attained and my women have neither suffered nor starved.' She had fed her party on the provisions she had brought from Scutari and cooked on her own stoves.

Two of the hospitals were five miles up the country; the Monastery Hospital was five miles the other way; there were no roads, only rough tracks. Now the weather had become bitterly cold, and snowstorms were continuous. Soyer, who accompanied her, wrote: 'The extraordinary exertions Miss Nightingale imposed on herself ... would have been perfectly incredible if not witnessed by many and well ascertained. ... I have seen that lady stand for hours at the top of a bleak rocky mountain near the hospitals, giving her instructions while the snow was falling heavily.' The long hours in the saddle without food proving too much for her weakened health, a mule cart was procured, but one night it overturned on one of the rough tracks. Colonel McMurdo then presented her with a springless, hooded baggage cart, which gave some protection from the weather. This was the Crimean carriage in which she said she henceforward 'lived', and which is still preserved at St Thomas's Hospital, London.

Sir John Hall's ingenuity in petty persecution was inexhaustible. On March 28, Mother Bridgeman and her nurses having left the General Hospital, Miss Nightingale went down to take over and found the nurses' huts locked. Sir John Hall had given the keys to Mr FitzGerald, who had locked the doors and taken the keys away. A message was sent to Mr FitzGerald. Could Miss Nightingale have the keys, please – she was outside the huts and would wait there until the keys came. It was late in the afternoon and snowing. She waited, an hour and another hour. Darkness fell. Still she waited with the snow thick on her, and at last the keys came.

The General Hospital was filthy. 'The patients were grimed with dirt, infested with vermin, with bed sores like Lazarus (Mother Bridgeman I suppose thought it holy),' wrote Miss Nightingale on April 17, 1856. The Bermondsey nuns were horrified. After two days had been spent in washing, scrubbing, and disinfecting, and three days in cleansing the patients and their bedding – one man was such a mass of bed sores that it took six hours daily to dress him – Sir John Hall paid a visit of inspection and at once wrote an angry letter. He was 'disgusted at the state of the hospital and ordered it all to be put back in the admirable order it was in previously', and he desired the Principal Medical Officer in charge of the hospital 'not to interfere with the Purveyor Mr FitzGerald's admirable arrangements'.

It was a letter designed for the official file, to stand as a useful piece of evidence if the state of the Balaclava General Hospital under Mother Bridgeman was ever called in question. Miss Nightingale did not answer it; she was sickened.

The only 'diversion' was provided by Miss Weare, who had been sent to the Monastery Hospital. Though over seventy, Miss Weare was causing a scandal. 'She spent,' wrote Miss Nightingale, 'much of her time cooking good things, eating, drinking, and gossiping with an old Medical Officer until the small hours of the morning. I don't think I ever felt in such a ridiculous position in my life, as when I was called upon by the authorities to put a stop to the midnight gossipings as causing "scandal" and I had to speak to these two old fogies each of whom was twice as old as myself. Both were over 70.'

On April 29 peace was proclaimed to the Allied Armies, but she felt no exultation; she looked forward with a sense of doom. 'Believe me when I say that everything in the Army (in point of routine versus system) is just where it was eighteen months ago. . . . "Nous n'avons rien oublié ni rien appris." . . . In six months all these sufferings will be forgotten.'

English and Russian soldiers were fraternizing and getting

197

drunk together. English officers were getting up steeplechases and breaking their necks. Interpreters were in request to arrange for collections of Crimean crocuses and hyacinths to be sent home to English gardens, and Lord Panmure was writing to Sir William Codrington on the importance of bringing the army home without beards.

The nurses began to go home by detachments, and one of the first was Jane Evans, the old farm-worker, 'made happy by the possession of a buffalo calf she had reared, to which beast, with herself, a free passage was granted'. Mrs Shaw Stewart went home, the persistent martyr who had been the prop and mainstay of Crimean nursing. Rev. Mother Bermondsey was invalided from Scutari – 'What you have done for the work no one can ever say,' wrote Miss Nightingale. 'My love and gratitude will be yours dearest Reverend Mother wherever you go. ... I do not presume to give you any other tribute.' Miss Tebbutt, one of Mary Stanley's 'ladies' who had proved an excellent nurse, went to Embley to rest, and Miss Nightingale wrote, 'as she has only a mother at home it would give great pleasure if the mother were invited too.' Miss Noble, another of Miss Stanley's 'ladies', went to take up a post Miss Nightingale had procured for her. 'She has been one of our best, kindest and most skilful surgical nurses, I feel a real attachment for her,' she wrote to Lady Cranworth. Every nurse was to be provided for. No one was to be 'thrown off like an old shoe'. 'That they remained with me I consider proof that I considered them, on the whole, useful to the work and worthy of having a part in it,' she wrote. Those she did not feel she could ask the Government to assist, she helped out of her private pocket. A nurse who had been drunken, but had pulled herself together and done well was to be met when she arrived in London, given money if she needed it, and have a place found for her. A Miss Laxton had been disgraced for receiving visits from an officer; Miss Nightingale thought she had been too severely treated and asked W. E. N. to meet her and supply her with money. One thing only she implored – that her party should keep out of print. 'If I do not conclude our campaign without saving all my ladies and nurses from expressing themselves in print (Oh that mine enemy would write a book!) I shall think myself quite out generalled.'

At the end of June she returned to Scutari, where the camp was empty, the Inkerman Café deserted, only a few convalescents lingering where once lines of dying men had lain on the bare floor. Father Michael Cuffe went home, who once had compared her to Herod but now, she wrote, 'ate out of her hand'. The Sellonite sisters, the 'ancient dames in black serge' who had proved among the best of her nurses, departed in tears.

On July 16, 1856, the last patient left the Barrack Hospital, and her task was ended.

Once more Fanny and Parthe began to hope. Surely she must be satisfied at last; surely now she would come and live at home, repose on her laurels and enjoy them. Would she accept an official reception, or should they meet her privately at Aix? They wrote Aunt Mai a great many letters. What were Florence's plans?

Aunt Mai answered that she mentioned no plans: she seemed in high spirits, in great good looks, but they must make no mistake, her health was seriously shaken. She was painfully thin and, when alone, deeply depressed. She did not enjoy her fame; she was afraid of it. Her reputation stood so high that whatever she did must disappoint expectations. As to her agreeing to settle down at home, of that Aunt Mai, hastening tactfully to agree that nothing could be more desirable, held out no hopes whatsoever.

The nation passionately desired to honour her. She had emerged from the War with the only great reputation on the British side. The Government offered a man-of-war to take her home in state, and Parthe wrote that 'the whole regiments of the Coldstreams, the Grenadiers and the Fusiliers would like to meet her, or failing that they would like to send their bands to play her home wherever she might arrive, by day or night'.

The Mayors of Folkestone and Dover desired Mr Augustus Stafford to 'find out privately where Miss Nightingale would first touch English ground in order to rouse the whole community'. There was a rumour that she would go to Lea Hurst. Committees met; triumphal arches were planned; there were to be bands, processions, addresses from the parish, and a carriage drawn by the neighbourhood to take her home. She rejected everything. She was bereaved; a haunted woman. She began to write private notes again: 'Oh my poor men; I am a bad mother to come home and leave you in your Crimean graves – 73 per cent in 8 regiments in 6 months from disease alone – who thinks of that now?' At night Aunt Mai heard her pacing endlessly up and down.

On July 28 she embarked at Constantinople for Marseilles, travelling incognito with Aunt Mai as 'Mrs and Miss Smith'. A

Queen's Messenger travelled in the same boat to attend to formalities. There preceded her what she called her 'Spoils of War': a one-legged sailor boy, a Russian orphan, a large Crimean puppy, and a cat had already arrived at Embley.

From Marseilles she went to Paris, where she left Aunt Mai and walked in unexpectedly at 120 rue du Bac. M. Mohl was at home, but Clarkey was in England. She stayed the night and next day went on alone to England. At eight in the morning she rang the bell at the Convent of the Bermondsey Nuns. It was the first day of their annual retreat, and she spent the morning in prayer and meditation with Rev. Mother. In the afternoon she took the train north, still alone, and in the evening walked up from the station to Lea Hurst.

Parthe, Fanny, and W. E. N. were in the drawing-room, but Mrs Watson, the housekeeper, was sitting in her room in front of the house. She looked up, saw a lady in black walking alone up the drive, looked again, shrieked, burst into tears and ran out to meet her.

Two figures emerged from the Crimea as heroic, the soldier and the nurse. In each case a transformation in public estimation took place, and in each case the transformation was due to Miss Nightingale. Never again was the British soldier to be ranked as a drunken brute, the scum of the earth. He was now a symbol of courage, loyalty, and endurance, not a disgrace but a source of pride. 'She taught officers and officials to treat the soldiers as Christian men.' Never again would the picture of a nurse be a tipsy, promiscuous harridan. Miss Nightingale had stamped the profession of nurse with her own image. Jane Evans and her buffalo calf, Mother Bridgeman and her proselytizing, Mary Stanley's ladies and their gentility, the hired nurses and their gin have faded from history. The nurse who emerged from the Crimea, strong and pitiful, controlled in the face of suffering, unself-seeking, superior to considerations of class or sex, was Miss Nightingale herself. She ended the Crimean War obsessed by a sense of failure. In fact, in the midst of the muddle and the filth, the agony and the defeats, she had brought about a revolution.

12

SHE said she had seen Hell, and because she had seen Hell she was set apart. Between her and every normal human pleasure, every normal human enjoyment, must stand the memory of the wards at Scutari. She could never forget. She wrote the words again and again, in private notes, on the margins of letters, on scraps of blotting-paper; whenever her hand lay idle the phrase formed itself – 'I can never forget.'

She was a haunted woman, but she was pursued not by ghosts but by facts, the facts of preventable disease. Blood was calling to her from the ground; the blood of the ghastly army of vermin-devoured skeletons who had died before her eyes in the hospitals of Scutari, but their blood called 'not for vengeance but for mercy on the survivors'.

The mortality of the Crimean disaster, 73 per cent in six months from diseases alone, was the ghastly fruit not of war but of the system which controlled the health administration of the British Army. The system was in operation still. Every day, every hour, wherever the British Army had barracks and hospitals, the system was murdering men as surely as it had murdered them in Scutari. The Crimean tragedy cried aloud not for revenge but for reform. She, and she alone it seemed, had discerned this self-evident truth. The summons to save the British private soldier had come to her.

She recoiled. It was too much. Must she pass her life struggling with the forces which had defeated her in the Crimea? Must she now renounce all human contacts for the aridity of official correspondence, the compiling of statistics, the drafting of regulations, the formality of official interviews?

There were midnight agonies, tears, prayers. Fanny, Parthe, W. E. N., hearing her pace her room, thought she was struggling with fearful memories; but she was struggling with herself. She did not find it easy to submit. But the voices of ten thousand of her children spoke to her from their forgotten graves. 'I stand at the altar of the murdered men,' she wrote in a private note of August 1856, 'and while I live I fight their cause.'

Early in September she wrote a message to the parishioners of East Wellow: 'We can do no more for those who have suffered and died in their country's service. They need our help no longer; their spirits are with God who gave them. It remains for us to strive that their sufferings may not have been endured in vain – to endeavour so to learn from experience as to lessen such sufferings in the future by forethought and management.'

She obeyed the summons. She, a woman, ill, alone, exhausted, a voice, she said, crying in the wilderness, prepared to undertake the gigantic task of reforming the health administration of the British Army – but she resented her fate. She wept for herself. No one appreciated what she was being forced to renounce for the sake of the work. She grew angry and the characteristics which had been so marked in her youth, the benevolence, the patience, the quality which Clarkey described as 'Flo's extraordinary bonté' faded. Her astonishing mind developed; her penetration, infinite capacity for taking pains, persistence, iron will to work, scrupulous sense of fair play became still more extraordinary, but the woman of her early years gradually ceased to exist.

She confided in no one. Her family and her friends were bewildered, but she would not enlighten them. The time when she had ached for sympathy was past; she revelled now in the consciousness that she was alone. The urgency of the situation drove her. Action must be taken now, within the next few months, while the Crimean disaster was still fresh in the nation's mind. The iron was hot and must be struck. How was she to strike it? London was empty. Politicians and administrators were taking their summer holidays. The war was over; it had been discreditable, and there was a universal wish to forget it.

Early in August she wrote to Lord Panmure announcing her arrival in England and asking for an official interview. Lord Panmure was in Scotland shooting grouse. He replied through his secretary on August 13, that later he would be delighted, as always, to hear Miss Nightingale's views. Meanwhile, 'it will be more pleasant for you to rest a little while'.

She wrote to Sidney Herbert. He was fishing salmon on his estate in Ireland. She drove herself to write him letter after letter, lying on her sofa sick and exhausted, her fingers hardly able to hold the pen, entreating him, imploring him, commanding him to take action for the army now, at once, before it was too late.

On August 16 he told her candidly that he thought her letters over-wrought. She should follow the excellent prescription of his doctor in Carlsbad, '*Ni lire, ni écrire, ni réfléchir*.' On August 26 he wrote to Mr Sam Smith that Florence's state of mind was causing him concern; complete rest was badly needed, but since he realized, having regard to her temperament, that this was almost impossible, he advised her relations to plan a life for her of 'some, tho' very limited and moderate, occupation'. He did not suggest a meeting; indeed he avoided her.

She became frantic. Her whole being cried out for action. 'If I could only carry *one* point which would prevent *one* part of the recurrence of the colossal calamity, then I should be true to the brave dead,' she wrote in a private note of August 1856. What could she do? Sidney Herbert had failed her; Panmure evaded her. She was so ill it seemed madness to contemplate work. She found difficulty in breathing, suffered from palpitations, and was overcome by nausea at the sight of food. W. E. N., unable to contemplate her condition, left Lea Hurst and retreated to the peace and the shadows of the library at Embley.

If only she would rest: her family, her friends, the whole world in an international chorus implored her to rest. A host of unknown admirers from every country in Europe, from America, from Asia, in letters, newspaper articles, poems, songs, implored her to repose on her laurels. She could not. She was driven by the certainty that delay was fatal. And yet – if delay was fatal, a false step was fatal too. On August 25, in a long letter to Colonel Lefroy, she explained her dilemma.

Special difficulties, she wrote, confronted her. The first was that she was a woman – that was very bad; the second, that she was a popular heroine, which was worse. The two together formed a pill which officialdom would never swallow. Any scheme known to emanate from her would instantly be rejected because it came from her. Sir Benjamin Hawes had written inviting her to put forward suggestions for improvements in the Army Medical Department. She had reason to believe the invitation was given with the object of creating an opportunity for registering an official rejection of her proposals, and she had refused. Dr Pincoffs had asked her to be patroness and organizer of a scheme to provide treatment for discharged wounded men, and she had told him that if he used her name the authorities would see to it that the scheme failed, 'so great is the detestation with which I am regarded by the officials'. Frankly, she continued, she did not know how to proceed. She might begin to work in the military hospitals at home as she had worked in Scutari and gradually reorganize the whole system. The Queen had written to her, and the Queen would certainly grant female nursing departments in all military hospitals. Again the difficulty was her position as a national heroine. It was nothing but an embarrassment. 'The buzfuz about my name,' she wrote contemptuously, 'has done infinite harm.'

Suddenly she scribbled a postscript: 'If I could only find a mouthpiece.' She was convinced that she herself would shortly die – if only she could find someone to carry on the work! 'If I could leave one man behind me,' she wrote in a private note; and she returned to the idea again and again. 'If I could leave one man behind me, if I fall out on the march, who would work the question of reform I should be satisfied, because he would do it better than I.' She needed a man who would be acceptable to the official world, who would carry weight in official circles, but who would be ready to submit himself to her and be taught by her. Where could such a man be found? She did not think of Sidney Herbert.

*

After Miss Nightingale's return from the Crimea she never made a public appearance, never attended a public function, never issued a public statement. Within a year or two most people assumed she was dead. She destroyed her fame deliberately as a matter of policy. The authorities expected that on her return she would make revelations. She neither revealed, nor attacked, nor justified herself. She wrote nothing; she made no speeches; she was not even seen. Instead, with infinite patience and self-effacement, she set out to win the authorities over to her side. She was laying aside a powerful weapon; at the moment adoration of her had reached an extraordinary pitch. 'She may truly be called the voice of the people of the present,' wrote Dr Pincoffs to Fanny. In Sheffield a lady who resembled Miss Nightingale found herself surrounded by a large but respectful crowd, who pressed round her asking for permission 'just to touch her shawl'. Society wished to lionize her, and she was inundated with invitations. The Duke of Devonshire, who had formed a collection of Press cuttings relating to her work, presented her with a model of Athena, her pet owl, in silver, and wished to give a reception in her honour at

Chatsworth. She refused. She refused interviews, receptions, presentations. She refused to go out to dine. She refused to be painted. 'The publicity and talk there have been about this work have injured it more than anything else,' she wrote in a private note of August 1856, 'and in no way, I am determined, will I contribute by making a show of myself.'

Her post-bag was enormous, but she would barely glance at it. 'As to her indifference to praise it is quite extraordinary,' wrote Parthe. Congratulatory letters arrived in 'hail storms'. Unknown admirers showered gifts, poems, songs, illuminated addresses, and proposals of marriage. Begging letters came 'in shoals'. An unknown gentlewoman asked to be provided with a post, 'but nothing derogatory because I am an Irish lady of good family'; one gentleman requested her to get his jewellery out of pawn and another asked for the gift of a donkey; a young lady wrote: 'I have had a passion for soldiers all my life and now wish to get my bread by it.' 'How would you construe this?' Miss Nightingale scribbled in the margin.

Parthe wrote the acknowledgements. Miss Nightingale herself wrote no letters, signed no autographs, granted no interviews. The few who did receive a personal reply were humble people – the parishioners of East Wellow, the working-cutlers of Sheffield who presented her with a canteen of cutlery, 1,800 working-men of Newcastle-on-Tyne who sent her an address. As far as the rest of the world was concerned, Miss Nightingale had disappeared.

At the end of August Sidney Herbert returned from Ireland; they met briefly at the Bracebridges' house, and he was 'lukewarm about army reform'. Then he retired to Wilton. She was in despair when suddenly she was given an opportunity of the most dazzling and unexpected nature.

Early in September her old friend Sir James Clark invited her to stay at his house in the Highlands. He wrote at Queen Victoria's desire: the Queen wished to hear the story of Miss Nightingale's experiences with the army, not only officially but privately. Sir James's house, Birk Hall, was only a mile or two from Balmoral. She would be commanded to Balmoral for an official interview, and in addition the Queen intended to have private conversations with her at Birk Hall.

She rose from her sofa and flew to work, determined to do much more than relate the story of her experiences. She would seize the opportunity to convince the Queen and, equally important, the Prince Consort, of the urgent necessity for immediate army reform. Her case must be presented in a way to impress the Prince, who took a keen interest in army affairs and whose meticulous accuracy and love of detail were notorious. Since she must be ready with facts and figures, with tables and statistical comparisons, she plunged into correspondence with experts she had known in the Crimea, Sir John McNeill, Colonel Tulloch and Colonel Lefroy. Her plan was to ask for a Royal Commission to examine the sanitary condition, administration, and organization of barracks and military hospitals and the organization, education, and administration of the Army Medical Department. For the first time in history the living conditions of the private soldier in peace and war, his diet and his treatment in health and sickness would, she hoped, be scientifically investigated. In addition to the Royal Commission she intended, on Colonel Lefroy's advice, to ask permission to address a Confidential Report to Lord Panmure which would be a frank account of her own experiences. 'In some form or other we have almost a right to ask at your hands an account of the trials you have gone through,' wrote Colonel Lefroy on August 28, 'the difficulties you have encountered, the evils you have observed ... no other person ever was, or can be, in such a position to give it.'

On September 15, accompanied by W. E. N., she arrived at Sir John McNeill's house in Edinburgh and was joined there by Colonel Tulloch for four days' furious and concentrated work. Her condition was causing grave anxiety; she was weak, emaciated, and still experiencing nausea at the sight of food; nevertheless, she was able to work day and night sorting, digesting, and collating the vast mass of figures and facts which had been collected in the course of the McNeill and Tulloch enquiry. Her few free hours were spent in visiting and inspecting barracks, hospitals, and institutions.

Sidney Herbert played no part in getting up the case. He did not even advise. Writing to Miss Nightingale from Wilton on September 9, he was affectionate but detached. 'I hope your Highland foray will do you good. I am sure it will if you find help and encouragement for your plans.' It seemed he did not take her seriously.

On September 19 Miss Nightingale left Edinburgh with W. E. N. for Birk Hall; and on September 21 she was commanded to Balmoral for an afternoon's talk with the Queen and the Prince Consort.

The meeting, an informal one, lasted for more than two hours and was a triumphant success. 'She put before us,' wrote the Prince in his diary that night, 'all the defects of our present military hospital system and the reforms that are needed. We are much pleased with her; she is extremely modest.' 'I wish we had her at the War Office,' wrote the Queen to the Duke of Cambridge, the Commander-in-Chief.

She was commanded to Balmoral again and yet again. She conversed with the Prince Consort on metaphysics and religion and went with the royal party to church. On several occasions she dined informally. Most important of all, the Queen, as she had indicated, paid her private visits. One day she appeared suddenly quite alone, driving herself in a little pony carriage, and took Miss Nightingale off for a long walk. Another day she came over alone and unannounced, spent the afternoon, stayed to tea, and there was 'great talk'. Parthe reported Lord Clarendon as having said the Queen was 'enchanted with her'.

The first step had been successfully taken, but Miss Nightingale was aware that it was only the first step. Under the British Constitution the Queen and the Prince had no power to initiate action; that was exclusively in the hands of the Ministers of the Crown. On September 26 Miss Nightingale wrote to Uncle Sam: 'Everything is most satisfactory. Satisfactory that is as far as their *will*, not their *power*, is concerned.' By their good-will and eagerness, she said soberly, her hopes were 'somewhat raised'.

Before the warrant for the Royal Commission could be issued, the Queen must be advised to do so by the Secretary of State for War. Lord Panmure must be convinced of the necessity for army reform.

The next week Lord Panmure was to be in attendance at Balmoral, and the Queen, almost too anxious to be of use, insisted that Miss Nightingale come to Balmoral to meet him: 'The Queen has wished me to remain to see Lord Panmure here rather than in London,' she wrote to Fanny on September 25, 'because she thinks it more likely something might be done with him here with her to back me. I don't but I am obliged to succumb.'

Lord Panmure was a difficult subject whose personal appearance was surprising; he had an enormous head, crowned with thick upstanding tufts of hair, and a habit of swaying it slowly from side to side which had earned him the nickname of 'the Bison'. Detail he hated, nor did he appreciate system. His position as Secretary of State for War involved an immense amount of work which, as Sidney Herbert said, he 'found easy through the simple process of never attempting to do it'. He detested 'bothers' and had been infuriated by Sir John McNeill's and Colonel Tulloch's revelations in their Report on the Supplies for the British Army in the Crimea, though he himself had appointed them and given them their instructions. He had gone through their report before its official publication with the avowed object of cutting out anything that seemed unpleasant, and he had desisted only because he found that the only way to render the report innocuous was to rewrite the evidence. He had been heard to say that he wished both Sir John McNeill and Colonel Tulloch at the devil.

The greatest difficulty in dealing with him arose out of his habit of procrastination. He would not take action, for he had discovered that if action is avoided consequences are avoided too.

Yet despite these defects he was a man of character. When he was a boy of fifteen, his father had quarrelled with his mother, and he was informed that he must choose between his parents. If he broke completely with his mother, he would enjoy all the privileges of an eldest son with a large income and a safe seat in the House of Commons; if he persisted in seeing his mother he would be cut off with £100 sterling a year and a commission in the army. He refused to be separated from his mother and never saw his father again. After twelve years in the 79th Highlanders he retired with the rank of Captain and entered politics, living, in the words of his biographer, 'as best he could on a yearly income of £1,000 and £10,000 down which he raised on a post-obit from the Jews'. In 1852, after thirty-six years of estrangement, his father died, and at the age of fifty-one he succeeded to vast estates and a large income. Now at Brechin Castle, in Forfarshire, he lived as a feudal chieftain. He was an enthusiastic sportsman, and his deer forests, his salmon rivers and his grouse moors were the best in Scotland; one of his chief interests were the promotion of measures for the protection and improvement of game. Good stories were told of him. When he was introducing the deputation on the Tweed Acts for the protection of salmon the Home Secretary was Sir George Lewis, who was a scholar rather than a sportsman. After listening attentively to the chairman's speech for some minutes Sir George enquired: 'Do I understand you to say, sir, that the salmon sometimes visits the sea?' Panmure brought his stick down on the floor with a loud crash. 'Good God,' he burst out, 'with how little wisdom is this country governed?'

In 1855 his favourite nephew, Captain Dowbiggin, was serving before Sebastopol; and to an official telegram to General Simpson, then commanding the British forces, he added this cautious postscript: 'I recommend Dowbiggin to your notice, should you have a vacancy and if he is fit.' In those early days of telegraphy operators were accustomed to condense, at their own discretion, messages that seemed verbose, and the postscript reached headquarters in the simple form, 'Look after Dowb'.

Panmure was a religious man, keenly interested in the organization of the Free Church of Scotland and an active member of the assembly. During his life generosity was not considered one of his characteristics; but after his death among his papers was found a long list of persons to whom he had been paying annuities in secret.

To win over Panmure was a formidable task, and the ground was carefully prepared. The Queen wrote informing him of the proposed meeting – 'Lord Panmure will be much gratified and struck with Miss Nightingale – her powerful clear head and simple modest manner.' The Queen also fell in with a stratagem designed to prevent Panmure from evading the main issue. Miss Nightingale wrote her a letter outlining her suggestions for army reform; this the Queen accepted 'with great grace', and a copy of it was then sent to Panmure with the information that the Queen had accepted the original. By this means it was hoped that the main lines of the discussion would be defined.

Sidney Herbert was pessimistic. Miss Nightingale had written to him before the Queen's invitation asking him to arrange for her to see Panmure. He had, at one point, promised to meet her if he were in London for a 'combined attack on the Bison'. But he had evaded her by never being in London. Before her interview with Panmure she wrote from Birk Hall to remind him of this 'very important promise'. It had been impossible to refuse the Queen's invitation, but 'I would rather have seen Panmure with you'. His reply was discouraging. There was no harm in her trying to see what she could do with Panmure, but 'I am not sanguine, for tho' he has plenty of shrewd sense there is a *vis inertiae* in his resistance which is very difficult to overcome.'

On October 5 the interview took place, and Miss Nightingale's success exceeded all expectations. Lord Panmure succumbed to the spell which drunken orderlies, recalcitrant nurses, and suspicious officials had been powerless to resist. 'You may like to know,' wrote Mr John Clark, Sir James's son, 'that you fairly overcame Pan. We found him with his mane absolutely silky; and a loving sadness pervading his whole being.' On November 2 Sidney Herbert wrote: 'I forget whether I told you that the Bison wrote to me very much pleased with his interview with you. He says that he was very much surprised at your physical appearance, as I think you must have been at his. God bless you.'

When she returned to Birk Hall, Panmure, like the Queen, came to see her privately, and it seemed that she had obtained everything. There was to be a Royal Commission, and the instructions were to be drawn up in accordance with her suggestions. She was to be invited to make a 'Confidential Report'; and the request was to come jointly from Lord Panmure as Secretary of State for War and Lord Palmerston as Prime Minister. Netley, the first general military hospital to be built in the country, was in process of construction: Lord Panmure volunteered to send her the plans and invited her to make observations, declaring himself to be at her service to discuss details as soon as they were both in London at the end of the month. When she left Birk Hall to go south, the prospect was rosy.

Some years later Clarkey wrote down what Miss Nightingale had told her of her impressions at Balmoral. 'She was struck with the difference between the minds of the Queen and Prince Albert and the fine folk about them and how little the latter were capable of appreciating them. For instance, the Duchess of Wellington and Lady Someone did nothing but complain what a dull place it was and how tiresome – they seemed occupied with nothing but trifles, but the Queen and Prince Albert's whole thoughts were about Europe, the Crimea War, etc. etc. etc. – all things of importance. She says that the Queen is a remarkably conscientious person, but so mistrustful of herself, so afraid of not doing her best, that her spirits are lowered by it.'

She spent a few days in Edinburgh with Sir John McNeill; stayed for a fortnight at Lea Hurst establishing contact with 'Crimean confederates'; and on November 1, accompanied by

Fanny, Parthe, and W. E. N., went to London to the Burlington Hotel, the 'dingy old Burlington'.

She had drawn up a list of names for the Commissioners. Civilians and military men were to be equally balanced. On the civilian side she put forward Sir James Clark, Sir James Martin, Dr Sutherland, and Dr Farr.

Dr Farr was of special importance. He was a pioneer in the science of statistics, then in its infancy, and she intended, with his help, to make a statistical comparison of the rate of sickness and death in barracks with the rate in civilian life. It would, she believed, be startling.

Dr John Sutherland was one of the leading sanitary authorities of the day. He had been head of the Sanitary Commission of 1855, had become Miss Nightingale's intimate friend, and was her personal physician. He was an invaluable worker but was prone to a flippancy she found intensely irritating. On November 12, 1856, he wrote: 'I am led to believe there must be a foundation of truth under the old myth about the Amazon women somewhere to the East. All I can say is that if you had been Queen of that respectable body in old days Alexander the Great would have had rather a bad chance. Your project has developed far better than I expected.'

On the military side she intended to press for Sir Henry Storks, who had worked 'hand in hand' with her in army welfare work at Scutari, Colonel Lefroy, Dr Balfour, assistant surgeon to the Grenadier Guards and an able statistician, and, most important of all, Dr Alexander. 'Get ALEXANDER,' Dr Sutherland had written. 'Nobody else if you cannot. He is our man.'

Sidney Herbert called Alexander 'unquestionably the ablest man in the British Medical service'. He had been first-class Staff Surgeon to the Light Division and had spent the Crimean campaign in the fighting line. William Howard Russell described him at Calamita Bay as 'a gentle giant of a Scotchman, sitting on the beach with a man's leg in his lap,' and pouring out the vials of his wrath on Sir John Hall for landing the army without medical supplies. Throughout the war he was distinguished by his skill, his powers of organization, and his fearless independence. His achievements did not endear him to his superiors, and at the close of the campaign he was relegated to a second-class appointment in Canada. If he were to be a member of the Commission, it would be necessary for the authorities to reconsider their decision and recall him. It was impossible to put forward the names of Sir John McNeill and Colonel Tulloch owing to their dispute with Lord Panmure.

The stage was set for the play, but the principal player lingered: Sidney Herbert still hesitated. He would not come to London. He remained at Wilton, and Miss Nightingale's letters became impatient. 'If you come to London during the next fortnight will you have the goodness to let me know you are there,' she wrote on October 31. '... I should have much to tell you about my "Pan", could I only see you.' In the first week of November he did come to London and called to see her; and when he left he had agreed to accept the chairmanship of the Commission. The spell cast by her presence coupled with his own sense of duty had been too powerful for him. He was far from well, easily tired, easily depressed – the fact was, he already carried within him the seeds of incurable disease. His term at the War Office during the Crimean campaign had half broken his heart; he viewed the future with gloom.

When Lord Panmure made an appointment to call at the Burlington Hotel on November 16, expectation ran high. 'I long to hear what results you obtain from the Bison,' wrote Sidney Herbert. During the morning Sir James Clark sent her a note by hand: 'I think it would be well, when you see Lord Panmure to make him understand that the enquiry is intended as ... an investigation into everything regarding the health of the Army.' Sir James Clark had opportunities of observing Panmure at Court, and he realized that he had no idea of the revolutionary and explosive ideas which lay concealed under Miss Nightingale's quiet, modest manner. Panmure had come to his first meeting fearing that 'with so strong a hold on the feelings of the nation, she is not unlikely to use it for personal ambition'. He had found a charming, well-bred woman, completely altruistic, and he acknowledged he had misjudged her. He was destined to discover that her altruism was more troublesome than most women's ambition.

The interview was a very long one. Once more her extraordinary personal charm worked its spell; once more she triumphed. The main point was achieved. The Commission was to go forward, and its scope was to be 'general and comprehensive, comprising the whole Army Medical Department and the health of the army at home and abroad'. As soon as Panmure had left, she

212

sat down and scribbled Sidney Herbert a long, gay, disconnected note: 'My "Pan" here for three hours. ... Won't bring back Alexander, will have three Army Doctors. So like a sensible General in retreat I named Brown, Surgeon Major Grenadier Guards ... an old Peninsular and Reformer. ... He [Panmure] had generously struck out Milton.' (Mr Milton had been sent to Scutari to straighten out the purveying, and her verdict had been that he dealt only in 'official whitewash'.) 'Seeing him in such a "coming-on disposition" I was so good as to leave him Dr Smith the more so as I could not help it.

'Have a tough fight of it; Dr Balfour as secretary. Pan amazed at my condescension in naming a Military Doctor; so I concealed the fact of the man being a dangerous animal and obstinate innovator.

'Failed in one point. Unfairly. Pan told Sir J. Clark he was to be on. Won't have him now. Sir J. Clark has become interested. Agreeable to the Queen to have him – just as well to have Her on our side. ... Besides things Ld P. finds convenient to forget, has really an inconveniently bad memory as to names, facts, dates and numbers. ... Does not wish it to be supposed he takes suggestions from me, a crime indeed very unjust to impute to him.'

Outside the Commission a point of first-class importance had been won. Dr Andrew Smith, Director-General of the Army Medical Department, was very shortly due to retire; and Lord Panmure had pledged his word that Sir John Hall should not be made Director-General as long as he was in office.

She had done a great deal, but the pressure necessary for complete success could not be applied by her. Sidney Herbert must do that. He must be made to understand that everything depended on him; and she finished her note with these words: 'You must drag it through. If not you, no one else.' A few days later an official letter from Panmure invited Sidney Herbert to accept the chairmanship of the Commission. He accepted, subject to certain conditions, chief of which was Alexander's recall from Canada.

And then, inexplicably, nothing further happened. The official announcement of the issue of the Royal Warrant to set up the Commission, which should have been made within the next few weeks, never came. Sidney Herbert's letter asking for Alexander's recall was not discussed. Instead, Sidney Herbert received a friendly note from Lord Panmure regretting he was unable to write as he had gout in both his hands. 'Gout is a very *handy* thing; and Lord Panmure always has it in his hands when he is called upon to do anything,' Miss Nightingale wrote to Sir John McNeill on December 15.

Unfortunately at the moment Panmure was being put to the greatest possible trouble by Miss Nightingale's friends, Sir John McNeill and Colonel Tulloch. The storm raised by the inconvenient revelations made in their report was still raging. He had appointed them out of a genuine sense of duty, and horrid bothers had resulted. It seemed only too probable that the Commission on the Health of the Army would result in even greater bothers, and there was certainly a strong party against it at the War Office. Miss Nightingale, who was so charming and could be so very persuasive, pushed him one way, but when he left her and returned to the War Office the permanent officials pushed him the other.

The solution was to do nothing, to be friendly and pleasant, because unfriendliness and unpleasantness also led to bothers, but to take no action whatsoever. Without apparent reason the Commission appeared to be fading away.

'Do not allow yourself to be discouraged by delays,' wrote Sir John McNeill on December 19; but to Miss Nightingale, whose overwrought mind burned with unearthly brilliance in her sick body, delay was intolerable. She suffered torments of frustration, pouring herself out once more in private notes. 'My God, my God why hast thou forsaken me,' she wrote at Christmas, 1856, 'We are tired of hearing of the Crimean Catastrophe. We don't want to know any more about the trenches cold and damp, the starved and frozen camp, the deficient rations, the stores which might have served the great army of the dead lying unused. ... Generals who, looking at dead dogs polluting the atmosphere where men lay, said "You are spoiling the brutes". G.H.Q. feeding their horses on the biscuits the men could not eat; and saying that anyway they kept the horses fat. ... Words were given in plenty to the great Crimean Catastrophe, but the real tragedy began when it was over.' Hour after hour her pen rushed on; hour after hour she paced her room sleepless, raging against the indifference, the forgetfulness of the world. In letter after letter she incited her fellow workers to action. Her shrewd eye had penetrated the Bison's secret. 'My Lord is,' she wrote, 'as I have often found the most bullyable of mortals.' She entreated that he should be bullied. She besought Sidney Herbert to write threatening to resign the chairmanship publicly unless the Royal Warrant for the Commission was issued forthwith. She vowed, 'I will never leave Panmure alone until it is done.' However, her personal relations with him remained pleasant, and their correspondence was conducted with arch playfulness. 'Here is that bothering woman again,' she wrote on January 22, 1857, 'just to remind you I am in London awaiting your decision.' Panmure jestingly called her 'a turbulent fellow' and sent her presents of game.

But he could not evade her entirely. As well as being involved together on the subject of the Royal Commission, they were waging what she called a private campaign on the subject of Netley Hospital. When he had volunteered to send her the plans for her observations, he had wished to pay her a compliment. Before he could take breath, she had fallen upon the opportunity with relentless thoroughness, obtaining leave to 'report confidentially' with the assistance of Dr Sutherland. She inspected; she consulted authorities; she drew up exhaustive reports bristling with statistics derived from sources both at home and in Europe. She prepared additional memoranda, dealing in detail with certain aspects of the case; she drew up alternative suggestions. Finally, she condemned the Netley plans root and branch and sent the whole result to Lord Panmure.

He felt he had accidentally released a genie from a bottle. The accumulated experience of fourteen years was suddenly put at his disposal; the fruit of her researches in France, Germany, Italy, London, and Switzerland; of the endless miles she had tramped down the corridors of hospitals, prisons, asylums, orphanages; of the endless questions she had asked; the endless figures she had tabulated. He was taken aback. Moreover, since issuing the invitation, a fact which he had failed to take into account had been brought to his notice. Building had progressed so far that radical alterations were impossible. He wrote her a soothing letter. Her objections were no doubt sound, but there were 'susceptibilities' to be considered. But she was not to be put off; she appealed to the Prime Minister, her old friend Lord Palmerston, and during the Christmas holidays of 1856 went over to his house, Broadlands, to dine and sleep and open his eyes to the truth about Netley.

On January 17 Lord Palmerston wrote Panmure a sharp letter. Miss Nightingale had left on his mind a conviction that the plan was fundamentally wrong; and that it would be better to pull down everything that had been built and start again. 'It seems to me,' wrote the Prime Minister, 'that at Netley all consideration of what would best tend to the comfort and recovery of the patients has been sacrificed to the vanity of the architect, whose sole object has been to make a building which should cut a dash when looked at from the Southampton river. ... Pray therefore stop all progress in the work till the matter can be duly considered.'

Lord Panmure was aghast. Vistas of bother, of explanations to be made, letters to be written, answers to questions in the House, unrolled themselves before him. There would be 'rupture of extensive contracts', 'reflections cast on all concerned in the planning of the designs'. In addition there were 70,000 reasons which were completely unanswerable – it would cost £70,000 to pull down the partially constructed building and start again. The vision of himself attempting to explain away a loss of £70,000 to the House was a nightmare he refused to contemplate. Work at Netley went on.

Lord Palmerston wrote again: He continued to feel very anxious about Netley Hospital, and he would rather pay for throwing it brick by brick into Southampton Water than construct a building which should be a charnel house rather than a hospital.

Still Panmure would not be moved. He offered to incorporate improvements, but reconstruction was impossible. Miss Nightingale refused to give up hope. Were not her criticisms admitted to be justified, her new plans to be infinitely superior to the old? She argued, cajoled, threatened, but she was defeated. The 70,000 reasons conquered, and Netley was constructed on the existing plans.

Time has proved her to be correct. Throughout its history Netley has been a difficult, depressing, and unsatisfactory hospital. The design she put forward was the 'pavilion' design, separating the building into blocks. 'The object sought,' she wrote, 'is that the atmosphere of no one pavilion or ward should diffuse itself to any other pavilion or ward, but should escape into the open air as speedily as possible, while its place is supplied by the purest obtainable air from outside.' Each pavilion was to form a separate detached hospital with an administration in common. Though the bacteriological nature of infection had not yet been discovered, Miss Nightingale deduced from her own observation the fact that, keeping patients isolated in small groups, decreased the incidence of hospital diseases. The pavilion design is to be seen in the Herbert Hospital at Woolwich and, until it was damaged by air bombardment, in St Thomas's Hospital, Lambeth, both of which were built under her supervision. Netley, with its large central block, its long corridors, its impressive façade, is a typical eighteenth-century hospital built fifty years too late.

Once defeat was a fact she accepted it and laboured with good-will to introduce improvements which should make the original plans tolerable. Her correspondence with Lord Panmure was more than usually playful. Netley was christened 'the patient', and she advised him of the progress made in letters written in the form of bulletins on the patient's condition. Nevertheless, she had been defeated, and by the spring of 1857, with the issue on Netley lost and the Commission still delayed, her spirits were at their lowest. 'I am very miserable,' she wrote to Mrs Bracebridge in February. 'I think he [Panmure] means to shelve me.'

Frustration had its inevitable effects; she could not sleep, could not eat, spent the nights pacing her room or writing private notes. 'No one,' she wrote in a private note of February 9, 1857, 'can feel for the Army as I do. These people who talk to us have all fed their children on the fat of the land and dressed them in velvet and silk while we have been away. I have had to see my children dressed in a dirty blanket and an old pair of regimental trousers, and to see them fed on raw salt meat, and nine thousand of my children are lying, from causes which might have been prevented, in their forgotten graves.'

Disappointment piled on disappointment. Early in 1857 the exasperating affair of the McNeill and Tulloch Report came to a head. The Report of the Chelsea Board set up with the avowed intention of white-washing the officers concerned was published, all blame was removed from individuals, and the gigantic misfortunes endured by the army were attributed to the non-arrival of a certain single consignment of pressed hay. Lord Panmure accepted it, disowning the Commissioners he had himself appointed, and the McNeill and Tulloch Report was set aside.

By March 1 Miss Nightingale had reached complete despair. 'Lord Panmure has broken all his promises,' she wrote to Sir John McNeill, 'defeated the Army Reformers at every point, simply by the principle of passive resistance, the most difficult of all resistances to overcome, the easiest of all games to play. I think our cause is lost. . . . Mr Herbert is ill and going abroad and so ends all chance of a Commission to enquire into the Sanitary State of the Army, of which he was to have been chairman.'

In fact, the first hopeful signs of change had already appeared. On February 18, after six months' delay, Panmure had written from the War Office with an official request for her Confidential Report. 'Your personal experience and observation, during the late War, must have furnished you with much important information relating not only to the medical care and treatment of the sick and wounded, but also to the sanitary requirements of the Army generally. I now have the honour to ask you to favour me with the results of that experience . . . should you do so . . . I shall endeavour to further as far as lies in my power, the large and generous views which you entertain on this important subject.'

She was not elated. She had no faith in Panmure's intentions of furthering her 'large and generous views'. She believed he meant to shelve the Commission, and she regarded the report as a sop to her, something to keep her quiet, which in due course would also be shelved. Nevertheless, she at once set to work. If he failed her, she had a larger audience. She had put the weapon of publicity aside, but the weapon still lay ready to her hand. At the end of February she wrote to Sidney Herbert: 'All that Lord Panmure has hitherto done (and it is just six months since I came home) has been to gain time. . . . He has broken his most solemn promise to Dr Sutherland, to me and to the Crimea Commission. And three months from this day I publish my experience of the Crimea Campaign and my suggestions for improvement, unless there has been a fair and tangible pledge by that time for reform.'

It was a threat which could not fail to make the Bison uneasy, and there were other indications that public opinion was turning in the direction of Army Reform. The whitewashing done by the Chelsea Board had by no means settled the matter of Sir John McNeill and Colonel Tulloch. Meetings of protest had been held in many large towns; addresses of sympathy and support from citizens and municipalities had been presented to them. Feeling in the country became so strong that Panmure was forced to act, and he attempted to buy off the Commissioners by offering them each £1,000 cash down, on the understanding that the matter was to be considered closed. They indignantly refused, and Miss Nightingale pressed Sidney Herbert to raise the matter in the House. On March 12 he moved a humble address to the Crown, amid loud applause, praying that Her Majesty might be pleased to confer some signal mark of favour upon Sir John McNeill and Colonel Tulloch. The atmosphere of the House was such that Lord Palmerston accepted the motion without a division. 'Victory!' scribbled Miss Nightingale that night. 'Milnes came in to tell us.' Colonel Tulloch was created a K.C.B. and Sir John McNeill, already a G.C.B., was created a Privy Councillor. 'They have been borne to triumph on the arms of the people,' she wrote.

The tide was setting toward reform; and once the tide had turned, Panmure was not the man to resist.

On April 27 he paid another official call at the Burlington Hotel. So extraordinary was Miss Nightingale's position, so clearly was it recognized that she was the leader of the Reform party, that Panmure brought the official Draft of the Instructions for the Royal Commission to her before submitting it to the Queen. It was a long and difficult interview. The forces of reaction were in retreat, but they were by no means conquered; and War Office officials had provided Panmure with a list of Commissioners which in Miss Nightingale's opinion would have nullified the inquiry. 'Every one of the members of the Commission was carried by force of will against Dr Andrew Smith,' she wrote to Sidney Herbert that evening.

She won all along the line. The Bison's capitulation was, for the moment, complete. Dr Alexander was to be recalled from Canada. Colonel Lefroy could not be spared from his work in the War Office, but in his place she secured her old admirer and Crimean confederate, Mr Augustus Stafford, M.P. Sir Henry Storks was to sit, so was Sir James Clark. Dr Sutherland sat as sanitary expert. Sidney Herbert was, of course, chairman. 'I could do nothing without him,' she scribbled on the edge of a document. Only one member of the old gang was included – Dr Andrew Smith inevitably sat as Director-General of the Army Medical Department. The Instructions, the official directions indicating the ground the Commission was to cover, were drawn up by Miss Nightingale herself and accepted without alteration by Panmure. On May 5 the Royal Warrant was issued, and the following week the Commission began to sit.

13

THE strain on her was enormous. Three months before she had been an invalid; now, as well as working night and day on the Commission, she was turning her Confidential Report into an important work covering the whole field of army medical and hospital administration in the recent war, in previous wars, and in peace. This work emerged six months later as *Notes on Matters affecting the Health, Efficiency and Hospital Administration of the British Army*, a volume of nearly 1,000 closely printed pages, crammed with figures, facts, tables, and statistical comparisons. The strain was intensified by the petty irritations, the tensions, the emotional conflicts, inseparable from the Nightingale family life.

In November 1856 Fanny, Parthe, and W. E. N. went to the Burlington Hotel. W. E. N. soon fled to the peace and shadows of the library at Embley, but Fanny and Parthe remained to extract the fullest possible enjoyment from their position as the mother and sister of Miss Florence Nightingale. 'No one,' she wrote in a private note, 'has enjoyed my reputation more than my own people.' Strangers stopped Fanny and Parthe in the hotel and in the streets to enquire after Miss Nightingale. Letters were written to Fanny. 'I hardly know how to express myself about your daughter's delicate health,' an unknown admirer wrote. 'She has the sympathy of two continents (one might say of all humanity).' Parthe struck the Crimean note in her conversation, christened Aunt Mai 'My Scutari Aunt', and assumed the role of the privileged guardian of her sister's shrine. 'I cannot conceive of anything more beautiful than her state of mind. She is so calm, so holy,' she told a cousin in the autumn of 1856. 'I cannot believe she will live long,' she wrote to Clarkey.

Fanny and Parthe were profuse in fine phrases and expressions of affection. In practice they behaved with total want of considera-tion. The only place, besides her bedroom, where Miss Nightin-gale could work was a little inner drawing-room opening off the outer drawing-room. She sat in the inner room working at a table piled with papers, immersed in figures or talking in low tones to

Dr Sutherland or Dr Farr, while in the outer drawing-room Fanny and Parthe entertained their friends, interrupting her whenever the fancy took them. A carriage was provided for Fanny and Parthe, but it was not put at her disposal. She used cabs, or, if a cab was not available, travelled in the public omnibus, an unusual proceeding for a woman of her social position in the eighteen-fifties. Several times she mentions having been stranded, unable to get a cab or omnibus and forced to walk in wind and rain. In November 1856 Fanny told W. E. N., who had fled to the peace of the library at Embley: 'Yesterday Flo went with Sir John Liddell and her good angel Hilary to Chatham, setting off at $9\frac{1}{2}$ o'clock and not returning until $9\frac{1}{2}$ at night; 30 miles to Chatham by rail, several miles in cabs and, Sir John says, up to 20 miles walking about the 3 hospitals.' The next morning she walked about the wards of St Mary's, Paddington, of which she had been made an Honorary Life Governor; in the afternoon she went down to Bermondsey and walked about the wards of a hospital there and came back late to the Burlington and sat up until the small hours working with Dr Sutherland and Dr Farr.

It would have been hard work for a woman in good health; that Miss Nightingale could perform it in her physical condition was unbelievable. Statistical science was in its infancy. Statistics relating to the British Army were almost non-existent. She was doing the work of a pioneer, visiting civil and military institutions, barracks, infirmaries, asylums, and prisons. After a day of toil she returned, almost fainting with exhaustion, to be scolded by Fanny and Parthe because she would not go to parties.

Again the old painful story dragged itself out. But it was impossible for Miss Nightingale not to perceive that Fanny and Parthe coveted not her companionship and affection but the reflected glory from her celebrity. 'I was the same person who went to Harley Street and who went to the Crimea,' she wrote in a private note of November 1856. 'Nothing was different except my popularity. Yet the person who went to Harley Street was to be cursed and the other blessed ... this false popularity has made all the difference in the feelings of my family towards me.' 'Nothing has been learnt from their former experience,' she wrote in another note, 'but the *world* thinks of me differently.'

After her return from the Crimea she took her furniture out of the rooms she had used at Harley Street. In a pretty little scene before a gathering of admirers and friends Parthe begged to be allowed to keep the blue teacups 'as a remembrance'. In a private note Miss Nightingale commented that Parthe might also keep as a remembrance the fact that she never crossed the threshold of Harley Street without having hysterics.

Financial arrangements were a constant source of irritation. Fanny never ceased to resent Miss Nightingale's allowance of £500 sterling a year and behaved as if £500 a year was immense wealth. She made a practice of sending in accounts for a proportion of the hotel bill, for any expenditure that could possibly be construed as having been incurred for or through her younger daughter.

The problem was insoluble. The truth was that where money was concerned Miss Nightingale was Fanny's daughter. She was high-handed, not squandering money but disregarding it, and she was incurably generous. Almost every January 1 she wrote in a private note, 'This year I *must* retrench,' but she never succeeded in living on £500 a year.

When the Commission began to sit, pressure on her steadily increased. Early in May Fanny wrote that Florence had spent the morning in Belgrave Square 'coaching up' Sidney Herbert for the sitting of the Commission next day and 'the afternoon at Highgate performing the same office for Dr Sutherland', returning to work far into the night. Next day she set off for Highgate at 9 a.m., worked there until after dark, then went on to work with Dr Farr and did not get back to the Burlington until very late. The day after that she started for Highgate at 8.30 a.m., worked there until 7.30 p.m., came back to the Burlington to find a message from Sidney Herbert, went straight off again to Belgrave Square and did not get back until after 11 p.m.

Meanwhile Fanny and Parthe gave breakfasts and dinners, drove in the Park, received callers, and paid calls. Through a mistake, one of their bedrooms was let. Fanny and Parthe stayed in the hotel but Miss Nightingale turned out, sleeping at an annex in Albemarle Street and coming into the Burlington to eat and work. By June Parthe and Fanny were weary, but they would not go home. The season ended. London became, wrote Parthe, 'dismal as for a funeral', but Florence was not going away and, until she went, Fanny and Parthe were determined not to go either.

Alarming reports reached W. E. N. at Embley. He wrote to Aunt Mai and asked her to call at the Burlington and see what could be done – Fanny was unwell, Parthe was unwell, and Florence was apparently dying. Aunt Mai hurried to London to find she had stirred up a hornet's nest. She hastily retreated with a burst of apologies: 'Dearest Parthe. I am anxious to prevent two things being thought which may APPEAR other than they are. 1, that there was any want of consideration for your dear mother either on F. N.'s part or mine. 2, that I was interfering in any way. When I came to the Burlington on Monday not only did I not know what would be best to do. I only knew that each *wished* to do what was best for all, and that it was very difficult to do what was desirable for each.' And so on for five, six, seven pages.

The summer wore on and became a nightmare. The weather was heavy, close, without sun but with great heat. The rooms at the Burlington were dark and airless, the sky perpetually grey. Water-carts sprinkled the streets to lay the dust, but the water they used was putrid and produced a horrible smell. Still Florence had not finished her work; still Fanny and Parthe refused to leave her. 'The days draw on and bring each their burden,' wrote Parthe to Aunt Julia on August 17 and signed herself 'yours wearily'.

Ten years later Miss Nightingale described to Clarkey what she had endured from Fanny and Parthe in the summer of 1857. 'The whole occupation of Parthe and Mama was to lie on two sofas and tell one another not to get tired by putting flowers into water. ... I cannot describe to you the impression it made on me.' She was in a fever of fatigue. Every day brought more than could possibly be accomplished in a year. She longed, she wrote, for rest as a man dying of thirst longs for water. In this state she reached the Burlington one evening to be told that the Duke of Newcastle had called to see her. As she passed through the drawing-room, Fanny and Parthe were lounging on the sofas. 'You lead a very amusing life,' they said to her. 'It is a scene worthy of Molière,' she wrote, 'where two people in tolerable and even perfect health, lie on the sofa all day, doing absolutely nothing and persuade themselves and others that they are the victims of their self-devotion for another who is dying of overwork.'

Her position was extraordinary. She was, as the men round her delighted to call her, the Commander-in-Chief. She collected the facts, she collated and verified them, she drew the conclusions, she put the conclusions down on paper, and, finally, she taught them to the men who were her mouthpieces. As each witness came up for examination, she prepared a memorandum on the facts to be elicited from him and coached Sidney Herbert before each sitting of the Commission. 'These men seem to make her opinions their law,' wrote Fanny to W. E. N. in June 1857.

'By Sidney Herbert's desire,' Miss Nightingale wrote in a reminiscence, 'I saw everyone of the witnesses myself and reported to him what each could tell him as a witness in public.' Notes from him reached her two or three times a day while the Commission was sitting. 'Let me know what you think,' 'Give me your notes on this,' 'What are we to do?' 'What shall we say?' 'This is Hebrew to me, will you look it over.' 'Your report is excellent.' 'I am at a stand still until I see you.'

'Sidney is again in despair for you,' wrote Mrs Herbert in the summer of 1857. 'Can you come? You will say "*Bless* that man, why can't he leave me in peace!" But I am only obeying orders in begging for you.'

'She is the mainspring of the work,' burst out Dr Sutherland to Aunt Mai in May 1857. 'Nobody who has not worked with her daily could know her, could have any idea of her strength and clearness of mind, her extraordinary powers joined with her benevolence of spirit. She is one of the most gifted creatures God ever made.'

The Reformers christened themselves the 'band of brothers', the Burlington Hotel was 'the little War Office', and the meals they shared were 'our mess'. Within this circle was an inner council of three – Miss Nightingale, Sidney Herbert, and Dr Sutherland. Though Dr Sutherland had sacrificed his life to Miss Nightingale, he still irritated her. In 1855, when he was appointed to the Sanitary Commission, he was at the opening of a distinguished career. He was Sanitary Adviser to the Government Board of Health and received £1,500 a year for part-time duties. He met Miss Nightingale at Scutari, became her slave, and his career was at an end. He worked with her throughout the Sanitary Commission on the Health of the Army without remuneration. She never thanked him and seldom had a good word for him. Once in February 1857 when they were working together on the plans for Netley, she wrote to him: 'As for Sanitary matters – Lord help you I'm only a humbug. I know nothing about them except what I have learnt from you. But you would never have found a more practical pupil.' It was the only acknowledgement she ever made.

He was untidy. He left his papers scattered over the sofa at the Burlington. He was unpunctual. He lost documents. He took a great deal of care of his health. Most exasperating of all, he was deaf. The more annoyed she became, the less he appeared to hear. She threatened to buy him an ear-trumpet, and he sent his wife to explain that his was no ordinary deafness but a peculiar nervous affliction. When she was pleased with him, Miss Nightingale laughed at him and called him the 'baby'; when she was annoyed, she belaboured him with the full force of her invective, and he became 'my pet aversion'. Dr Sutherland described himself to her as 'one of your wives'.

Dr Sutherland made his wife write for him when he feared Miss Nightingale was angry, and during the immense labours of the summer of 1857 she had to write frequently. In one note sent down from Highgate by hand, she begged Miss Nightingale to forgive her husband for not keeping an appointment. 'The rain is so tremendous that he would be drenched in five minutes so he hopes the Commander in Chief will excuse him for this once.' He was not excused. First thing in the morning a messenger arrived with an angry letter. 'I *hope* you will have seen Dr Sutherland before the return of your messenger,' wrote Mrs Sutherland, 'I am so sorry he could not go to you last night. I am afraid it worried you.' 'My dear Lady,' wrote Dr Sutherland on May 22, 1857, 'do not be unreasonable. ... I would have been with you yesterday had I been able but alas my will was stronger than my legs.' One evening at the Burlington Hotel he was late with an urgent report. After a scene he consented to stay and finish it. When Miss Nightingale read it through, she was not satisfied and sent up to Highgate telling him to come back at once and go through it with her again. Dr Sutherland lost his temper and refused, upon which she collapsed and fell into an 'agitated half fainting state'. Aunt Mai hurried up to Highgate and implored Dr Sutherland to come or he would kill her. He came immediately and expressed 'great sorrow and penitence'.

Her demands were fantastic; yet once within her orbit it was impossible not to be fascinated. The tremendous vitality, the passion of feeling she poured into her work made the rest of the world colourless. Mrs Sutherland was devoted to her husband, devoted to her domestic life; Miss Nightingale broke up the Sutherlands' home. Dr Sutherland was a strict Sabbatarian; she made him work on Sundays. He complained he was ill; he complained he was overworking; she abused him. Yet Mrs Sutherland adored her. She became 'Miss Nightingale's fag', shopping for her, running errands, buying oranges, pencils, new-laid eggs, dark curtains. She called her 'My dear dear friend', 'My dear and ever kind friend'. It was impossible to know Miss Nightingale without recognizing that she possessed qualities which allowed her to transcend the ordinary rules governing human behaviour.

She pressed hard on Dr Sutherland, but she pressed even harder on Sidney Herbert. The phrase she had scribbled against his name when the Commission was first set up she repeated again and again. 'Without him I can do nothing.' His standing, his prestige, his power with the House of Commons, were the means by which the Commission was raised to first-class importance. His powers were incomparable – if only she could get him to devote them to the work. But he was still hanging back. He complained of his health; he suffered from lassitude, fits of depression, and general malaise. They were the first symptoms of mortal disease, but to Miss Nightingale working, as she was convinced, under sentence of death, driving herself on by sheer force of will, fainting from exhaustion, forcing herself to get up and grind on again, Sidney Herbert's complaints were 'fancies'. She spoke of them contemptuously as 'fancies'. She grumbled about him. Speaking of a difficult negotiation she said: 'Mr Herbert and no one else can do it. If I can only bring him up to the scratch.'

Working together they were unequalled. Her industry, energy, and passion for facts, his incomparable talents as a negotiator, were a combination impossible to resist. 'He was a man of the quickest and the most accurate perception I have ever known,' she wrote. 'Also he was the most sympathetic. His very manner engaged the most sulky and the most recalcitrant witnesses. He never made an enemy or a quarrel in the Commission. He used to say "It takes two to make a quarrel and I won't be one".'

As the Commission proceeded, it became evident that the Reformers were succeeding beyond all their hopes. The long hours of close work, the exhausting journeys, the interviews, were bearing fruit, and Miss Nightingale's case for reforming the living conditions of the British Army was proving unanswerable.

In May Dr Andrew Smith was examined. He cut a poor figure. 'Never was man so shown up as Smith,' wrote Dr Sutherland. He gave no further trouble and allowed himself, wrote Sir John McNeill in June, 'to slip quietly into the current of reform.'

In the middle of June Sir John Hall gave evidence. Miss Nightingale had endured insolence at Sir John Hall's hands; he had been able to humiliate her, flout her authority and finally defeat her. Now the tables were turned, and she had him at her mercy, but she wrote: 'We do not want to badger the old man in his examination, which would do us no good and him harm. But we want to make the best out of him for our case.'

In July came the turn of the most important witness of all, Miss Nightingale herself. How much dared she and should she say? Sidney Herbert wished her to make no reference whatever to her Crimean experiences. He did not want her to 'make bad blood by reviving controversial issues'. His plan was for her to appear personally before the Commissioners but to be asked only questions on hospital construction. Miss Nightingale disagreed. Sidney Herbert's plan combined the two worst possible policies for her. She would appear in person and provoke the personal attention – the 'buz fuz' – which she was convinced did the work harm and yet say nothing of importance to compensate. 'I am quite as well aware as he can be,' she wrote to Sir John McNeill on July 7, 1857, 'that it is inexpedient and even unprincipled to bring up past delinquencies, but it would be untrue and unconscientious for me to give evidence upon an indifferent matter like that of hospital construction, leaving untouched the great matters which affect (and have affected) our sick more than any mere architecture could do. ... It would be treachery to the memory of my dead.' She decided not to give evidence at all. 'Let me entreat you to reconsider your determination,' wrote Mr Augustus Stafford, M.P., on June 11. 'The absence of your name from our list of witnesses will diminish the weight of our Report, and will give rise to unfounded rumours. It will be said either, that we were afraid of your evidence and did not invite you to tender it, or, that you made suggestions the responsibility for which you were reluctant to incur in public.'

A compromise was reached: she did not appear in person; she submitted written answers to questions, but she did not confine herself to hospital construction. Her evidence was read by the Commissioners 'with the greatest eagerness and admiration' and was agreed to be conclusive. 'It must,' wrote Sir James Martin, one of the Commissioners, '... prove of the most vital importance to the British soldier for ages to come.'

Her evidence is of great length, occupying thirty closely printed

pages of the report of the Commission, and is a verbatim reproduction of part of that great work which she completed in the
same month, *Notes on Matters affecting the Health, Efficiency and
Hospital Administration of the British Army.*

The enormous volume of the *Notes* was written by Miss Nightingale in six months at the same time as she was working day and
night on the Commission. It is a work on the grand scale. The
canvas is immense; great masses of detail, vast quantities of facts
and figures are handled with admirable lucidity; yet detail never
obscures the main theme. The huge volume written at white-hot
speed burns with an urgency which still strikes the reader with a
physical shock. She uses the Crimean Campaign as a test case, a
gigantic experiment in military hygiene. 'It is a complete example
(history does not afford its equal) of an army after falling to the
lowest ebb of disease and disaster from neglects committed,
rising again to the highest state of health and efficiency from
remedies applied. It is the whole experiment on a colossal scale.'

In six long and detailed sections she examines the causes; the
course and the cure of the Crimean disaster, quoting facts and
figures, giving tables, plans, diet sheets, proving conclusively that
the hospital was more fatal than the battlefield, that bad food, inadequate clothing, insanitary conditions led inevitably to defeat,
while good food, proper clothing, tolerable conditions restored
discipline and efficiency.

Let the past, she pleads, be buried, but alter the system so that
the soldier is more humanely treated in future. 'It would be useless as injudicious to select individual instances or persons as the
objects of animadversion. The system which placed them where
they were is the point to be considered. ... Let us try to see
whether such a system cannot be invented as men of ordinary
calibre can work in, to the preservation and not to the destruction of an Army. It has been said by officers enthusiastic in their
profession that there are three causes which make a soldier enlist,
viz. being out of work, in a state of intoxication, or, jilted by his
sweetheart. Yet the incentives to enlistment, which we desire to
multiply, can hardly be put by Englishmen of the nineteenth
century in this form, viz. more poverty, more drink, more faith-
less sweethearts.'

The most important part of the book follows. She had collected
figures which proved living conditions in the barracks of the
British Army in time of peace to be so bad that the rate of mortal-

ity in the army was always double, and in some cases more than double, the rate of mortality of the civilian population outside. For instance, in the parish of St Pancras the civil rate of mortality was 2·2 per 1,000. In the barracks of the Life Guards, situated in St Pancras, the rate was 10·4. In the borough of Kensington the civil rate of mortality was 3·3. In the Knightsbridge barracks, situated in the borough of Kensington, the rate was 17·5. Yet the men in the Army were all young strong men who had been subjected to a medical examination to guarantee their physical fitness, while the civilian population included old people, infants, and the physically unfit. 'The Army are *picked* lives,' she wrote. 'The inferior lives are thrown back into the mass of the population. The civil population has all the loss, the Army has all the gain. Yet, with all this, the Army from which the injured lives are subtracted dies at twice the rate of mortality of the general population, 1,500 good soldiers are as certainly killed by these neglects yearly as if they were drawn up on Salisbury Plain and shot.' In a phrase which became the battle-cry of the Reformers she declared: 'Our soldiers enlist to death in the barracks.'

Here was something very different from the dry bones of administrative reform, very different from jobbing back to old disasters, old grievances in the Crimea. The Crimean disasters had faded from the public mind, but the event which had obscured them, the Indian Mutiny, had focused attention on the army once more. 'Our soldiers enlist to death in the barracks' was a challenge no Government could afford to ignore.

There were two channels through which her disclosures could reach the public, through the Commission and through publication of the *Notes*. In July, the month in which she completed the *Notes*, it became evident that, owing to the white-hot speed at which she had driven the Commission, its Report would be written in August. She decided to put the *Notes* on one side – it was a confidential report addressed only to Lord Panmure and easily shelved by him, while facts stated before a Royal Commission and included in its report could not be suppressed. The enormous volume, representing such agonizing effort, such almost incredible toil, was sacrificed. Lord Panmure was not presented with *Notes on Matters Affecting the Health, Efficiency and Hospital Administration of the British Army* until after he had received the report of the Commission in November 1857. The public never had the opportunity of seeing it at all. It was never published and has remained unread. 'It is an old story now,' Miss Nightingale wrote of the *Notes* in December 1858. She had a few copies privately printed at her own expense which from time to time she gave away, and in these alone this monumental work survives.

In July she began to write the report of the first Royal Sanitary Commission on the Health of the Army. She lists the Report as 'one of my works'. Success had been achieved, but at the price of her health. 'Most people in her state,' wrote Aunt Mai to W. E. N., 'would be in bed not attempting to work but ... if she can keep up for this time her object will be gained.' She had set herself a goal: to finish the report. Everything was to be sacrificed, everything was to be endured until it was completed. When that was done she would rest.

But one day at the end of July she scribbled a sentence on the margin of a draft: 'Reports are not self executive.' She wrote the sentence again and again, in a private note, in a letter, on scraps of paper. 'Reports are not self executive.' She had realized that her task would end only when the recommendations of the report were put into force. Panmure had nearly succeeded in shelving the Commission; he would certainly try to evade the bothers involved in carrying out its recommendations. Once more the Bison must be bullied.

On August 7, 1857, Sidney Herbert wrote to Lord Panmure and communicated the outstanding points which would emerge from the report. In suave terms he pointed out that the disclosures as to the living conditions of the army were sensational, that public attention would certainly be arrested, and the Government would be attacked. He suggested the Government should protect itself by taking measures to remedy the worst of the abuses before the report came before the House. 'The simultaneous publication of the recommendations of the Commission and of new orders and regulations already introduced by the Government to remedy the abuses the Report disclosed, will give the Government the prestige which promptitude always carries with it.'

He then outlined a plan drawn up by Miss Nightingale which put the reorganization of the health administration of the army into the Reformers' hands. Four sub-Commissions were to be appointed at once by Lord Panmure, and Sidney Herbert was to be chairman of each. The four Commissions would begin to put the recommendations of the report into practical operation at once. Each Commission would have executive powers, and finance would be provided by an interim grant from the Treasury.

The four sub-Commissions would:

(1) put the Barracks in sanitary order;
(2) found a Statistical Department for the Army;
(3) institute an Army Medical School;
(4) completely reconstruct the Army Medical Department, revise the Hospital Regulations, and draw up a new Warrant for the Promotion of Medical Officers.

The fourth sub-Commission was christened the 'wiping Commission' by Miss Nightingale, because its wide scope enabled the Reformers to wipe the slate clean and start afresh.

On August 9 Sidney Herbert told her that 'Panmure writes fairly enough but he has gone to shoot grouse'. A week later Panmure was forced to come south on urgent business, was caught 'on the wing' at the War Office, and after a long discussion agreed to the four sub-Commissions 'in general terms'. Sidney Herbert then left for Ireland to fish for a month, writing on August 14 that he went 'with a lighter heart after seeing Pan. But I am not easy about you .– Why can't you who do man's work take man's exercise in some shape?'

Miss Nightingale was still at the Burlington, still toiling in the stuffiness and heat, still plagued by Fanny and Parthe. The report was completed, but the four sub-Commissions must be prepared. It must be decided in what places and from what persons evidence should be taken and what questions should be asked. In addition, she was unexpectedly overwhelmed with work in connexion with the Nightingale Fund. On August 11 she had a complete collapse.

'I must be alone, quite alone,' she suddenly broke out to Parthe. 'I have not been alone for 4 years.' She was at her last gasp. She had eaten no solid food for four weeks and had lived on tea. Parthe forgot her grievances and was frightened. 'It was very affecting poor dear,' she wrote to Aunt Mai. When Miss Nightingale was calmer, she told Parthe: 'I who required more time alone than anybody, who could not live without silence and solitude, have never had one moment to myself since I went to Harley Street. I don't call writing being alone. It is by far the greatest sacrifice I have made.' She refused to go to Embley; she refused to be nursed. She must go away by herself. She admitted she was ill and consented to take a cure at Malvern, but she must be quite alone. 'She took,' wrote Fanny to W. E. N., 'a sudden resolution to go to Malvern. Nothing would induce her to take anyone but George [a footman]. It makes us very unhappy to think of her so forlorn and comfortless.'

She was very ill, so ill that it was generally thought she must die, and in London Harriet Martineau brought her obituary notice up to date for the *Daily News*. Dr Sutherland wrote imploring her to stay on at Malvern, pointing out as her physician that the Burlington Hotel, dark, stuffy, and in the centre of London was the worst possible place for a person in her state of health. 'The day you left town,' he wrote at the end of August, 'it appeared as if all your blood wanted renewing and that cannot be done in a week. You must have new blood or you can't work and new blood can't be made out of tea at least so far as I know. ...'

Ill as she was, she seized her pen and wrote him an enormous angry letter. Her brain wandered; her sentences were incoherent; her writing straggled over the page, but she still had strength to be irritated by him. '... Had I lived anywhere but handy would Mr Herbert have used me? Had I not been ever at hand *could* he have used me? ... Now had I lost the Report what would the health I should have "saved" have profited me, or what would the ten years of my life have "advantaged" me exchanged for the ten weeks this summer? Yes, you say, you might have walked, or driven, or eaten meat. Well ... let me tell you Doctor, that after any walk or drive I sat up all night with palpitation. And the sight of animal food increased the sickness. ... Now I have written myself into a palpitation. ... I have been greatly harassed by seeing my poor owl lately without her head, without her life, without her talons, lying in the cage of your canary ... and the little villain pecking at her. Now, that's me. I am lying without my head, without my claws and you all peck at me. It is *de rigueur, d'obligation*, like the saying something to one's hat when one goes into church to say to me all that has been said to me 110 times a day during the last 3 months. It is the obbligato on the violin, and the twelve violins all practise it together, like the clocks striking twelve o'clock at night all over London, till I say like Xavier Le Maistre "*Assez, je le sais, je ne le sais que trop*".'

Dr Sutherland replied on September 7 that she was decidedly wrong in passing herself off for a dead owl. He himself had got all the pecking 'and your little beak is of the sharpest'. Nevertheless, he loved her. 'You little know what daily anxiety it has caused me to see you dying by inches in doing work fit only for the strongest constitutions. When I think of it all I can hardly bear the sight of a report. .... One thing is quite clear that women can do what men could not do, and that women will dare suffering knowingly where men would shrink.'

His affection and concern only provoked her. Impatience was gaining on her. Everything was to be wiped out of her life but work. Work loomed always before her, mountains of it, endless labour, endless toil which somehow must be struggled through. Mr Herbert went away to fish; Dr Sutherland prated of rest. Did they not understand, could they not understand that the only way anything could be accomplished was by unremitting effort, unceasing toil? Did they think she had brought herself to the verge of the grave for her own amusement?

She wrote Dr Sutherland a cross, snubbing note. He was to cease this nonsense about being afraid of her and this nonsense about her taking a rest. Far from taking a rest, she ordered him to come to Malvern on Monday, when she intended to start work again. Everything, figures, facts, plans for the four new sub-Commissions must be ready for Mr Herbert when he came back to London.

'I have your note Caratina Mia,' Dr Sutherland wrote on September 11, 'and write to say how sorry I am that I should seem to be afraid of your biting me. .... But what are Mr Herbert and I to do when you are buried? How is the play to go on without Hamlet? .... The daughters of Sermiah be too much for me. I'll take the veil. I'll retire from the world. There's no help for it then but my coming down on Monday.'

He arrived to find her apparently at death's door. For once her iron will was defeated, and she was forced to stay in Malvern for over a month. Her pulse raced, and she was given two cold-water packs a day to bring it down. In spite of her physical condition she obstinately continued to work.

Fanny and Parthe, both unwell themselves and sobered by the frightening spectacle of her collapse in London, made only half-hearted attempts to join her. W. E. N. came to Malvern and insisted on seeing her for a few minutes. He was horrified. 'Her days may be numbered,' he wrote to Fanny. 'Her breathing betrays her moments of distress, her power to take food fails her if excited, her nights are sleepless in consequence. .... 'Tis a sad tale. I'm not able to say more. Adieu, "W. E. N." I've said too much.'

Once more Aunt Mai was called in. In the middle of September she left her husband and family and went to Malvern. She was expected to return in a week or two but she did not return. Husband and family were put on one side, and when Miss Nightingale returned to the Burlington at the end of September, Aunt Mai went with her.

The collapse of August 1857 was the beginning of Miss Nightingale's retirement as an invalid. For the first year after her return from the Crimea, though she had gone into strict retirement as a public figure, she had led her normal life. She had had strength, she said, to 'rush about'. Though she had refused to attend functions, she had seen her friends. 'You should know Lord Stanley ... come and dine with him here on Sunday,' Richard Monckton Milnes had written in February 1857. After August 1857 she had strength only to work. 'It is an intolerable life she is leading,' wrote Parthe to Clarkey in December, 'lying down between whiles to enable her just to go on.' After any prolonged exertion her exhaustion was frightening, and she often fainted.

She not only became an invalid; she began to exploit her ill health. From the summer of 1857 she used her illness as a weapon to protect herself from her family. The summer had discouraged Fanny. She had been ill, and she announced that her health would not permit her to 'attempt the Burlington' in the winter. Parthe, however, wrote that she proposed coming to London. Aunt Mai was told to write and tell Parthe not to come. Parthe was furious. Did Florence think her own sister was not capable of doing what was wanted for her? She insisted on coming.

Miss Nightingale's reply was to have an 'attack'. Aunt Mai wrote in the greatest agitation. After reading Parthe's letter Florence had been ill all night. Dr Sutherland had been much alarmed and had said he could not sleep for thinking of her. 'It was excessive hurried breathing with pain in the head and the heart.' As a result, Parthe did not come to London. W. E. N. then insisted on coming; he must see Florence and discuss her future plans. She had another 'attack', and he retreated.

It was evident that in the present state of her health, while her life, as Aunt Mai wrote, 'hung by a thread', it was too much for her to see her family. They must keep away from her.

This unpalatable news was conveyed to the Nightingales by Aunt Mai in a series of letters of immense length, every page criss-crossed, every statement wrapped in layer upon layer of explanation, withdrawal, and apology. They were forced to give way.

By January 1858 a point had been reached at which Aunt Mai could suggest that Parthe and Fanny had better give up coming to London at all. 'My fear would be,' she wrote to 'dear friends at Embley', 'that if you were staying in London and she trying to think day after day *when* would she see you, thus would be caused the agitation we so dread.' In an undated note of 1858 Fanny wrote to Parthe: 'My dear – Florence thinks we are staying in town for her sake, so we must go on Friday.' Parthe made one further effort. In February 1858 she wrote that she was coming up to London with Fanny for the season and intended to stay at the Burlington. This announcement brought on another 'attack'. Again Parthe and Fanny gave way. They must come to London, but they would go to another hotel. Miss Nightingale's condition improved instantly, and Aunt Mai wrote to W. E. N.: 'Thank God all seems relieved.'

Fortunately a new interest was occupying Parthe's mind. In the previous summer Fanny, writing to W. E. N. from the Burlington, described a visit from Sir Harry Verney, 'an old guardsman, very much interested in Flo's work, his wife died last year and left as her earnest request that her daughters should become acquainted with F. but F. had not time for young ladies and she would not see Sir H.'

Sir Harry Verney, head of the Verney family and owner of the historic mansion of Claydon House, in Buckinghamshire, was fifty-six years old and one of the handsomest men in England. He was immensely tall, his features were aristocratic and aquiline, and he possessed an air of nobility so extraordinary that people turned to look after him in the street as if he were a visitant from another world. He had originally held a commission in the Life Guards and intended to make a career in the army. When he inherited the family estates in Buckinghamshire, he was so horrified by the miserable condition of his land and his tenants that he gave up the army and devoted himself to becoming a model landlord. Agriculture and the agricultural labourer, owing to a long depression, were shamefully neglected. Sir Harry became a pioneer in rural housing and administration, drained and reclaimed land, built model cottages, founded schools and was active in the administration of the poor law. When the cholera broke out in Aylesbury Sir Harry worked among the sick, and subsequently

235

collected funds to build the county hospital at Aylesbury. At the age of thirty he went to Cambridge, where his abilities won him the friendship of the philosopher Whewell. He became Liberal member for Buckinghamshire in 1832 and held the seat for periods amounting to thirty-one years. Members of the Verney family had represented Buckinghamshire, either the shire or the borough, at frequent intervals since the reign of Edward VI. Sir Harry was an intimate friend of Lord Shaftesbury, an Evangelical, an active member of the Bible Society, the Church Missionary Society, and the Evangelical Alliance, but he never obtruded his opinions or 'lost the tone of good society'. Clarkey, who met him with a strong prejudice because she said she 'detested psalm singers', declared he was ' a jewel'.

Before the end of the summer of 1857 Sir Harry was a frequent caller at the Burlington. He fell in love with Miss Nightingale and asked her to marry him, and she refused him. During the winter he stayed at Embley, and it began to be evident that he was becoming attached to Parthe. The engagement was announced in April, and the marriage took place quietly at Embley in June 1858.

The prospect of becoming Lady Verney occupied Parthe's mind; the business of marrying a daughter delighted and distracted Fanny. Hilary Bonham Carter told Clarkey in May 1858 that she had been seeing Fanny and Parthe in London and spent 'some horrible long days fussing and shopping with Aunt Fanny', but though Fanny and Parthe were staying in a hotel close to Florence, 'they did not wish to interfere with her in the slightest'. At last she was left alone.

14

A HUSH fell over her life. She moved to new rooms in an annexe of the Burlington – three bedrooms and a dressing-room on one floor, a double sitting-room on a floor below. The street was quiet; the house had none of the bustle of the hotel. The atmosphere was heavy with solemnity. Voices were lowered; feet trod lightly as her fellow workers were shown into the drawing-room where she lay prostrate on the sofa. She was convinced, everyone round her was convinced, that she had at most a few months to live.

Aunt Mai broke up her family life. She shut up her house, her husband and girls went to stay at Embley, and she came to the Burlington to make Florence's last months on earth easy. Her son-in-law, Arthur Hugh Clough, became Florence's slave. He came to the Burlington every day, wrote notes, delivered reports, fetched letters, tied up parcels, and was content, she wrote, 'to do the work of a cab horse'. The Nightingales, cowed, remained at a distance. Aunt Mai and Clough became the twin guardians of a shrine.

It was a strange, hot-house existence led under the shadow of impending death. One day Miss Nightingale had a long conversation with Clough in which she arranged all the details of her funeral. She wrote many letters 'to be sent when I am dead'. On November 26, 1857, Sidney Herbert was given as a sacred trust the task of carrying out the reforms recommended in the report of the Commission. He was not to regret the manner of her death. 'You have sometimes said you were sorry you employed me. I assure you that it has kept me alive. .... I hope you will have no chivalrous ideas of what is "due" to my "memory". The only thing that can be "due" to me is what is good for the troops.'

On December 11, 1857, she gave Parthe directions for her burial. Her love for the troops and her association with the men had made her feel what she never expected to feel – a superstition. She had a yearning to be buried in the Crimea, 'absurd as I know it to be. For they are not there.'

In November 1857 she made a will in which the property she would one day inherit from her father and mother was to be used to erect a model barrack, not forgetting the wives but having a kind of Model Lodging House for the married men.

Personal remembrances were to be sent to Mrs Herbert, Dr Sutherland, and Sir John McNeill, 'after I am dead', and Parthe was instructed to bring them from Embley.

It was an atmosphere in which emotion ran riot, and the exalted affection between Miss Nightingale and Aunt Mai burst into strange flower. Wounded by the behaviour of Fanny and Parthe during the summer, she conceived an idea which she called the 'Virgin Mother', to explain the love and sympathy she felt in Aunt Mai, the indifference she felt in Fanny. 'Probably there is not a word of truth in the story of the Virgin Mary,' she wrote in a private note of 1857. 'But the deepest truth lies in the idea of the Virgin Mother. The real fathers and mothers of the human race are NOT the fathers and mothers according to the flesh. I don't know why it should be so. It "did not ought to be so". But it is. Perhaps it had better not be said at all. What is "Motherhood in the Flesh"? A pretty girl meets a man and they are married. Is there any thought of the children? The children come without their consent even having been asked because it can't be helped. ... For every one of my 18,000 children, for every one of these poor tiresome Harley Street creatures I have expended more motherly feeling and action in a week than my mother has expended for me in 37 years.'

In another note she wrote: 'I have had a spiritual mother without whom I could have done nothing, who has been all along a "Holy Ghost" to me and lately has lived the life of a porter's wife for me.' On Aunt Mai's side affection passed into worship. 'My child, my friend, my guide and uplifter, my dearest one on earth or in heaven,' she wrote. Recalling this period to Clarkey, Miss Nightingale said: 'We were like two lovers.'

Her family had agreed not to expect letters from her; her energy must be preserved for work. Aunt Mai was to send reports. And now, lying on the sofa in the drawing-room, seldom sitting up and almost never going out, she proceeded to toil as she had never toiled before. In November 1857 Aunt Mai wrote: 'Mr Herbert for 3 hours in the morning, Dr Sutherland for 4 hours in the afternoon, Dr Balfour, Dr Farr, Dr Alexander interspersed.' A little later: 'Flo is working double tides, labouring day after day until she is almost fainting.' The task differed from the task of a year ago when the material for the Commission was being collected. 'There is now the most important work of all to be done namely to gain the fruit of the Report in working the Reforms which have been its purpose – they have now not only to work but to fight.' 'Dr Sutherland quite admits,' wrote Aunt Mai, 'that ... completion depends on her. She alone can give facts which no one else hardly possesses ... she alone has both the smallest details at her finger ends and the great general view of the whole. He has been saying all this while at his luncheon, now he is at his work, and I only hope he won't too soon say "Will you tell her I am at a standstill until I see her," for she is now resting, and I am always afraid to hear those words which don't at all the less come because he begins by saying, "I don't want her at all, I only want her to rest".'

Lord Panmure was behaving over the four sub-Commissions exactly as he had behaved over the Commission itself. He had been frightened by Sidney Herbert into agreement, but, under pressure from within the War Office, he lost his nerve and once more retreated to the Highlands where he shot grouse, left letters unanswered, and refused to come to London.

'Pan is *still* shooting,' wrote Sidney Herbert on September 28. 'In future you must defend the Bison, for I won't.' Miss Nightingale sent letters to Brechin Castle; she got wind of a flying visit Panmure was secretly paying to London and sent him round a note by hand. She declared in a letter to Sir John McNeill on October 10: 'I shall not leave P. alone till this is done.' Her personal relations with Panmure continued playful. Grim determination was masked by compliments and jokes. She wrote him humorous notes; he sent her grouse. In private she raged against his 'unmanly and brutal indifference'. 'I have been three years serving the War Department,' she wrote to Sidney Herbert in November 1857. 'When I began there was incapacity but not indifference. Now there is incapacity and indifference.'

Lord Panmure was being rent apart. The Reformers were powerful, they had public opinion with them, and public opinion was to be dreaded; but the reactionary party within the War Office was powerful and to be dreaded as well. On the administrative side Sir Benjamin Hawes was capable of causing infinite trouble, while Dr Andrew Smith was fighting furiously against the 'wiping' sub-Commission, which wiped the slate clean for the complete reorganization of the Army Medical Department. In November 1857 pressure from the War Office was so great that Panmure revoked the 'wiping' sub-Commission. The Reformers were appalled, Sidney Herbert forced Panmure to see him, and after a long and stormy interview this sub-Commission was reinstated.

It became clear to Miss Nightingale that the issue turned on the ability of each side to frighten Panmure. Whichever side could frighten him most thoroughly would be victorious.

She devised a new idea. Public opinion was the Reformers' strongest weapon. She would instruct public opinion and at the same time put pressure on Panmure through a Press campaign which should tell the nation the truth about conditions in the army.

The outlines, the facts, even the headings for all articles were supplied to contributors by her. 'I enclose a sketch, add to it, take away from it, alter it,' she wrote to Edwin Chadwick in August 1858. Lord Stanley, Sidney Herbert, Edwin Chadwick, who had sat on the first Poor Law Commission of 1834 and instigated the first Sanitary Commission of 1839, wrote for her. She refused to sign anything herself. Her contribution was an unsigned pamphlet, *Mortality in the British Army*, which was one of the earliest, if not the earliest, instance of the presentation of statistical facts by means of pictorial charts – Miss Nightingale believed she invented this method. The facts that the majority of deaths in the Crimean mortality were due not to war but to preventable disease, and that mortality in barracks was double the mortality in civil life, were driven home at a glance by means of coloured circles and wedges. The text was an explanation of the diagrams based on the Reformers' battle-cry, 'Our soldiers enlist to death in the barracks'. Sidney Herbert printed the charts and the memorandum as an appendix to the report of the Commission, and the pamphlet was published as 'Re-printed from the Report of the Royal Commission'. Miss Nightingale had 2,000 copies printed at her own expense, and sent them to the Queen, the Commander-in-Chief, Members of both Houses, Commanding Officers and doctors, with a personal letter designed to provoke the recipient's curiosity. 'It is always more gratifying to people to have a thing which they think other people have not got,' she wrote to Sidney Herbert on Christmas Day, 1857. One letter runs: 'It is confidential and *I* have no right to give away copies.'

'This is for your own reading,' she writes in another. And, 'Please do not leave this about.' Three copies of the charts were handsomely framed and sent to the War Office, the Horse Guards, and the Army Medical Department. She never discovered if they were hung.

By the end of 1857 Panmure had given way. The four sub-Commissions were granted and set up in December. 'With such ample instructions as you may guess them to be, when I tell you they were written by me,' wrote Miss Nightingale to Sir John McNeill.

She did not go to Embley for Christmas, but stayed in London at the Burlington with Aunt Mai and Clough. The year 1858 dawned brightly for the Reformers, and as the summer came they gained success after success; even though in February 1858 Lord Palmerston's government fell. Lord Panmure went out of office, and for a moment it appeared as if his exit might be fatal. Dr Andrew Smith had at last retired, and the next man in order of seniority was Sir John Hall. Lord Panmure had pledged his word that as long as he held office Sir John Hall should never be Director-General, but at this crucial moment Panmure was succeeded by General Peel. Miss Nightingale was in an agony. The appointment hung in the balance, and she raged at the fate which had decreed the Bison's disappearance at the one moment when he could be of use. However, Sidney Herbert, with his matchless talent for negotiation, approached General Peel. On February 27 Peel promised he would make no appointment until he had conferred with Sidney Herbert, but would not commit himself further. Sidney Herbert continued to put himself in Peel's way, not speaking directly on the appointment but 'throwing in a little praise of Alexander when talking or writing on other subjects'. His persuasiveness was effectual. On May 25 he wrote that he had won the day. Sir John Hall was passed over, and Alexander was appointed Director-General of the Medical Department of the British Army and officially gazetted on June 11.

A citadel had fallen. Co-operation would replace obstruction from within the Medical Department at the War Office. 'Smith is really gone,' wrote Sidney Herbert. 'It is no use trying to realize the enormous importance of such a fact.' Yet another victory followed. On May 11 in the House of Commons Lord Ebrington moved a series of resolutions on the health of the army founded on the report of the Royal Sanitary Commission. He called atten-

241

tion to the figures published in the report revealing the high mortality in barracks compared to the mortality in civil life. 'Improvements,' he concluded, 'are imperatively called for not less by good policy and true economy than by justice and humanity.' There were deafening cheers. After reading the account of this debate, Sir John McNeill wrote to Miss Nightingale on May 13: 'To you more than to any other man or woman alive will henceforth be due the welfare and efficiency of the British Army. I thank God that I have lived to see your success.'

Throughout the summer of 1858 she was at the Burlington, going out of London only twice for a week's cure of 'fresh air and water packs' at Malvern. She travelled by railway in an invalid carriage attended by Aunt Mai as 'dragon' and Clough as courier. Bystanders were struck with awe. She was carried in a chair, and usually her bearers were old soldiers, who carried her as if she were a divinity. A space was cleared on the platform, curious onlookers were pushed back, voices were hushed, and the station-master and his staff stood bareheaded as she was carried into the carriage. She was already becoming a legend.

Though she cut herself from the world, her rooms at the Burlington, the 'little War Office', were a hive of industry. She had made her rooms cheerful with new carpets and curtains at her own expense. Lady Ashburton, who before her marriage had been Miss Louisa Stewart Mackenzie, 'beloved Zoë', sent flowers and plants from Melchett Court every week; Fanny supplemented the catering of the Burlington with game, hot-house fruit, eggs, and cream. Miss Nightingale was very ready to provide her fellow workers with breakfasts, luncheons, and dinners. 'If you will come and talk ought it not to be with dinner?' she wrote to Dr Farr in 1859. 'Please come here you shall have dinner at 7,' she wrote to Dr Balfour in 1859. She offered the delegates to the Statistical Congress of 1860, 'a room, breakfast, dinner and a place to work at any time – a better dinner with notice.' She seldom joined these parties herself, but from her couch she kept a hand on detail. 'Take care of your cream for the breakfast – it is quite turned.' 'Put Dr Balfour's big book back where he can see it while drinking his tea,' she told Aunt Mai.

She visited no one, but eminent visitors came to her. The Queen of Holland, the Crown Princess of Prussia, the Duke of Cambridge called on her regularly. Kinglake consulted her when he was writing his *Invasion of the Crimea*. She was not impressed.

'I found him exceedingly courteous and agreeable,' she wrote to Edwin Chadwick, 'looking upon the war as a work of art and emotion – and upon me as one of the figures in the picture ... upon figures (arithmetical) as worthless – upon assertion as proof. He was utterly and self *sufficiently* in the dark as to all the real causes of Crimean Mortality.' Kinglake's well-known description of herself and her work she dismissed shortly: 'He meant it to be kind, but it was fulsome.'

Manning visited her, and she wrote to him in February 1860 as 'one of those whom I *know* to be friendly to me'. She told Sidney Herbert that Manning had always treated her fairly; he advised her on the special needs of the Catholic regiments of the army.

In the spring of 1858 she had begun an important friendship. Captain Douglas Galton, a brilliant young Royal Engineer, was the army's leading expert on barrack construction, ventilation, heating, water supply, and drainage. He held an important War Office appointment, had been appointed referee for the consideration of plans for the drainage of London, and was a member of the Barrack sub-Commission. He was also a family connexion, for in 1851 he had married the beautiful Marianne Nicholson. In 1857 Marianne and the Nightingales were reconciled, and Clarkey was asked to meet her at a breakfast given by Fanny and Parthe at the Burlington. 'In comes Marianne by invitation,' wrote Clarkey to Mrs Bonham Carter on July 18; 'we are all loving, she is always as pretty and much improved in character, she takes an interest in amendments and don't never flirt no more. Let people talk against matrimony!'

After the Barrack Commission was set up, Miss Nightingale and Douglas Galton met and corresponded almost daily. She absorbed him. She was too busy to see Marianne, but she wrote her kind letters and was godmother to one of the Galton children. Otherwise Marianne's name was not mentioned. Once at the foot of a long letter dealing with the construction of a hospital at Woolwich, he scribbled in pencil: 'Marianne had a boy this morning.'

The support of Douglas Galton, the appointment of Alexander as Director-General of the Army Medical Department, the appointment of Dr Balfour to establish a statistical department within the War Office strengthened Miss Nightingale's hand, and in the autumn of 1858 she judged the time was ripe for another

Press campaign. The only way to influence Ministers, she wrote, was through the public.

Notes on Matters affecting the Health, Efficiency and Hospital Administration of the British Army was lying unused. Copies were sent to the Queen, the Commander-in-Chief, Members of the Cabinet, War Office officials, and well-known public figures with a covering letter. 'This is an advance copy of a Confidential Report,' she wrote. 'May I ask you not to mention to anyone that you have this Report.' A copy was sent to Harriet Martineau with a letter calculated to provoke her interest by warning her, 'this Report is in no sense public property'.

Harriet Martineau was a leader-writer on the *Daily News*. The daughter of an unsuccessful sugar-refiner, deaf, sickly, physically unattractive, and born without the sense of taste or smell – she said she had only once in her life been able to taste a leg of mutton and found it delicious – she had become a political power through her writings. She had hit on the idea of conveying knowledge in the form of fiction, and in an enormously successful series of tales had illustrated the facts of political economy, taxation, and the poor law. Her prestige was very great, she dined out every day except Sunday, Cabinet Ministers asked her advice, she had been offered and had refused a Government pension, and Richard Monckton Milnes, Kingsley, Gladstone, Bulwer Lytton, Sydney Smith, Carlyle and Lord Brougham were her friends.

Like Miss Nightingale, she suffered from bad health, and her character was, she wrote, gloomy, jealous, and morbid. In 1839 she was pronounced to be suffering from an incurable disease and had spent five years in bed. She recovered, cured by mesmerism. In 1855 she was announced to be dying of heart disease, but in fact she did not die until twenty years later.

She was a passionate supporter of the movement for 'Women's Rights', unlike Miss Nightingale who, though she did more to open new worlds to women than perhaps any other woman, was not a feminist. Miss Nightingale dedicated herself to the cause of the unfortunate, the weak, the suffering, and the defenceless, and it was a matter of indifference to her whether they happened to be women or men.

She forbade Harriet Martineau to use the *Notes* as the text for a sermon on pioneer women. 'I have a great horror of its being made use of after my death by "*Women's Missionaries*" and those kinds of people. I am brutally indifferent to the wrongs or the rights of my sex,' she wrote on November 30, 1858. Miss Martineau used information from the *Notes* for a series of articles on the army which were published in the *Daily News* and successfully reprinted in 1859 in book form under the title *England and her Soldiers*.

In December 1858 Miss Nightingale herself wrote a short popular version of the facts contained in the *Notes* and in the report of the Royal Commission; it was published anonymously under the title, *A Contribution to the Sanitary History of the late War with Russia*, and Sir John McNeill described it as 'complete and unanswerable'.

It was a hopeful period. The four sub-Commissions, in spite of official opposition, were accomplishing a great deal. Barracks were being inspected and plans laid for rebuilding and reconditioning. Alexander was hard at work on the new regulations which were to transform the Army Medical Department. For the future, developments of the greatest importance were taking shape.

In the summer of 1857 the Indian Mutiny had broken out. Miss Nightingale longed to leave her desk and go out to the troops, but Sidney Herbert prevented her. 'I may tell you in confidence,' she wrote to Dr Pattinson Walker in 1865, 'that in 1857, that dreadful year for India, I offered to go out to India in the same way as to the Crimea. But Sidney Herbert . . . put a stop to it. He said that I had undertaken this work, caused him to undertake it and that I must stay and help him.' She consoled herself with the reflection that by her work for the army in England she was saving more lives than by going to India. 'What are the murders committed by these miserable Bengalese compared to the murders committed by the insouciance of educated cultivated Englishmen?' she wrote to Sidney Herbert in September 1857. As a result of the Mutiny India passed from the government of the East India Company to the government of the Crown, and the welfare of the troops in India became the responsibility of the British Government. In the course of the Royal Commission appalling reports were received of sanitary conditions in India, and for six months Miss Nightingale had been asking for a second Royal Sanitary Commission to deal with the health of the army in India. Now she seemed likely to succeed. The first Secretary of State for India to be appointed after the passage of the Control of India Bill in 1858 was Lord Stanley, her admirer and friend, and there was every probability that the Commission would be set up in the near future.

All too soon the sky darkened. In August 1858 Alexis Soyer died. At the time of his death he was collaborating with her on the Barracks' Commission, and one of his last acts was to open, on July 28, his model kitchen at the Wellington Barracks. On August 28 she wrote to Douglas Galton: 'Soyer's death is a great disaster. My only comfort is that you were imbued before his death with his notions.'

But the disaster of Soyer's death was as nothing compared with the next disaster that threatened – the breakdown of Sidney Herbert's health. He had never been robust. After his term of office during the Crimea his health had broken down and he had gone abroad for a cure, and during the Royal Commission he had broken down again. Two months of fishing in Ireland, and riding and shooting at Wilton during the long vacation, partially restored him, but when he had returned to London in the autumn of 1857 he had to grapple with the enormous tasks of setting up the four sub-Commissions and then of administering them, since he was chairman of each of the four.

The work was crushing, physically. Inspection of barracks meant constant travelling; there was opposition; commanding officers were insolent. Facilities for inspection were refused, and the Commissioners were kept waiting on barrack squares in cold and wind. 'The Big-Wigs were surly,' he wrote after a visit to Aldershot in March 1858. Physical exertions were succeeded by the close and gruelling labour of drafting and revising regulations, for three of the sub-Commissions dealt with administration. (Miss Nightingale's rough agenda for one week's work ran: 'Draft Instructions and Regulations defining the duties of etc. etc., and revising the Queen's Q.M.Gs, Barrack and Hospital Regulations, a new Constitution for the Army Medical Board and a Warrant for Promotion.') The strain was too great. From the beginning of 1858 a marked deterioration in his health began.

In January of that year he was suffering from acute neuralgia and tic in the temples. Miss Nightingale recommended 'saturating a small piece of cotton with chloroform and camphor, putting it up the nose and inhaling strongly'. He followed this prescription, did it too frequently, and made himself sick. At the end of the month he wrote: 'My head is very shaky in the neuralgic way.'

On February 2 he was apologizing for missing a conference – 'I am fairly broken down, but will be up again directly.' He got up but felt so ill that he had to go back to bed. Three days later he wrote that he was suffering tortures from headache and was unable to work. On March 15 he was in bed again – 'Here I am idling away my time in bed. I have been heartily ashamed of myself these last few days.'

Never did a man receive less sympathy. Miss Nightingale, working as she believed on her deathbed, had small consideration for lesser ailments. What was a headache, a feeling of wretchedness compared with what she was enduring? It was no new thing for her to complain of him. They had differed when she was in the Crimea. He had evaded her on the subject of Army Reform when she came home. 'Ten years have I been endeavouring to obtain an expression of opinion from him and have never succeeded yet,' she wrote to Sir John McNeill in November 1857. Nevertheless, she finished the letter with the phrase she used so often, 'without him I could do nothing'.

She drove him. That was her function. He did not shrink from her white-hot energy, her implacability – he needed its vitalizing warmth. The fundamental differences between their two characters balanced each of them and gave their collaboration its immense value. But because they were so different, complications ensued. They irritated each other. Miss Nightingale lavished no admiration on Sidney Herbert while he was alive; her eulogies were written after his death. She was impatient with him; she hunted him; she grumbled at him; and Sidney Herbert, renowned for his urbanity and gentleness, scolded her. He told her she was irritable, exacting, impatient, that she exaggerated and was too fond of justifying herself. He never broke into the panegyrics commonly indulged in by her fellow workers. Only the words used by him at the end of every note he wrote her, of every interview they had together, 'God bless you,' spoke of the affection between them. The tie which united them was so strong that it did not need support. 'We were identified,' she wrote to Clarkey in 1861. 'No other acknowledgement was needed.'

She was not alone in thinking Sidney Herbert took his ailments too seriously. 'How can a man like you get ill?' wrote Delane, editor of *The Times*, on February 6, 1858. 'If I could lead your life I would live 1,000 years, and never have a headache.'

While on one hand Miss Nightingale drove him; on the other

he was urged on no less relentlessly by his wife. Far from resenting his work with Miss Nightingale, Liz encouraged it; work with Florence was the part of her husband's life which she most thoroughly shared.

Before his marriage Sidney Herbert had been the close friend of the beautiful and unhappy Caroline Norton, one of three lovely sisters nicknamed the Three Graces. The grand-daughters of Sheridan, they had inherited his wit and captivating charm. One sister became Lady Dufferin, one became Duchess of Somerset, and Caroline herself married the Hon. George Chappel Norton, younger brother of Lord Grantley. It was an unhappy match; George Norton, an unsuccessful barrister, had a violent temper, and the Nortons were incessantly in financial difficulties. Caroline became a professional author and achieved considerable success. She had wit and brilliant beauty, and the parties she gave in her little drawing-room in Storeys Gate became famous. One of her most intimate friends was Lord Melbourne. George Norton, though insanely jealous, was willing to profit from his wife's friendships, and in 1831 he was made a Metropolitan Police Magistrate by Lord Melbourne. Five years later he brought proceedings charging Lord Melbourne with committing adultery with Caroline. The action failed, the jury dismissed the case without even calling upon the defence, but as a result the Nortons separated and the tragedy of Caroline's life began. As a woman living apart from her husband, she had no rights either over her income or her children. Her children, whom she adored, were removed, and George Norton not only refused to make her an allowance but brought an action demanding the money she made from her books. She did not think her children were well treated; her youngest boy, after being taken away, died at the age of nine as the result of a fall from a pony. The miseries she endured were instrumental in bringing about an improvement in the laws relating to women.

In the early forties Sidney Herbert, 'beautiful as an angel', chivalrous, brilliantly clever, immensely rich, was always to be seen at Caroline Norton's house. Caroline was not only beautiful, she was powerful. Motley the historian described her face as 'most dangerous, terrible, beautiful', hair 'violet black ... eyes very large with dark lashes, and black as death; the nose straight; the mouth flexible and changing; with teeth which in themselves would make the fortune of any ordinary face.' She had force and she had altruistic passion. In 1836 she had written *A Voice from the Factories*, a plea for the sweated worker. Her attachment to Sidney Herbert was well known, and Percy Dacier, the hero of *Diana of the Crossways*, was said to have been drawn from him.

Sidney Herbert was now first in the succession to Wilton. His father's eldest son by his first marriage had, in 1814, contracted a disastrous marriage with a Sicilian pseudo-countess, which the family had tried in vain to annul, and in 1827 had succeeded to the title as the twelfth Earl of Pembroke. It was evident that his health was hopelessly impaired, and, as he had no children, it had become increasingly desirable that Sidney Herbert should marry. In 1846 he married Miss Elizabeth à Court, who had been devoted to him since childhood, who was beautiful, well born, and devoutly religious; they had been married for eighteen months when Miss Nightingale met them in Rome. Caroline Norton disappeared from his life. In 1847 Fanny Allen met one of Sidney Herbert's 'intimates' who 'detailed the course of his [Sidney Herbert's] marriage and the loosening of the tie between him and Mrs Norton, who behaved very well on the occasion and assured him when he married she would never cross his path.'

Liz Herbert's devotion to her husband was possessive. Miss Nightingale, meeting her in Rome, had noticed her almost unbalanced affection for Sidney, her eagerness to be everything to him, to share his every thought. She was insecure. 'You know all I have to bear more than anyone else,' she wrote to Miss Nightingale after Sidney's death. 'It is strange but I think his whole family believe *he did not love me*.' Far from remonstrating with Florence for driving Sidney too hard, Liz supported her. She clung to Florence – she had clung to her from the first moment they met in Rome in 1849, because through her she drew nearer to Sidney.

Thus urged, goaded, driven, Sidney Herbert struggled through 1858, and immeasurably greater demands were made on him the next year. Early in 1859 it became evident that what Miss Nightingale described to Harriet Martineau as 'eight months importunate widowing of Lord Stanley' was to be successful. A Royal Sanitary Commission was to be set up to do for the army in India what the Royal Sanitary Commission of 1857 had done for the army at home, and Sidney Herbert was invited to be chairman. It was a hideously laborious prospect. The work would be gigantic; the state of India was inextricably confused, the opposition obstinate; the distance from which data must be collected was an enormous complication. His health was steadily deteriorating, and he still had to devote long hours to the four sub-Commissions. Nevertheless, he felt himself bound to accept. A month later an even greater task was thrust on him. The Government had fallen in March, and in the general election which followed Lord Palmerston was returned to power. He invited Sidney Herbert to become Secretary of State for War. It was, on the face of it, a triumph. What could not Sidney Herbert do for Army Reform in the place of Panmure? But his first sensation was one of despair. On June 13 he wrote to Miss Nightingale: 'I must write you a line to tell you I have undertaken the Ministry of War. I have undertaken it because I believe that in certain branches of administration I can be of use, but I do not disguise from myself the severity of the task, nor the probability of my proving unequal to it. But I know you will be pleased at my being there. I will try and ride down to you tomorrow afternoon. God bless you.'

She received the news soberly. 'Dearest I hardly can congratulate you,' she wrote to Liz on June 14, 'but you know what I think about it. I don't expect even him to turn apes into wise men and lions into Unas. I am afraid he has inefficient servants, a disorganized department and a silly C-in-C to deal with. But for all that, (and all the more because of all that), it is a great national benefit his undertaking it. And certainly he is the only man who could do it. ... I don't believe there exists a more disorganized office than the War Office.'

The Reformers seemed now to be in a strong position. Sidney Herbert was Secretary of State for War; Alexander was Director-General of the Army Medical Department. The Royal Sanitary Commission on the Health of the Army of 1857 was being put into operation, and a new Commission on the Health of the Army in India was being set up. It seemed there was every reason for optimism; but there was no optimism. Instead there was depression. The Reformers felt that the future was dark.

Only now, when so much progress had been made, did the almost insuperable difficulties confronting them emerge. The basic difficulty was the administrative system of the War Office itself. In November 1859 Miss Nightingale summed up her experience. 'The War Office is a very slow office, an enormously expensive office, and one in which the Minister's intentions can be entirely negatived by all his sub-departments and those of each of the sub-departments by every other.'

Sidney Herbert had just had a demonstration of War Office power in the matter of the Army Medical School. Miss Nightingale had set out the necessity for an Army Medical School in *Notes on Matters affecting the Health, Efficiency and Hospital Administration of the British Army*. 'Young men were formerly sent to attend sick and wounded soldiers who *perhaps* had never dressed a serious wound ... who certainly had never been instructed in the most ordinary sanitary knowledge, although one of their most important functions was hereafter to be the prevention of disease in climates and under circumstances where *prevention* is everything.' The school was designed to provide training in military hygiene and military surgery. Miss Nightingale drew up the regulations in consultation with Sir James Clark, and the nomination of its professors was left entirely in her hands. The third of the four sub-Commissions which Sidney Herbert extracted from Lord Panmure was concerned solely with it.

The nominations were made in 1857. Dr Parkes, the great military sanitarian, was to be Professor of Hygiene and Dr William Aitken, later Sir William Aitken, to be in charge. Panmure could not be got to the point of making the appointments; he would not actually appoint anyone 'even if the Angel Gabriel had offered himself, St Michael and all angels to fill the different chairs', wrote Sidney Herbert. Panmure went out of office and General Peel succeeded him, but still nothing was done. Then General Peel was succeeded by Sidney Herbert, who wrote that something should be done about the Army Medical School 'at last'. He converted nominations into appointments, but delay continued; the officials in the War Office were not yet defeated. Premises were selected at Fort Pitt, Chatham, yet work on them did not begin; the professors were appointed, but their salaries were not paid; requisitions were sent in for instruments and equipment, but they were not filled. Month added itself to month, it was a year, it was two years, it was nearly three years since the original authority for the establishment of the school had been given, and still nothing had been accomplished. In 1860 Sidney Herbert insisted on fixing a date for the opening of the school. Three letters sent by Miss Nightingale to Douglas Galton, in August of that year, relate what occurred. The first from Dr Aitken, marked 'Wail no. 1', states: 'No work even begun.' The second also from Dr Aitken, marked 'Wail no. 2', states: 'No money for instruments.' The third from Miss Nightingale herself, dated September 3, 1860, marked 'Wail no. 3', relates the 'disaster of the opening day'. 'On Saturday I had a letter from the Professors of the Medical School quite desperate . . . the authority for the instruments and the money had not yet come. Ten of the students arrived. They stared at the bare walls and in the absence of all arrangements for their work concluded the school was a hoax.'

It was one of a hundred petty triumphs scored by a jealous and strongly entrenched bureaucracy. A new issue had become clear. Progress was impossible with the existing machinery. Before reforms could be carried through the War Office itself must be reformed.

Sidney Herbert must nerve himself to yet another gigantic task. The amount of work exacted by the office of Secretary of State for War was in itself enormous; General Peel had been 'overpowered' by it, Panmure had surmounted it 'by the simple process of never attempting to do it'. Sidney Herbert was attempting far more than either of his predecessors. He nominally resigned in June 1859 the chairmanship of the Commission on the Health of the Army in India, but he continued to be chairman of the four sub-Commissions. And the Chancellor of the Exchequer was Mr Gladstone, whose two ruling passions were national economy and anti-militarism. Sidney Herbert came into conflict with both. Mr Gladstone's passion for economy produced interminable wrangles on the Army Estimates and the Estimates for Defence; his anti-militarism made him unwilling to provide amenities for the British soldier. Liz sent Miss Nightingale a riddle: 'Why is Gladstone like a lobster? Because he is so good but he disagrees with everybody!'

Yet once again Sidney Herbert felt he had no choice. In consultation with Miss Nightingale a scheme was prepared. Its objects, she wrote, were 'to simplify procedure, to abolish divided responsibility, to define clearly the duties of each head of a department and of each class of office; to hold heads responsible for their respective departments with direct communication with the Secretary of State.'

She approached this new task with a determination so grim that it was almost despair. The enthusiasm, the exhilaration with which she had approached the first Royal Commission of 1857 were gone for ever. She was being crushed, as Sidney Herbert was being crushed, by the weight of her labours. 'I am being worked on the tread-mill,' she wrote.

Miss Nightingale had no secretary. The compilation of statistics, the noting down of columns of figures, the laborious comparisons were done by herself. The innumerable letters, the immensely long reports were written by her own hand. The physical effort of writing down the enormous number of words she produced each day was staggering. The only method of duplicating was to have the text set by a printer and copies struck off. She had this done at her own expense, recording that from 1857–60 she spent £700 sterling out of her own private income on printing.

As she toiled her sense of resentment grew fiercer. Was ever suffering like mine? Was ever self-sacrifice like mine? she constantly asked herself. Sidney Herbert with his health, Dr Sutherland with his desire for holidays, Dr Alexander, even Aunt Mai, even Clough, all aggravated her; all were inadequate.

At the end of the summer of 1859 she had another collapse, with the familiar symptoms of fainting, breathlessness, weakness, and inability to digest food. It was impossible for her to leave London, impossible for her to pass another summer at the Burlington. She compromised by taking rooms at Hampstead, a custom she continued for many years. Dr Sutherland and Clough came daily and stayed all day. Aunt Mai took the opportunity of going home to see her family, and Hilary Bonham Carter took her place as 'dragon'. Sidney Herbert, also kept in London by his work, rode out nearly every evening from Belgrave Square.

For a short interval she was quiet. Hilary, writing a bulletin to Embley, described her 'lying on her couch, wrapt up in her delicate blanket, her little head resting on the pillow peeping from the blanket gives her quite an infantine appearance'. In September Fanny came to Hampstead – she had not seen Florence for nearly six months. 'She received me as if we had only just been parted, very affectionately, but her manner was nervous as if she feared to touch upon exciting subjects.' She told Parthe, now Lady Verney, 'She would have made a beautiful sketch, lying there reclining upon pillows in a blue drifting gown, her hair so picturesquely arranged, her expression most trusting, hardly harmonizing with the trenchant things she sometimes says, her sweet little hands lying there ready for action.' There were, Fanny noticed, 'several pussy cats' in the room; one was lying on Florence's shoulder.

She was now in bed or on her couch continuously. She never walked; she seldom went out. Fanny noticed that her face was flushed, her hands hot, and that talking seemed an effort. Nevertheless, ill though she might be, the attitude of her circle toward her physical condition was changing. Time had passed. She still spoke as if she were on her deathbed, her life was still described as hanging by a thread, but – it had been hanging by a thread for two years.

In June 1859 Aunt Mai wrote to Embley that she did not now have the 'daily fears' she had at one time. Dr Sutherland told Fanny in September that he did not now think Miss Nightingale's life would be shortened, as she herself expected it would be. There was even a family unpleasantness over her invalidism. The Verneys came to London and she said she was not well enough to see Parthe. Meanwhile her maid told Sir Harry that she was fairly well. Sir Harry was offended, which brought on one of her 'attacks' of 'agitated breathing'. Aunt Mai explained in a long apologetic letter that 'it is now not her life but her getting *worse* which hangs upon a thread'.

Aunt Mai's family became impatient. Two years had passed since Aunt Mai, who was greatly beloved at home, had gone to the Burlington. In the early summer of 1859 Uncle Sam wrote to Fanny complaining – not to Florence; she had established herself in a position where none of the family dared write to her direct. His grievance did not stop at Aunt Mai – there was Clough. His daughter Blanche's life had been broken up by Florence's absorption of Clough. For the past year Blanche had been living with her children in her father's house, while Clough stayed in London. Clough was delicate, his health was causing grave anxiety, and it was felt Florence asked too much of him.

At the end of September 1859 when Miss Nightingale left Hampstead and went back to the Burlington, Uncle Sam did his best to persuade Aunt Mai to leave her. Miss Nightingale was very angry. To her, Aunt Mai's problem was not a personal problem. Aunt Mai's presence in London was essential to the work. Since she was the instrument chosen to do the work, if she suffered the work must suffer. She insisted that Aunt Mai must return.

In October, back in the Burlington once more, she had Aunt

Mai and Clough to slave for her and cherish her, the familiar round of conferences and interviews, the inevitable burden of crushing work. But though outwardly everything was the same, inwardly nothing was the same. Difficulties were piling up. She was anxious over Clough's health and Aunt Mai's troubles with her family. The task of War Office reform became daily more complicated, more hopelessly involved, more infinitely laborious. Above all, there was the constant menace of Sidney Herbert's failing health. Everywhere she turned she saw threats of disaster.

15

To contemplate the work which Miss Nightingale performed for the army produces a sensation of weariness. It is too much. No one person should have driven herself to accomplish all this. What must not these mountains of paper, these innumerable reports and memoranda, these countless letters, have cost in fatigue, in strain, in endless hours of application, in the sacrifice of every pleasure? And yet by 1859 work for the army was only a part of her labours. From military hospitals and military nursing she had passed to civilian hospitals and civilian nursing; from working for the army she had passed to working for the nation and the world.

She returned from the Crimea with the intention of devoting her life to the British Army. It was impossible. Her knowledge, her genius, and her experience were such that she could not be allowed to limit herself to military affairs.

In her evidence before the Royal Sanitary Commission of 1857 she was asked: 'Have you devoted attention to the organization of civil and military hospitals?' She replied: 'Yes, for thirteen years I have visited all the hospitals in London, Dublin and Edinburgh, many county hospitals, some of the Naval and Military hospitals in England; all the hospitals in Paris and studied with the 'Sœurs de Charité'; the Institution of Protestant Deaconesses at Kaiserswerth on the Rhine, where I was twice in training as a nurse, the hospitals at Berlin and many others in Germany, at Lyons, Rome, Alexandria, Constantinople, Brussels; also the War Hospitals of the French and the Sardinians.'

The Commissioners were startled. It was an experience such as no other person in Europe possessed; and it was impossible that its benefits should be restricted to the British Army.

In a letter to Dr Farr, written during the autumn of 1859, Miss Nightingale described her feelings when she became aware of the deplorable state of civil hospitals. She came back from the Crimea, she said, suffering from a delusion. She knew military hospitals were in administrative confusion, but she imagined civil hospitals to be much better; and her state of mind when she discovered civil hospitals to be 'just as bad or worse' was 'indescribable'.

During the long, losing fight over the construction of Netley Hospital she learned that abysmal ignorance of the first principles of hospital construction existed even among educated and liberal-minded people. She suffered a notable defeat because no single person concerned had had the faintest idea that any special importance ought to be attached to the way in which a hospital was designed. Steps must be taken to educate public opinion, and through Netley she entered the field of public health. In October 1858 Lord Shaftesbury arranged that two papers written by her on Hospital Construction should be read at the annual meeting of the Social Science Congress. They were received 'with enthusiasm', and she expanded them into a book which was published in 1859 under the title *Notes on Hospitals*.

It was her revolutionary thesis that the high rate of mortality, then invariable in large hospitals, was preventable and unnecessary.

It may seem a strange principle to enunciate as the very first requirement in a Hospital that it should do the sick no harm. It is quite necessary nevertheless to lay down such a principle, because the actual mortality in hospitals, especially those of large crowded cities, is very much higher than any calculation founded on the mortality of the same class of patient treated *out* of hospital would lead us to expect.

Notes on Hospitals draws an alarming picture of contemporary hospital conditions; walls streaming with damp and often covered with fungus, dirty floors, dirty beds, overcrowded wards, insufficient food, and inadequate nursing. The answer to hospital mortality was neither prayer nor self-sacrifice but better ventilation, better drainage, and a higher standard of cleanliness.

Notes on Hospitals was a success; it went into three editions, and after its publication she was constantly asked for advice on Hospital Construction.

The plans for the Birkenhead Hospital, the Edinburgh Infirmary, the Chorlton Infirmary, the Coventry Hospital, the Infirmary at Leeds, the Royal Hospital for Incurables at Putney, the Staffordshire Infirmary, and the Swansea Infirmary were submitted to her. The county of Buckinghamshire rebuilt its Infirmary at Aylesbury 'in accordance with the requirements specified in Miss Nightingale's *Notes on Hospitals*'. At her instigation a pitched battle was fought in Winchester, which resulted in a new county hospital being built on a new site instead of the old hospital in the city, where an erysipelas epidemic had raged, being patched up. The Government of India officially consulted her on the plans for the new General Hospital at Madras. The Crown Princess of Prussia and the Queen of Holland submitted hospital plans. The King of Portugal asked her to design a hospital in Lisbon. She did so, the plans were accepted, and she then learned that the hospital was intended not for adults, but for children; the King of Portugal waved aside her protests – it did not matter, the children would have all the more room.

She had to deal with an enormous mass of practical detail. The organization and equipment of hospital kitchens, hospital laundries, and hospital wash-houses involved the choice of baths, taps, sinks, and basins. The piping of water was novel, and each choice was in some degree an experiment. She wrote hundreds of letters to ironmongers, engineers, builders, and architects. Huge bundles of these survive, though notes attached to the bundles state 'Many destroyed'. An immense series of letters was written to Douglas Galton, dealing with such subjects as the superiority of plate-glass for windows, the proper person to carve in the kitchen, the necessity for bathrooms to be securely locked when not in use, the preference of sick men for washing at a table with a basin on a slab (white enamel showing dirt, it is therefore in this case preferable to slate, which does not), the construction of shutes for soiled linen, the best variety of saucepans, the popularity of maps in recreation rooms. She did not like the dark green walls which were becoming popular in hospitals and wished to have 'the palest possible pink'. She forwarded one long report with the title 'A treatise on sinks'.

In 1859 each hospital followed its own method of naming and classifying diseases. Miss Nightingale drafted model hospital statistical forms which would, she wrote, 'enable us to ascertain the relative mortality of different hospitals, as well as of different diseases and injuries at the same and at different ages, the relative frequency of different diseases and injuries among the classes which enter hospitals in different countries, and in different districts of the same countries.' The model statistical forms were well received. St Mary's Hospital, Paddington, St Thomas's, St Bartholomew's, and University College Hospital agreed to use them at once. A year later representatives of Guy's, St Bartholomew's, the London Hospital, St Thomas's, King's College Hospital, the Middlesex, and St Mary's, Paddington, met and passed a resolution that they would adopt a uniform system of registration of patients and publish their statistics annually, 'using as far as possible Miss Florence Nightingale's Model Forms'.

She found statistics 'more enlivening than a novel' and loved to 'bite on a hard fact'. Dr Farr wrote in January 1860: 'I have a New Year's Gift for you. It is in the shape of Tables.' 'I am exceedingly anxious to see your charming Gift,' she replied, 'especially those returns showing the Deaths, Admissions, Diseases.' Hilary Bonham Carter wrote that however exhausted Florence might be the sight of long columns of figures was 'perfectly reviving' to her.

In the spring of 1859 St Thomas's Hospital found itself in a dilemma. The South Eastern railway was about to build a line from London Bridge to Charing Cross, and St Thomas's lay directly in the path proposed. The Governors of the Hospital were unable to agree on a policy. Some wished the railway company to acquire the whole site and move the hospital to a new district; others wished for only a part of the site to be sold so that the hospital could remain in its ancient place. A deadlock was reached, and in February 1859 Mr Whitfield, the Resident Medical Officer, called on Miss Nightingale and asked her to help. In his opinion the whole site should be sold and the hospital rebuilt in another district. Would she influence the Prince Consort, who was a Governor, to adopt this point of view?

She would not allow herself to be easily convinced. She went into the matter thoroughly, studied figures, interviewed railway and hospital officials, and came to the conclusion that Mr Whitfield was correct. She then sent a memorandum to the Prince Consort. He read her memorandum and was converted.

However, the battle was not yet won, for a new point was raised. Financially it might be preferable to sell the whole site and rebuild elsewhere, but there was surely an ethical consideration. Ought the hospital to leave its ancient position among the people it had served for centuries? Again an appeal was made to Miss Nightingale. She collected statistics of patients treated in the hospital and was able to prove that the largest number of patients treated at the hospital did not come from the immediate neighbourhood but from districts further away. A memorandum containing this evidence was drawn up by her and submitted to the Governors and the Prince Consort, and as a result it was agreed that the whole site should be sold and the hospital moved.

Yet another crisis followed. The Governors, becoming greedy, decided to ask the railway company the then very large sum of £750,000. Miss Nightingale, called in once more, pointed out that if the demand was persisted in, the company would go to arbitration; and the sum awarded would almost certainly be smaller than the present offer. Her advice prevailed; the entire site in the Borough was sold by agreement, and the hospital moved to its present position in Lambeth.

These negotiations produced a close association. Mr Whitfield became devoted to her, and Mrs Wardroper the matron was already a close friend. She became identified with St Thomas's and in time held a position in the hospital which was almost that of patron saint.

Her interest in nursing and nursing reform had never diminished, though her work for the army had pushed it into second place. She had the Nightingale Fund of £45,000 at her disposal to found a Training School for Nurses, but there had been great difficulty in finding suitable connexions for the school. In 1859, when she became concerned with the affairs of St Thomas's, she began work on a scheme to establish the school there.

While she worked on this scheme for training the professional nurse, she wrote a little book on nursing for the use of the ordinary woman, which became the most popular of her works. *Notes on Nursing* was intended to make the millions of women who had charge of the health of their children and their households 'think how to nurse'. 'It was,' she wrote, 'by no means intended ... as a manual to teach nurses to nurse.' It is a book of great charm, sympathetic, sensible, intimate, full of witty and pungent sayings, and possessing a remarkable freshness. 'It is as fresh as if nobody had ever before spoken of nursing,' wrote Harriet Martineau, and Sidney Herbert wrote in January 1860 that it was 'more interesting than a novel'. Neither its good sense nor its wit has dated, and *Notes on Nursing* can be read to-day with enjoyment.

When the book was published in December 1859 it caused a mild sensation. Habits of hygiene now taken for granted were then startling innovations. Mothers of families were shocked when Miss Nightingale attacked the education of the mid-Victorian girl, to whom 'the coxcombries of education' were taught, while she was left in ignorance of the physical laws which governed her own body.

The book was not cheap – the price was 5s. – but 15,000 copies were sold within a month; and it was reprinted at 2s. and later at 7d. Thousands of copies were distributed in factories, villages, and schools, and it was translated into French, German, and Italian. 'There is not one word in it written for the sake of writing but only forced out of me by much experience in human suffering,' she wrote to Sir John McNeill on July 29.

It is impossible to doubt, after reading it, that Miss Nightingale was a gentle and sympathetic nurse. She understood that the sick suffer almost as much mental as bodily pain. 'Apprehension, uncertainty, waiting, expectation, fear of surprise, do a patient more harm than any exertion. Remember he is face to face with his enemy all the time, internally wrestling with him, having long imaginary conversations with him.' 'Do not cheer the sick by making light of their danger.' 'Do not forget that patients are shy of asking.' 'It is commonly supposed a nurse is there to save physical exertion. She ought to be there to save (the patient) taking thought.'

She spoke of the 'acute suffering' caused a sick person by being so placed that it is impossible to see out of the window; of the 'rapture' brought to an invalid by a bunch of brightly coloured flowers; of the intense irritation caused to an invalid by a noise such as the constant rustling of a nurse's dress. She understood the relief afforded a sick person by being taken out of himself. 'A small pet animal is an excellent companion for the sick. A pet bird in a cage is sometimes the only pleasure of an invalid confined to the same room for years.' She loved babies and recommended visits from the very young. 'No better society than babies and sick people for each other.'

It will be recalled that when she was convalescing in Mr Sabin's house at Scutari she had become fond of a certain Sergeant Brownlow's baby; and when the 7d. edition of *Notes on Nursing* was published she added a chapter on 'Minding Baby' inspired by this child. 'And now girls I have a word for you,' the chapter opens. 'You and I have all had a great deal to do with "minding baby", though "Baby" was not our own baby. And we would all of us do a great deal for baby which we would not do for ourselves.' Jowett of Balliol said that a world of morality was contained in the parenthesis 'though "baby" was not our own baby'. She received letters telling her that this chapter was most fruitful in results. Unhappily, Sergeant Brownlow's baby died shortly after the return of the troops to England from the Crimea, owing, she said, to the insanitary condition of the barracks in which the father's regiment was quartered.

She attacked 'invalid food'. As a result of invalid diet thousands of patients are annually starved, she declared. 'Give 100 spoonfuls of jelly and you have given one spoonful of gelatine which has no nutritive power whatever. Give a pint of beef tea and you have given barely a teaspoonful of nourishment. Bulk is not nourishment.' Milk, in her opinion, was the best of all invalid foods, and she urged also that people who are ill should not be deprived of vegetables. Tea 'admittedly has no nourishing qualities but there is nothing yet discovered which is a substitute to the English patient for his cup of tea'. Invalid food must be carefully served: Do not give too much, do not leave any food by the patient's bed. 'Take care nothing is spilt in the saucer.'

In a series of pungent paragraphs she cut to pieces the current idea of a nurse. 'No *man*, not even a doctor, ever gives any other definition of what a nurse should be than this – "devoted and obedient". This definition would do just as well for a porter. It might even do for a horse. It would not do for a policeman.' 'It seems a commonly received idea among men, and even among women themselves, that it requires nothing but a disappointment in love, or incapacity in other things, to turn a woman into a good nurse.'

Though she denounced the education of the Victorian girl, and advocated the training of women, *Notes on Nursing* ends with a vigorous attack on the 'jargon about the rights of women'. 'Keep clear of both the jargons now current everywhere,' she wrote; '... of the jargon, namely about the "rights" of women, which urges women to do all that men do including the medical and other professions, merely because men do it, and without regard to whether this *is* the best that women can do; and of the jargon which urges women to do nothing men do, merely because they are women, and should be "re-called to a sense of their duty as women" and because "this is women's work" and "that is men's" and "these are things which women should not do" which is all assertion and nothing more. ... You do not want the effect of your good things to be "How wonderful for a *woman*!"; nor

262

would you be deterred from good things by hearing it said "Yes, but she ought not to have done this, because it is not suitable for a woman." But you want to do the thing that is good whether it is suitable for a woman or not.' To praise women for doing what men did habitually and easily, she wrote, was to reduce them to the status of Dr Howe's idiots whom, after two years of ceaseless labour, he succeeded in teaching to eat with a knife and fork.

Six months after the publication of *Notes on Nursing*, the scheme for establishing a Training School for Nurses, endowed with the proceeds of the Nightingale Fund, was at last carried through. For the past three years she had been 'continuously deluged' with suggestions for spending it. On January 31, 1856, while still at Scutari, she had written that she must decline considering any scheme for the present; she was too much overwhelmed with work to plan, and she was afraid of failure. 'People's expectations are highly wrought,' she wrote. 'They think some great thing will be accomplished in six months, although experience shows that it is essentially the labour of centuries – they will be disappointed to see no apparent change, and at the end of a twelve-month will feel as flat about it as people do on a wedding day at 3 o'clock when breakfast is over.' In March 1856 she had written: 'The first fruits of a long series (as I expect) of the brick and mortar plans of needy or philanthropic adventurers who wish to get hold of the "Nightingale Fund" have already come in upon me. ... I take at random those which first present themselves. One is a magnificent elevation with my statue on the top to be called the "Nightingale Hospital" ... another is a Home for Nurses with no hospital at all. One advises me to admit none but gratuitous services. This includes a threat if the obnoxious word "Sister" is allowed, and a terrible warning as to the "cut of our aprons" which are to be "Large and White", and a caution as to "Celibacy", which I was not aware before came into the question. We are also solemnly assured that the "Apostles received a salary" (How much was it?) and that the Nurses must lead an "*ordinary*" life". I thought the object was they should not be ordinary nurses. One offers me a clergyman and his sons and insists upon a service every day in the week, probably a son for each day and the Father on Sunday. Another insists on no Clergyman at all and a strictly secular education. One desires to confine my operations to the Work Houses, another to the Hospitals and a third recommends the training of nurses for private families only. One wishes for an "Order", another for an "Asylum" for old age, and a third for high wages which will enable each to save for herself.'

She saw before her a fresh succession of the religious intrigues which had so nearly wrecked the nursing in the Crimea, and she refused to move. 'If I do *anything* at present I shall be smothered in the dust raised by these religious hoofs,' she wrote.

In March 1858 she wrote to Sidney Herbert asking to be released from the responsibility of conducting the Fund. He dissuaded her. The money was well invested and could accumulate; there was no need for immediate action. The subscribers expected that she personally would 'animate the work'; and he did not see how she could with propriety dissociate herself from it.

She unwillingly agreed but continued to feel that the Nightingale Fund was a millstone round her neck. She was afraid of private charities. She had been driven nearly frantic by the necessity of finally clearing up the frightful matter of the Free Gifts. In 1857, at her most over-driven, she had been forced to produce sixty-eight pages of *Statements exhibiting the Voluntary Contributions received by Miss Nightingale for the use of the British War Hospitals in the East with the Mode of the Distribution in 1854, 1855, 1856.* Her fingers had been burnt by private charities and she was unwilling to involve herself again. Moreover, she had been annoyed by an attempt by the Council of the Nightingale Fund to dictate to her what her activities should be. It was proposed that she should be asked for a pledge that she would work in civil hospitals and not in Military hospitals. 'I might go to the Opera and Races,' she wrote to Sidney Herbert on October 31, 1856, 'no pledge against amusing myself existing, but I might not take Government employment being pledged to work for Civil Hospitals by the Fund. ... I never can cease while I live, doing whatever falls in my way in the work I have mentioned above, viz. the Military Hospitals, which God and you have so singularly put into my hands.'

By 1859, however, it was evident that some action must be taken. She was now an invalid; it was most unlikely that she would ever be sufficiently recovered to become the active superintendent of a large institution; and in any case her achievements were so great that as the superintendent of an institution she would be wasted. The scheme of founding a training school for nurses was revived. She was to be the patroness and organizer;

and a sub-committee of the Nightingale Fund Council was appointed to hurry the scheme forward.

She hoped to work with Dr Elizabeth Blackwell, but the plan failed. 'During March and April in town,' Miss Nightingale wrote to Sidney Herbert on May 24, 1859, 'I saw and corresponded with pretty nearly all the hospital authorities and female superintendents in esse or in posse that could be applied to the Fund. I will not tell you in writing (tho' I could any day viva voce) all the pros and cons and the different plans I have successively tried to initiate. The most promising, that of the "London" qua hospital and of Miss Blackwell qua superintendent, has fallen through. I have talked over the matter at great length with Sir John McNeill. For some months past I have also discussed it with some of the authorities at St Thomas's Hospital. The Matron of that Hospital is the only one of *any existing* Hospital I would recommend to form a "school of instruction" for Nurses. It is not the *best conceivable* way of beginning. But it seems to me to be the *best possible*. It will be beginning in a very humble way. But at all events it will not be beginning with a failure, i.e. the possibility of upsetting a large Hospital – for she is a *tried* Matron.'

Mrs Wardroper, matron of St Thomas's Hospital, a gentlewoman by birth, had been left a widow with young children at the age of forty-two, and had taken up nursing. Training was unknown – she entered the wards of a hospital and learned what she could from experience. Miss Nightingale wrote, 'Her force of character was extraordinary, and she seemed to learn from intuition.' She was appointed matron of St Thomas's in 1853, a remarkable achievement for a woman of her upbringing and class. In 1854 Miss Nightingale had applied to Mrs Wardroper for nurses to go to the Crimea, and through her obtained her best nurse, Mrs Roberts. Mrs Wardroper's personal appearance was striking: she was tall, dark, and possessed a personal magnetism which enabled her to impose discipline on the most refractory nurses and patients; she had great powers of administration, was a tireless worker and an ardent reformer. When Miss Nightingale revisited St Thomas's after her two years in the East she found Mrs Wardroper had achieved astonishing results. 'She had already made her mark; she had weeded out the inefficient, morally and technically; she had obtained better women as nurses: she had put her finger on some of the most flagrant blots, such as night nursing ... but no training had been thought of.' Mrs Wardroper held the post of superintendent of the Nightingale Training School for twenty-seven years, and a great part of the success of the school was due to her energy and determination.

The scheme for a training school for nurses was not universally welcomed. A strong party in the medical world thought that nurses did very well as they were, and that training would merely result in their trespassing on the province of the doctors. Both Mrs Wardroper and Mr Whitfield, the Resident Medical Officer at St Thomas's, showed courage in committing their hospital to the scheme; and in May 1860 Mrs Wardroper wrote warning Miss Nightingale that they must be prepared for 'rather harsh criticism'.

Strong opposition came from within St Thomas's itself, led by the Senior Consulting Surgeon, Mr J. F. South. In 1857, when the scheme of a training school was first discussed, Mr South published a little book, *Facts relating to Hospital Nurses. Also Observations on Training Establishments for Hospitals*. He was at the top of his profession, President of the College of Surgeons and Hunterian Orator, besides being senior consulting surgeon at St Thomas's, and 'not at all disposed to allow that the nursing establishments of our hospitals are inefficient or that they are likely to be improved by any special Institution for Training'. He argued that the sisters learned by experience and could only learn by experience, that the nurses were subordinates 'in the position of house-maids' and needed only the simplest instruction, such as how to make a poultice. He asserted that the nursing at St Thomas's hospital was already on a very high level. 'That this proposed hospital nurse training scheme has not met with the approbation or support of the medical profession is beyond doubt,' wrote Mr South. 'The very small number of medical men whose names appear in the enormous list of subscribers to the (Nightingale) Fund cannot have passed unnoticed. Only three physicians and one surgeon from one London Hospital and one physician from a second are found among the supporters.' The Nightingale Training School was launched in an atmosphere of criticism. Its way was not to be made easy, and the probationers would be watched by unfriendly eyes.

The object of the school was to produce nurses capable of training others. The Nightingale nurses were not to undertake private nursing; they were to take posts in hospitals and public institutions and establish a higher standard. They were to be missionaries, and as such they must be above suspicion. If scandal centred upon a Nightingale nurse, an active opposition was eagerly waiting to fasten on it. One piece of indiscretion, one false step, and hopes of reforming the nursing profession and elevating its status might be set back for years. The future of nursing depended on how these young women behaved themselves. As a result, candidates to become Nightingale probationers were subjected to minute examination, and there was great difficulty in finding young women of suitable character.

In May 1860 advertisements appeared inviting applications for admission. The response was discouraging. However, fifteen candidates were chosen, and on July 9, 1860, the Nightingale School opened. No pupil was admitted without a certificate of good character, and the training was to last for one year – so long a period was hitherto unheard of. The Nightingale probationers lived in a nurses 'Home'; this was a novel idea originated by Miss Nightingale, and it was received with disapproval by the opposition. An upper floor of a wing of St Thomas's was fitted up so that each probationer had a bedroom to herself, there was a common sitting-room, and the sister in charge had a bedroom and sitting-room of her own. Books, maps, prints, and a supply of flowers from Embley were sent by Miss Nightingale, though Mrs Wardroper feared that flowers came dangerously near over-indulgence. The Nightingale probationers wore a brown uniform with a white apron and cap. Board, lodging, washing, and uniform were provided by the Fund. Each probationer was given £10 sterling for her personal expenses during training. At the end of the year's course the nurses who had satisfied the examiners were placed on the hospital register as 'Certificated Nurses'. A first-class cash gratuity of £5 and a second-class cash gratuity of £3 sterling were offered to nurses who were certified to have worked efficiently in a hospital for one year after completing their training.

It was a standard of life which nurses had never been offered before, and the opposition sneered at Miss Nightingale's 'lady nurses'. She did not believe nurses should be housemaids. 'The Nurses,' she wrote, 'should not scour; it is a waste of power.' The Nightingale probationers worked hard, attending daily lectures from the medical staff and sisters of St Thomas's Hospital and bi-weekly addresses from the chaplain. They were required to take notes, to be ready to submit their notebooks at any time for inspection and to pass examinations both written and oral; they acted as assistant nurses in the wards and received practical instruction from surgeons and sisters. What was required of them in work, however, was as nothing compared to what was required in behaviour. Every month a report entitled 'Personal Character and Acquirements' was filled in by Mrs Wardroper, who exercised the closest possible supervision over every probationer. The details of the report, planned by Miss Nightingale, were minutely comprehensive. Two main heads, 'Moral Record' and 'Technical Record', were further sub-divided; 'Moral Record' had six subdivisions – punctuality, quietness, trustworthiness, personal neatness, cleanliness, ward management and order. 'Technical Record' had fourteen subdivisions which were again subdivided, in some cases a dozen times. Mrs Wardroper wrote against each head 'excellent', 'good', 'moderate', 'imperfect', or 'o'. In addition she wrote confidential personal reports on each probationer.

Even this information was felt by Miss Nightingale to be insufficient. She was obsessed by the importance of these young women. The future of nursing hung on their behaviour. Their natures, their thoughts, might wreck or make the work. If only she could get inside their minds. She originated a new scheme by which each probationer was required to keep a daily diary which was read by her at the end of the month. 'I am sure,' wrote Mrs Wardroper, 'that your approbation will stimulate them to increased perseverance.' Miss Nightingale noticed that some of the probationers were weak in spelling and arranged for them to have spelling drill.

Flirtation was punished by instant dismissal – the girls selected and trained to redeem nursing must not allow themselves to be women; their mission was to prove that the woman can be sunk in the nurse. No Nightingale probationer was permitted to leave the Home alone; two must always go out together. 'Of course we always parted as soon as we got to the corner,' wrote one of the original probationers in a reminiscence.

The character and behaviour of each probationer was discussed by Miss Nightingale and Mrs Wardroper in anxious conferences and long letters. One set of letters considered in detail whether a certain young woman ought to be dismissed because she 'made eyes'. She was a competent nurse and her moral character was 'said to be unexceptionable', but she seemed unable to refrain from 'using her eyes unpleasantly'. Before she was dismissed, however, ought they not to consider whether she might not grow out of this objectionable habit as she became older?

Strictness was necessary. The Nightingale nurse must establish her character in a profession proverbial for immorality. Neat, lady-like, vestal, above suspicion, she must be the incarnate denial that a hospital nurse need be drunken, ignorant, and promiscuous. It soon became evident that the school was succeeding. Within a few months a flood of applications was being received to bespeak the services of Nightingale probationers as soon as their period of training was completed.

A second experiment, financed out of the Nightingale Fund at the end of 1861, was the establishment of a Training School for Midwives. With the co-operation of the authorities of King's College a maternity ward was equipped, and the physician accoucheurs of the hospital agreed to assist in giving a six months' training; it was a scheme which Miss Nightingale had had at heart since her Harley Street days. On September 24, 1861, she wrote to Harriet Martineau: 'In nearly every country but our own there is a Government School for Midwives. I trust that our school may lead the way towards supplying a long felt want in England.'

The school trained midwives, not only to work in hospitals, but to deliver women in their own homes. During their training at the hospital the candidates paid for board and lodging but received instruction free. A promising beginning was made, and a number of owners of large estates sent women at their own expense to be trained as village midwives. Unfortunately, after more than two years of success, an outbreak of puerperal sepsis brought the scheme to an abrupt end, but Miss Nightingale's interest in rural problems of health continued, and she was constantly consulted on the selection and training of village nurses.

Through these years of unremitting toil her sole recreation was theological and metaphysical speculation, and in the summer of 1858 she turned again to the philosophical manuscript she had written in 1851 designed to provide a new religion for intelligent artisans. She sent it to Dr Sutherland, who wrote on July 7 to Aunt Mai that he 'disagreed entirely and vehemently' with her theory, but 'I have preferred sending this to you because poor Florence is very unwell and in our own work we have enough of differences of opinion to make it desirable not to have more'. In spite of this discouragement Miss Nightingale did a considerable amount of work on the manuscript during the following year and at the end of 1859 had it privately printed under the title *Suggestions for Thought*.

She had a strong affection for the book; she believed it to be an important work, and she determined to obtain unbiased opinions. She sent out a number of copies anonymously, with a letter asking if, in the opinion of the recipient, the book should be offered to the general public. A copy was sent to Richard Monckton Milnes, who identified the author and wrote to her on January 21, 1860: 'I do not think the theory of omnipotent and implacable *Law* is any more satisfactory to the disturbed and distracted mind than that of a beneficent and benevolent Deity.' As to its suitability for the 'Artizans of England' he could not express an opinion as he had 'a morbid horror of touching on these subjects with what people call the "lower classes".' The book, he added, should be revised. The letter ended on a wistful note: 'My two little women are well and happy. I am as much of both as I believe is good for me.'

From John Stuart Mill she obtained unqualified approval. *Suggestions for Thought* was not sent to him anonymously – a copy was brought to him by Edwin Chadwick, who first sent him a copy of *Notes on Nursing*. 'I do not need it,' wrote Mill, 'to enable me to share the admiration which is felt towards Miss Nightingale more universally, I should imagine, than towards any other living person.' He read *Suggestions for Thought* with care, annotating the copy in the margin; and he wrote on September 23, 1860, giving his verdict in favour of publication.

An anonymous copy was sent to Benjamin Jowett by Clough. Jowett, at this time a tutor of Balliol, was Clough's close friend. His verdict was unfavourable: he had 'received the impress of a new mind', but the book must be rewritten. He thought that 'here and there I traced some degree of irritation in the tone, the book appears to me full of antagonisms – perhaps these could be softened.'

She also sent a copy to Sir John McNeill, and he, too, told her the book must be rewritten and replanned. She wrote that she had not time or energy to undertake it and regretfully laid the book aside, sending the manuscript to Richard Monckton Milnes for safe keeping. Yet she was not entirely convinced. More than ten years later she sent the manuscript to the historian, Froude. Once more the verdict was unfavourable, and this time she resigned herself. She wrote to him in July 1873: 'What you say about its "want of focus", want of "form", of its "bleating propensities", is of course felt by me more than anyone. I would re-write every word – if I could.'

Suggestions for Thought was a failure, but it had brought her friendship with Benjamin Jowett. 'I do so like Mr Jowett,' she scribbled on the margin of one of Jowett's first letters. In 1860 he was already a celebrated Oxford character. He was a fellow and tutor of Balliol and Regius Professor of Greek, and his personal appearance, wit, and eccentricities were University legends. In appearance he was short, cherubic, and strikingly handsome on a miniature scale; undergraduates nicknamed him the 'downy owl'. He spoke in a small piercing voice, 'that small sweet voice once heard never to be forgotten'.

Jowett, who had no private means, had been educated at St Paul's, and he had made his career by force of his attainments. In a University which was traditionally the stronghold of conservatism he represented a new order, and he had to endure opposition and disappointment. In 1854 he had failed to be made Master of Balliol; in 1855 he was made Regius Professor of Greek, but his opponents contrived that the appointment should be financially worthless. Four chairs, of which the Regius Professorship of Greek was one, had been founded by Henry VIII; the emoluments attached to the other three chairs were increased – but that held by Jowett was excluded. Until 1865 he received only £40 a year as Regius Professor of Greek. His enemies declared that his ambitions were more social than scholarly, and that he was fonder of the dinner-party than the library. Unquestionably he was worldly wise. It was said that he was entertaining a young cousin to dinner, and the youth was describing to Jowett how he made his way in the world. 'When a man insults me I always ask him to dinner,' he said. Jowett burst into delighted laughter, and rubbing his hands together exclaimed: 'You'll do, dear boy, you'll do.'

As Regius Professor Jowett did not confine himself to editing dictionaries and commentaries, but lectured on Plato and encouraged undergraduates to bring him Greek verses and Greek prose: an innovation which scandalized the conservative party in the University. He succeeded in having Plato read in public schools which had previously confined themselves to Aristotle and rhetoric. He was intensely interested in young people, saw his pupils constantly, advised them, and received their confidences. 'He wished,' wrote Miss Nightingale, 'to bring together the University life and the life of the world, so that the University life should be a direct preparation for the life of the world.' He had, she said, 'a genius for friendship'. He identified himself with his friends, forming friendships with their friends and relations and entering into their business. When a friend obtained a secretaryship in the Educational Department, the office of the Privy Council became of 'sustained interest' to Jowett. He loved children; he could be, he wrote, 'almost mad with fun', delighted in playing the fairy godfather and being 'patron in general to all poor devils'. When he became Master of Balliol in 1870 he gave entertainments so wildly assorted that his dinner-parties were known in the University as 'Jowett's Jumbles'. The butler was once heard to repress a groan as he took off a guest's coat. 'I hope you have no cause for anxiety at home,' enquired the guest. 'No, sir, it's nothing,' replied the butler, 'only the Master invited twelve to dinner tonight and you are the eighteenth who has come.'

Between Jowett and Miss Nightingale acquaintance quickly became intimacy. Like all the men who were fond of her, Jowett scolded her – she was not to exaggerate, not to fuss, not to be so hard on people; she was to try to be more cheerful, to look back on what she had accomplished and be proud of herself. Affection became devotion, and it was known to their friends that Jowett was pressing her to marry him. She refused, but their friendship was unaltered. They corresponded constantly, and she leaned on his devotion and advice. 'My darling Jowett,' she called him.

She needed friendship, for as 1860 drew to its close the structure for which she had sacrificed everything in life crashed round her ears in ruin – Sidney Herbert's health finally collapsed.

16

It was the worst possible moment for him to break down. Everything depended on War Office reorganization, and War Office reorganization could be pushed through by Sidney Herbert alone. 'One fight more, the last and the best,' wrote Miss Nightingale; let him nerve himself to this final task and he should be released. He should be allowed to resign his office. He should go away where he would, abroad, or to Wilton, or to Ireland. He should shoot, fish, hunt, and never be worked on the treadmill again. But he must not fail now. It was a possibility she refused to contemplate. Resentment against life, against fate, against Providence, blinded her. Why should not Sidney Herbert do as she had done? She had been declared to be at death's door and had driven herself to work day and night. She had proved that serious illness need not interfere with work; he must do the same. He could not be excused. The work was greater than the man. He must go on, he must be compelled to go on.

She had the habit of disregarding his complaints, and she shut her eyes to his physical condition. Indeed, his health varied. In November 1859 he insisted on going down to Wilton, spent the week fox hunting and wrote: 'I have been drenched to the skin every day and enjoyed myself very much.' Surely if he could hunt five days a week, he could find enough strength to carry through War Office reform. She admitted his health had deteriorated, but so had her own. In February she wrote to Manning: 'I am so much weaker that I do not sit up at all now.'

Sidney Herbert's health was not to be the sole catastrophe; from the beginning of 1860 blow after blow rained on her as if Fate was determined to discover how much she could be made to bear.

In February Dr Alexander suddenly died of a cerebral haemorrhage. Behind his death lay a history of obstruction and petty intrigue. The departmental machine had been strong enough to break him. Men he had trusted, who had been placed in their positions through his recommendations, had betrayed him. Disillusioned, snubbed, frustrated, he had laboured through the immense amount of work entailed by the Sanitary Commission and he was engaged in drafting the new Army Medical Regulations when he died. 'His loss undoes a great part of the work I have done,' Miss Nightingale wrote to Sir John McNeill in March 1860. 'I wish I had not lived to see it. . . .'

While she was still distracted by the loss of Alexander, another blow fell. Aunt Mai returned to her family. Since the previous autumn Aunt Mai's position had been intolerable. Uncle Sam refused to visit her when she was with Florence – he said he would be 'de trop'. Her second daughter was to be married that summer, and she implored her mother to come home. It was undeniable that Florence, who had been dying in 1857, was still alive. In the early summer of 1860 she decided it was her duty to return.

Her decision provoked intense bitterness. When Miss Nightingale realized that she was going to leave she refused to see her or speak to her. Aunt Mai wrote to Embley that she could not send her usual report on Florence's health because she had not seen her for over a week. Miss Nightingale did not forgive Aunt Mai for nearly twenty years; they never met, and the correspondence between them ceased. Uncle Sam was forgiven. She had shared no ideals with him and felt no sense of betrayal. She continued to consult him in her financial affairs and to write him friendly letters.

It was impossible for her to be left alone; and in June 1860 Hilary Bonham Carter came to the Burlington to take Aunt Mai's place. Clough remained faithful, calling daily and devoting every moment he could snatch from his office to the work, but his health, too, was causing anxiety, and on December 7, 1860, Miss Nightingale wrote a depressed letter about him to Uncle Sam: 'I have always felt that I have been a great drag on Arthur's health and spirits, a much greater one than I should have chosen to be, if I had not promised him to die sooner.'

Alexander was dead, Sidney Herbert fatally handicapped by ill health, and the burden on her increased. It was out of the question for her to leave London, and in August she went to Hampstead once more. She was very feeble, lying all day in bed by an open window in her rooms in South Hill Park, finding solace in the society of her cats and of children. She had Clough's children to stay with her, and on September 1, 1860, she wrote to Clough's wife describing a visit from the baby. ' "It" came in its flannel coat to see me. No one had ever prepared me for its Royalty. It sat quite upright, but would not say a word, good or bad. The cats jumped up upon it. It put out its hand with a kind of gracious dignity and caressed them, as if they were presenting Addresses, and they responded in a humble grateful way, quite cowed by infant majesty. Then it put out its little bare cold feet for me to warm, which when I did, it smiled. In about twenty minutes, it waved its hand to go away, still without speaking a word.'

Sidney Herbert rode out to see her every day. Forced to stay in London for the second summer in succession, he was weary, feverish, and dispirited. He felt 'a total inability to deal with business'. 'He shrank,' Miss Nightingale wrote to Sir John McNeill in July 1861, 'from the Herculean task of cleansing the Augean stable.'

Through July, August, and September he complained of biliousness, lassitude, and headaches. In September Liz asked Florence not to come back to the Burlington because the daily ride out to Hampstead did Sidney so much good; and she stayed in Hampstead with Hilary Bonham Carter through the autumn.

Sidney Herbert was enduring a martyrdom, while the two women closest to him shut their eyes and drove him on. More than a year before he had told his wife: 'Every day I keep the War Office with the House of Commons is *one day taken off my life.*' Since then, with Miss Nightingale urging him on one side and Liz on the other, he had forced himself to continue with both. Now he was nearly at the end of his powers. In October and November 1860 his health suddenly grew worse. Perhaps, wrote Florence, it was the London air. She conferred with Liz. He was so much better out of London – perhaps the solution was not to stay in Belgrave Square but to take a house in Hampstead. Liz agreed, and Mrs Sutherland was sent round the house agents. It was too late. No change of air could save Sidney Herbert now. Early in December he collapsed.

He was pronounced to be suffering from kidney disease, incurable and at an advanced stage: the amount of work he was doing must be cut down drastically and at once. It was practically a death warrant. On December 5 he rode out to Hampstead to see Miss Nightingale. 'He was not low, but awe struck,' she wrote to Uncle Sam on December 6; 'I shall always respect the man for having seen him so.'

He remained with her for several hours, his intention being to

275

consider what his future course of action should be. In fact it had already been settled for him. Liz had been to see Florence earlier, before Sidney was well enough to ride out, and they had agreed between themselves what he was to be persuaded to do. Three courses were open to him. He could retire altogether; he could give up the War Office, and keep his seat in the House of Commons; he could go to the House of Lords, give up the House of Commons and keep the War Office. For his own sake the first was the best. The doctors had enjoined complete and absolute rest as his only chance.

But that choice neither Miss Nightingale nor his wife would allow him to contemplate, nor indeed did he contemplate it himself. Work was his fate. He recognized that and inclined toward the second course. He would give up the War Office and keep his seat in the House of Commons. He was a House of Commons man; he had sat in the House for twenty-eight years and had been brilliantly successful there. He could 'do with the House what no one else could'. He was an orator and a matchless negotiator. The work of the War Office he frankly detested. It did not suit him. His talents were wasted. He had never had, as Miss Nightingale frequently told him, any genius for administration. He could not, drive himself as he would, master the enormous mass of intricate detail which War Office reorganization involved. His mind recoiled from it.

But Miss Nightingale's task was to persuade him to keep the War Office. For the sake of the work, for the sake of War Office reorganization, he must be persuaded to resign his seat in the Commons, keep the War Office, and go to the House of Lords.

The interview was long, very long, but she succeeded. For the sake of the work the War Office should be kept. 'A thousand thanks,' wrote Liz, 'for all you have said and done.' She had forced him to sign his death warrant.

She still refused to admit how fatally ill he was. How else could she justify what she had done? The shock of the doctors' verdict had been great, but presently she began to minimize it: doctors were often wrong; people with so-called fatal diseases often lived for years. 'I hope you will not judge too hardly of yourself from these doctors' opinions,' she wrote on December 8, '... it is *not* true that you cannot (sometimes) absolutely mend a damaged organ, almost always keep it comfortably going for many years, by giving Nature fair play. ... I know a very active intellectual

London man, now 65, whose albuminous symptoms were accompanied by one, the most advanced of all, which you have never had, but who by sleeping in the country etc., has given himself 15 years good life and may have 15 more. ... I am not going to bore you with a medical lecture. But I do hope you won't have any vain ideas that you can be spared out of the W.O. You said yourself that there was no one to take your place – and you must know *that* as well as everybody else. It is quite absurd to think Lord de Grey can do it. ... You cannot be the only person who does not know that you are necessary to the re-organizing of the W.O. It is more important to originate good measures than to defend them in the H. of C. ... I don't believe there is anything in your constitution which makes it evident that disease is getting the upper hand. On the contrary.'

He tried to follow her directions. He went down to Wilton and wrote cheerfully on the 12th: 'I went out hunting, had a lovely bright day and a good run, and I slept like a top for the first time for some days.' Early in January he was created a baron with the title of Lord Herbert of Lea, and took his seat in the House of Lords.

Miss Nightingale showed him no softness, and it almost seemed as if she regarded bad health as her personal monopoly. 'To him retaining office and giving up the H. of C. is like it was to me giving up men and taking to regulations,' she wrote to Harriet Martineau on January 1, 1861. Only once did she permit herself a quick horrified glance at the truth, when, on January 13, she wrote to Harriet Martineau: 'I see death written in the man's face. And, when I think of the possibility of my surviving him, I am glad to feel myself declining so fast.' She hid herself behind impatience, harshness; she blamed him. It was a relief to blame him and to close her eyes to the grim certainty advancing inexorably upon her.

Liz began to drag her husband on an anxious, melancholy round, consulting doctor after doctor, starting with eminent specialists, descending to fashionable quacks, trying treatment after treatment, each started in expectation of a miraculous cure, each only too soon discarded.

Throughout these months Miss Nightingale was plagued by domestic difficulties and disasters. The management of the Burlington she considered inefficient. She had advanced ideas on fresh air and open windows which she tried in vain to put into operation. On August 16, 1860, she drew up an agreement: 'If for one fortnight from this time I find all the doors shut and all the windows open, including those of the two water closets, which also must have their seats shut, I will give the servants a doctor's fee, viz – one guinea,' signed 'F. Nightingale.' To this agreement a slip is pinned, 'On going downstairs today I found every door open and every window shut.'

During this summer she suffered intense irritation from noise made by a water cistern immediately over her bedroom. Shortly afterwards the Burlington changed hands, and she refused to return unless the whole 'water apparatus' was removed from over her head; the new landlord gave his word of honour that it should be done. In the autumn Dr Sutherland and Hilary Bonham Carter went to inspect her rooms and reported that under the new proprietor there was 'a great improvement in house keeping and cleanliness'. But on January 21, 1861, with a 'loud bang', the cistern emptied itself on her head. Writing to Hilary Bonham Carter next day she described 'the ceiling saturated, water dripping at every pore into a wilderness of pots and pans, paper ruined and new carpets wrenched up'. She herself caught a severe chill. Sir Harry Verney, W. E. N., and Uncle Sam were summoned and the landlord was called in. After putting forward various excuses he was driven into a corner by Miss Nightingale and forced to admit that the cistern had never been removed at all. On a fragment of paper she dashed off an angry private note – she considered her male supporters had been ineffectual. 'I have had a larger responsibility of human life than ever man or woman had before. And I attribute my success to this. I NEVER GAVE OR TOOK AN EXCUSE.

'Yes, I do see the difference between me and other men. When a disaster happens, I act – and they make excuses.'

She had now been at the Burlington for two years, and it was both uncomfortable and expensive. Fanny and Parthe talked of buying a house, but W. E. N. was unwilling to sink a large sum of money in purchasing a house for Florence in her present state of health. However, the difficulty and expense of housing her steadily increased. For months she had two sets of accommodation, the rooms in Hampstead and in the Burlington. She slept at Hampstead and was driven in every day to the Burlington to work, sleeping odd nights at the hotel at her convenience.

In April 1861 Colonel Phipps, Private Secretary to the Prince

Consort, wrote to W. E. N. offering Miss Nightingale, on behalf of the Queen, an apartment in Kensington Palace. On April 20 she wrote to W. E. N. sharply declining: 'I have to see a great many people on a great variety of subjects and *no residence* would be any use to me which was not near enough the business centre of London to allow me to see these people *at a moment's notice*.' W. E. N. pressed her to reconsider her decision, and a few days later she wrote again with extreme irritation. She had discovered that, without her knowledge, the Board of Trade had been approached to find an apartment for her in one of the royal residences. 'Sir Harry Verney consulted by Phipps asks Parthe, who advises Kensington. ... They might as well dispose of me in marriage. What does Parthe know about my work? Where should I be now if I had taken Parthe's advice? ... I cannot see how Sir Harry having been Col. Phipps' school-fellow makes him competent as an adviser.'

She was not easy to live with. In a letter to Uncle Sam on June 2, 1861, she described her mental state. 'It is the morbid mind of a person who has *no* variety, *no* amusement, *no* gratification or change of any kind. ... No one ever understands ... the morbid *conscience* of anybody whose nervous system had been over-tasked. I am always thinking I might do more or I had better not have done what I did do.' Her physical condition was distressing. '*Every day*, from 5 or 6 p.m. to 8 or 9 the next morning, I am totally disqualified from *anything*, even reading an amusing book by that "grasping" action of the heart ... like being hung. ... In addition to this on an average of two days in every week I am totally unable to do anything. ... I can only lie quietly in bed ... if I tempt myself by getting up I am entirely laid aside for 3 or 4 hours by the exertion of dressing.' Trifles assumed immense importance; finding a servant a new situation and writing a character for her destroyed her peace of mind, she said, for four-teen hours.

When Hilary Bonham Carter joined her, a new difficulty arose. Hilary's family complained, first, that Hilary was being victim-ized, as Aunt Mai had been, and, secondly, that Florence was taking up all her time and preventing her from working at her drawing. Miss Nightingale blamed herself. Hilary must devote a certain number of hours each day to work, and to encourage her she agreed to sit for a statuette. It was an important concession, for she had a moral objection to having her likeness taken in any form. 'I do not wish to be remembered when I am gone,' she said.

For the bust by Steell, now in the Royal United Service Institution, she gave only two sittings; and then only because it was to be paid for out of the proceeds of a fund raised by small subscriptions from the non-commissioned officers and men of the British Army. Hilary had been given a unique opportunity. She began the statuette and at first it progressed well. But she did not finish it. Months went by, Hilary stayed with one sister while her husband was away, nursed another sister's children through measles; but she did not complete the statuette.

'Hi! It is now the seventh month since you told me that little horrid thing would be done "Next Monday" – since then 30 Mondays have elapsed,' Miss Nightingale wrote to her in July 1860. At the end of August Miss Nightingale refused to leave the Burlington for Hampstead until the statuette was done. Hilary was forced to finish it, but she was dissatisfied and began another. The second statuette was not satisfactory. Hilary took advice from everyone and was 'almost ill of trying to alter and improve to meet everyone's views and strictures'. The head of the first was united with the body of the second. The face was worked over again and again. Finally, she lost heart altogether.

The statuette was not finished until the spring of 1862. The sculptor Thomas Woolner was called in to give technical advice on the final version, and it was exhibited at the Royal Academy. Miss Nightingale's family did not like it. 'I have seen our F.,' wrote Fanny to Parthe, 'and am shocked at the poor little finnikin minnikin they call Florence.' Nevertheless, and in spite of the fact that it is slight and amateurish, a sketch rather than a finished portrait, the statuette is one of the best likenesses of Miss Nightingale. In 1866 Mrs Sutherland wrote: 'There are photographs of the statuette which (though it seems odd to say so) are more characteristic than the actual portraits, none of which ... give a real idea of what you were ten years ago.'

Unwillingly Miss Nightingale realized that with all her talent, all her charm, all her intelligence, Hilary was frittering her life away. She sat up half the night writing letters which Miss Nightingale did not want written. She exhausted herself over the housekeeping. She got up before seven when she might just as well have stayed in bed. She came home with a cold and would go out again to buy flowers. In January 1861 she was ill and continued to ail through February. In March she went to do a cure at Malvern, and while she was there Miss Nightingale wrote telling her not to come back. 'Dearest, I hope that you will have guessed that long before last year had ended, I had quite come to the conclusion that it would not be right for me any more to absorb your life in letter-writing and house-keeping. For this I gave you my reasons in five big conversations. ... If I could, if we were on that kind of terms together, I would go down on my knees and ask you to forgive me for having made such a use of your life. If you like to come back on the terms about the Atelier and the hours which we discussed by letter – as my guest and friend – oh my very best and dearest friend – but not as my letter-writer and housekeeper – let us now discuss what those hours shall be, you settling them before you come back. ...'

But Hilary would settle nothing, and she was not allowed to return, though sending her away, Miss Nightingale told Clarkey, was like amputating her own limb. A month later a fresh blow fell. Clough's health gave way completely. He was told his only chance of survival was complete rest in a warm climate, and in April 1861 he and his wife went abroad to Greece.

She was left entirely alone. In a year she had lost Alexander, Aunt Mai, Hilary, and Clough. She was in no condition to face the enormous burden of work involved in War Office reform combined with racking suspense over Sidney Herbert's health.

In January 1861 the scheme for War Office reorganization was launched. There was to be a pitched battle between the forces of reform and the forces of bureaucracy, Sidney Herbert on the one hand, Benjamin Hawes on the other. Miss Nightingale was not confident. Sidney Herbert was the pivot; Sidney Herbert was the essential; Sidney Herbert alone could carry the scheme through – and she feared Sidney Herbert was weakening. 'Our scheme of reorganization is at last launched at the War Office,' she wrote to Sir John McNeill on January 17, 'but I fear Hawes may make it fail. There is no strong hand over him.'

Again she hid fear behind impatience. She wrote to Douglas Galton in January that Sidney Herbert was the weak spot in the War Office reorganization scheme. 'No one appreciates as I do Mr H's great qualities. But no one feels more the defect in him of all administrative capacity in details.' By March she was frantic, declaring that Sidney Herbert was inefficient. She warned Douglas Galton. 'Though he says he will set about your committee as soon as ever you like, make haste, for he is like the son who said "I go and goeth not".' She railed at his delays; she complained of 'the difficulty of bringing him up to scratch', the impossibility of getting a decision from him, his 'total incapacity for attending to an administative question for a single hour'. She obstinately refused to recognize that he was a dying man. 'He is a great deal better of that there is no doubt,' she wrote to Liz in March, and added, 'he is a bad patient.' On May 14 she wrote: 'I am sure the Cid thinks "*Oh she does not know how weak I feel* and how much worse in general health" ... but I do. I see it every time I see him and sorrowfully perceive that he *is* weaker and thinner – and yet I don't think him worse in general health, not materially worse.' The opinion of the doctors was against her, but she dismissed their opinion. 'Almost all London physicians are quacks.'

She spared him nothing; she stood behind him insisting that his sick mind should flog itself on, his weary spirit brace itself for fresh struggle. No one intervened on his behalf; Liz seconded her demands, and he bore it all with 'angelic temper'. By the end of May it was useless for her to rail at him – what she demanded he no longer had the power to attempt. Disease was advancing with horrible swiftness. He spent the mornings on a sofa in Belgrave Square drinking gulps of brandy until he had the strength to crawl down to the War Office, where he arrived too exhausted to work. In fact, he was dying on his feet; an examination, made after his death two months later, showed disease so far advanced that it was a miracle he had been able to work at all for the past year.

The end came in June. In the first week of June he collapsed and on June 7 he wrote to Miss Nightingale telling her he could struggle no longer. He must resign the War Office and retire. It was the letter of a beaten man. 'As to organization I am at my wits' end. The real truth is that I do not understand it.' He suggested that he should remain in office for a few weeks longer so that he could carry through, as a last gesture, certain points of internal organization and military hospital reform.

On June 8 she replied, bitterly, contemptuously, and cruelly. He had failed her. She refused to accept his health as an excuse. 'I believe you have many years of usefulness before you. I have repeated so often my view of your case – and I never felt more sure of any physical fact in my life – that I will not trouble you with writing my letters all over again.' She told him what his failure meant. Their work was ruined. War Office reorganization was a general wreck. The reform of military hospitals was a general wreck. The suggestion that he should stay on in office just long enough to carry through certain essential reforms she contemptuously rejected. '*No* reform had better be done by anyone about to leave office. ... How perfectly ineffective is a reform unless the reformer remains long enough at the head to MAKE IT WORK.' For herself personally there was nothing he could do, but on Douglas Galton's behalf she asked him to establish and define Galton's position at the War Office, where his authority had been questioned. 'I consider your letter as quite final about the reorganization of the W.O. And I promise never to speak of it again. Many women will not trouble you by breaking their hearts about the organization of an office – that's one comfort. ... *Hawes has won.* If you will not think me profane I will say "Hell hath gotten the victory".'

She cut herself off from him. He was still at the War Office preparing to hand over to his successor. She would not see him or write to him. Uncle Sam remonstrated with her and was told, 'There is no uneasiness between me and Lord H. I am sure he does not at all realize what I feel about his failure, but thinks I do not see him or write to him because of my own health.' He did realize, and he could not bear it. Her anger he had always been able to bear, but he could not endure her unhappiness. 'Poor Florence'– he used the phrase so often. She gave up so much, and she was doomed to fail. He had always seen that she must fail, that the sacrifice must be in vain; from the beginning he had known that the task she insisted on undertaking was too difficult. But she had almost made him believe that faith could move mountains. Poor Florence.

It was not in his nature to leave her to eat her heart out; he had always said, 'It takes two to make a quarrel and I won't be one.' Ill and harried as he was, he went to face her. She was in the old familiar rooms at the Burlington. Here, where so much had been endured, so much had been hoped, so much had been sacrificed, a terrible interview took place. In September 1861, in a letter to Harriet Martineau, she described what passed between them. She was a woman possessed; she was consumed with grief and rage; she would not see that she had before her a dying man. By failing to endure – and who could be asked to endure more than she had

283

endured? – he was dooming the British Army. She felt no mor
pity for him than if he had in fact been an inanimate tool break
ing at the crucial moment in her hand, and she lashed him with
her tongue. 'A Sidney Herbert beaten by a Ben Hawes is a greater
humiliation than the disaster of Scutari,' she told him implacably
'No man in my day has thrown away so noble a game with all the
winning cards in his hands.' She said that he bore it all. He did no
justify himself. 'And his angelic temper with me I shall never
forget.'

Did she break his heart, or had he already passed beyond he
power to wound? The end had almost come. He had intended
when he resigned to remain at the War Office clearing up his
work for some weeks, but within a fortnight he had another and
even more serious collapse. He was ordered to give up work a
once and to go to Spa for a cure. On July 9 he came to the Burl
ington to say good-bye to her. They were not alone. He could no
longer walk easily, and he was brought in a carriage and assisted
up the stairs. She never saw him again.

He managed to reach Spa, but his condition was hopeless. He
wished to be at Wilton, and on the 25th he came home. It was
evident that he was dying. He reached Wilton, saw again the place
where he loved every spot as if it were a living person, and early
in the morning of August 2, 1861, he died. Liz kept notes of his
last hours and his last words for Florence. His last coherent
thought, she wrote, was of Douglas Galton's position in the War
Office which Florence had asked him to establish, his last murmur
'Poor Florence ... poor Florence, our joint work unfinished.'
'And these words he repeated twice.'

Miss Nightingale was in Hampstead when she heard the news
She was overwhelmed. Anguish, despair rushed in on her like the
bursting of a dam. She hurried down to the Burlington where she
collapsed, and was seriously ill for nearly four weeks. Uncle Sam
took charge of her affairs, informing all correspondents by her
orders that 'a great and overwhelming affliction entirely pre
cludes Miss Nightingale from attending to any business.'

The structure of her existence had been destroyed at a single
blow. 'He takes my life with him,' she told W. E. N. 'My work
the object of my life, the means to do it, all in one depart with
him. ... Now not one man remains (that I can call a man) of all
whom I began work with five years ago. And I alone of all men
"most deject and wretched" survive them all. I am sure I mean

to have died.' Successive frenzies of grief, of longing, of rebellion swept over her. She found it difficult not to blame God, who possessed, after all, absolute power, for letting Sidney Herbert die. Sidney Herbert's death 'involved the misfortune moral and physical of five hundred thousand men,' and it would have been 'but to set aside a few trifling physical laws to save him'.

There was no doubt of her agony, and yet – it was not quite what might have been expected. She felt grief, longing, the hopeless regret which follows bereavement, but she did not feel remorse. She felt she was justified. She knew that Sidney Herbert had found the burden of life almost too heavy to bear, that he had been overworked, harried, and subjected to unjust criticism, and she regretted it had been necessary for her to add to his burdens. But the necessity had existed; it had been her duty to act as she did; and she had nothing with which to reproach herself. In the letter to Harriet Martineau of September 9 she wrote: 'I too was hard on him.' It was the only admission she ever made that she had anything to regret. In the same letter in which she told Harriet Martineau of the 'angelic temper' with which he bore the hard things she said to him, she added implacably, 'at the same time he knew that what I said was true.'

He died with a broken heart, but she never admitted she had done anything toward breaking it. That had come about through a combination of the world's cruelty and his own weakness. She never felt she was to blame.

Yet now he was dead an extraordinary change took place in her. While Sidney Herbert was alive she had been the teacher, he the pupil; she had been the hand, he the instrument. Now he was dead she called him her 'Master'. It was the name by which she always referred to him. She who had criticized him never uttered now a word that was not praise, she who had lashed him with her tongue, now abased herself before him. She spoke of herself as his adoring disciple. 'I loved and served him as no one else,' she wrote to Sir John McNeill.

She developed an intense possessiveness about him. No other claim must equal hers; no knowledge of him could be compared with her knowledge. 'I understood him as no one else,' she wrote. It was the old story. What she felt, what she endured, must be unique. No illness was to be compared with her illness; no self-sacrifice was to be compared with her self-sacrifice; no grief could rival her grief. She would not even admit a wife could be more bereaved. 'How happy widows are,' she wrote to Uncle Sam on August 14, 1861, 'because people don't write them harassing letters in the first week of their widowhood and yet I know of no widow more desolate than I.'

Hand in hand with intense jealous emotion came resentment. The world did not understand; friends did not understand; they had wrong ideas, miserable misapprehensions.

In fact, the world was doing Sidney Herbert less than justice. His obituary notices were cold. The disasters in the Crimea were remembered, but it seemed that no one was aware of the benefits which had since been conferred on the British Army through him. His friends thought that something should be written which did justice to his achievements; and Miss Nightingale was approached by Gladstone to write a short memoir. On August 21, in a turmoil of grief, irritation, and misery, she wrote a vast letter to Sir John McNeill saying she knew nothing about what had appeared in newspapers. As far as she was concerned, she had stopped all newspapers from the day of Sidney Herbert's death because she could not bear to read one line about him. But 'before he was cold in his grave his wife, Mr Gladstone and the War Office have done nothing but harass me. ... Twice in the *first week* after his death I was written to for materials for his life. Mr Gladstone was one of these as you will guess. And he enclosed me a sketch written by *her*. There was not one word of truth in it from beginning to end!!! She represented him as having triumphed (and quoted words *of his* to this effect) in having effected the reorganization of the War Office, which he died of regret for not having done. I told Mr Gladstone a little of the real truth and wrote at his request a slight sketch of what he had done. (And the week was not out before *she* wrote to me for another.) ... This is just what I most dreaded and least asked. In fact I really would hide myself in the East of London not to do it.'

Grief, resentment, irritation, despair, discharged themselves in the floods of words, which were her safety valve. But when irritation and resentment were discharged, as in all the crises of her life, justice and generosity remained behind; and when fury at Liz's obtuseness had run its course, there emerged recognition of her qualities and rights, and Miss Nightingale became the support and consolation of Sidney Herbert's widow.

During the first fortnight after her husband's death, Liz wrote to her five times. 'You will say the children ought to be a comfort

286

to me,' she wrote on August 14, 1861, 'but I suppose I am not naturally fond of children – at any rate I have never been used to be much with them. He was *my all*. He is gone.' 'You know all I have to bear more than anyone else,' she wrote in November 1861.

She continued to include Miss Nightingale in her family life after Sidney Herbert's death. 'I cannot help repeating that there is a great "fond" of justice and magnanimity in her,' Miss Nightingale wrote to W. E. N. in May 1862. 'I am always first with her because I was first with him. My claim to be consulted, to be informed is always recognized.'

But she could not leave Sidney Herbert's reputation to the uninformed panegyrics of his wife or to the faint praises of men like General Sir John Burgoyne, who said at a memorial meeting: 'Lord Herbert's hobby was to promote the health and comfort of the soldier and his pet was Miss Nightingale who followed the same pursuit.' She reconsidered her decision and went to the Burlington, where she shut herself up for a fortnight, writing an account of Sidney Herbert's work for the army which she sent to Mr Gladstone.

On August 21 she had told Sir John McNeill that she had asked Gladstone to assume Sidney Herbert's mantle. 'I took advantage of my opportunity and told Mr Gladstone a little of what he [Sidney Herbert] had *not* done, asking him whether I should tell him the rest, and *asking* him whether I should *ask* him to help in it for S. Herbert's sake. The reply was truly Gladstonian – cautious, cold, complimentary yet eloquent – but evidently intending to do nothing.'

Her memoir was privately printed in 1861 and privately circulated under the title, *Private and Confidential. Sidney Herbert – on his Services to the Army*. In 1862 the memoir was enlarged, read as a paper at the London meeting of the Congrès de Bienfaisance in June, and subsequently published as a pamphlet under the title: *Army Sanitary Administration and its Reform under the late Lord Herbert*.

Sidney Herbert has left little impression. His term of office was a period of great and promising beginnings fated to come to almost nothing. Many years were to pass before the reforms for which he and Miss Nightingale laboured became realities. Much of what he did was undone by his successors; many improvements have necessarily been superseded and forgotten.

None the less, his term of office is marked by the first improvements in the health administration and living conditions of the British Army. When the Barrack and Hospital Commission began to inspect the barracks and hospitals of the British Army in 1857 it was found the troops were living in slum conditions. Barracks were frequently undrained and without water supply. Rooms were unlighted except by farthing dips – the allowance was two to a large room, and the resulting illumination was so dim that it was impossible to see to read. Cooking facilities were totally inadequate, and boiling was almost invariably the only method of cooking. Married couples lived, ate and slept together in low, dark, unventilated rooms where privacy was unknown. As many as thirty couples were quartered in one room, separated from each other only by blankets hung from the ceiling. Barrack rooms had no means of heating. The men had no facilities for keeping clean. One jack-towel, one piece of soap and one basin was the ordinary allowance for fifty men. There was no means of drying wet clothes. The men had no place where they could read or write. As in Scutari, the only facilities the authorities provided were facilities for drinking. Overcrowding, insufficient ventilation, defective drainage, in conjunction with bad food badly cooked, resulted in the prevalence of the 'pulmonary and zymotic complaints' which were responsible for raising the death-rate in barracks to double the figure of the death-rate in civil life. Conditions in military hospitals were equally deplorable.

Under Sidney Herbert a start was made on demolishing buildings condemned by the Commissioners and on reconditioning others. Barrack rooms and hospital wards were warmed and ventilated. Farthing dips were replaced by gas. Water was laid on. Kitchens were provided with ovens for roasting and baking. A training school for army cooks was established on lines laid down by Soyer. After his death much of this work was abandoned or shelved, with the result that thirty years later the nation was shocked by a series of barrack scandals. Nevertheless, what had been done was of importance. General principles laid down in the Barrack Commission's Report had actually been put into practice, and the British Army owes Sidney Herbert the first substantial improvements in its barracks and hospital accommodation.

At the end of her paper on Sidney Herbert's services to the army Miss Nightingale gave some striking figures. In the three years during which Sidney Herbert was Secretary of State for War only one-half of the men who entered the army died per annum (on home stations) as had previously died. The total mortality on home stations from all diseases was less than the former mortality from chest diseases and consumption alone. Such was the immediate and startling result of better food, more warmth, more sanitation and more fresh air. Sidney Herbert, she wrote, should be remembered 'as the first War Minister who ever seriously set himself to the task of saving life'.

It was his personal tragedy that fate called on him to expend his genius on a subject for which he was temperamentally unsuited. But he did not sacrifice himself in vain. The British Army was fortunate in finding such a champion. Without his influence, the prestige deriving from his high standing, and his altruism, the cause of the British soldier might well have languished for another half-century. He died before his work was done, and no outstanding reform is associated with his name, but he succeeded in making the health of the British soldier an issue of first-class importance, which no subsequent administration could ignore. The Royal Sanitary Commission to inquire into the Health of the Army in 1857 may not have done all that was anticipated in the first flush of hope. But that it should have sat at all was a triumph and marked the dawn of a new age.

When Miss Nightingale had finished writing her memoir, she left the Burlington for ever. It was haunted; she could never bring herself to go back there. She could see Sidney Herbert in the street. 'I could not bear to look down Burlington Street where I had seen him so often.' And Sidney Herbert was not the only ghost. She could see Alexander; Clough, now desperately ill; Aunt Mai who had deserted her; Hilary, the limb she had been forced to cut off. 'I have quite decided not to return to the Burlington where one by one my fellow workers whom I had so laboriously got together have been removed from me,' she wrote to W. E. N. on September 9, 1861. She retired to Hampstead, where she isolated herself completely. She would see no one, fellow workers, family, or friends. To overwhelming grief was added blank despair as report after report reached her of Sidney Herbert's work being undone. During the first week after his death three important decisions he had taken were reversed by the Duke of Cambridge, the Commander-in-Chief, 'who absolutely cringed to him when alive', she wrote to Sir John McNeill on August 21. 'On one of these occasions Lord de Grey, who *happened* to be in G. Lewis' room (everything happens – is not done – at that miserable place) said, "Sir, it is impossible; Ld Herbert decided it – and the House of Commons voted it" and walked out of the room. It was less wise than honest. But it had its effect for the time. G. Lewis was awed and the C-in-Chief silenced. *But only for the time*.'

On September 24 she wrote to Harriet Martineau: 'The Commander in Chief rides over the weak and learned Secretary of State (Sir G. Lewis) as if he were straw. Day rooms, Barrack Inspections, Hospitals, all the Sanitary Improvements, it is the same. Not one will they leave untouched.' 'As for me,' she wrote to Sir John McNeill on August 21, 'I feel like the Wandering Jew as if – I *could not* die.'

17

She was convinced that her life work was ended. The death of Sidney Herbert had closed the door through which she entered the official world. Great as her influence had been, intimately as she had been concerned in army administration, all had depended upon him. She had never had any official status; she could not force the War Office to use her. She had been inside because he was inside. Now he was gone, she would be shut out.

It was the general opinion. In a letter written by one of Aunt Mai's daughters in September 1861, Fanny asked Edwin Chadwick to suggest some new field of work for Florence, 'her own being now closed against her'. He agreed that her army work was ended but foresaw for her a long, useful life of labour for others in some different sphere. In the furnished rooms in Hampstead she was wretchedly ill. All the familiar symptoms of her collapses reappeared, fainting, extreme weakness, nausea at the sight of food. In addition, she suffered from nervous twitchings. She insisted on remaining alone.

In 1862 Miss Nightingale wrote of the widowed Queen Victoria: 'She always reminds me of the woman in the Greek chorus, with her hands clasped above her head, wailing out her inexpressible despair.' It is an apt description of her own behaviour after the death of Sidney Herbert. Shut away in the rooms in Hampstead, she wrote letter after letter wailing out ruin. 'My poor Master has been dead two months today,' she wrote to Dr Farr on October 2, 1861, 'too long a time for him not to be forgotten. ... The dogs have trampled on his dead body. Alas! seven years ago this month I have fought the good fight with the War Office. AND LOST IT.' 'Every day his decisions ... his judgements are over-thrown. ... We have lost the battle. Now all is over,' she wrote to Harriet Martineau in September. 'Would I could hide myself underground not to see what I do see,' she told Lady Herbert on August 17.

But as the weeks went by the sympathy of her fellow workers became tempered with irritation. As far as they were concerned, she might as usefully have been underground as shut away in Hampstead wailing out despair. At the end of September Douglas Galton wrote to her sharply: 'Notwithstanding what you say, Sidney Herbert *did* do a great deal, doubtless he left something still to be done. The Medical Department is in itself a great achievement. But perhaps your motto is "Nihil actum, si quid agendum".' (Nothing has been accomplished if there is still something to be done.) Army reform had not utterly perished with Sidney Herbert. The edifice had collapsed, but there were workers still among the ruins. Douglas Galton held his post as Inspector-General of Fortifications and controlled the erection and maintenance of barracks and hospitals. Lord de Grey, a convinced reformer and a disciple of Sidney Herbert, had been appointed Under-Secretary of State for War. Sir George Lewis, Sidney Herbert's successor, was not unfriendly. On October 21, 1861, Richard Monckton Milnes wrote, encouraging her to return to work: 'I should like you to know how you will find Lord de Grey willing to do all in his power to further your great and wise designs. You won't like Sir G. Lewis, but somewhere or other you ought to do so. .... I write this about de Grey because I was staying with him not long ago and he expressed himself on the subject with much earnestness.'

Even in the dark weeks immediately following Sidney Herbert's death, a few points had been gained. In September 1861 Douglas Galton's appointment as Inspector-General of Fortifications, which had been temporary, was confirmed. At the same time the scope of the Barrack and Hospital Commission, on which Dr Sutherland and Douglas Galton were the most active and influential members, was extended to take in the Mediterranean stations.

A considerable victory was won over the proposed construction of a new General Military Hospital at Woolwich. The Duke of Cambridge, the Commander-in-Chief, had steadily opposed the building of this hospital, and as soon as Sidney Herbert was dead he pressed for cancellation of the scheme. This was the occasion when Lord de Grey 'happened' to be in Sir George Lewis's room and said, with more honesty than wisdom: 'Sir, it is impossible. Lord Herbert decided it and the House of Commons voted it.' The building went forward and eventually, at Miss Nightingale's suggestion, was called the Herbert Hospital.

It was true something might still be done, but how woefully

little! A point here, a point there, might with infinite labour be carried, but all high hopes, all grand schemes had perished. 'It is really melancholy,' wrote Douglas Galton after Sidney Herbert had been dead a fortnight, 'to see the attempts made on all hands to pull down all that Sidney Herbert laboured to build up.'

Victory was no longer a possibility, but every inch of ground must be contested in retreat, to preserve something in the midst of disaster.

Miss Nightingale was drawn back, but the work was bitter to her now. She had dreamed of great achievements; there were to be no great achievements. All that was left was 'desperate guerilla warfare.' Heartbroken and weary, she revolted. 'It cannot last. I am worn out and cannot go on long,' she wrote to Harriet Martineau on September 14, 1861.

Seclusion proved impossible. As soon as she finished her paper on Sidney Herbert's services to the army she went back to Hampstead to shut herself up again, but at this point she received an appeal which it was impossible to ignore. In April 1861 civil war had begun between the Northern and Southern States of America. In October an appeal from the Secretary at War in Washington reached Miss Nightingale through the agency of Harriet Martineau, who had a channel of communication with the Northern States through her publisher in New York. She was asked for help in organizing hospitals and the care of the sick and wounded. On October 8 she told Dr Farr she had sent to Washington 'all our War Office Forms and Reports, Statistical and other. ... It appears that they, the Northern States, are quite puzzled by their lack of any Army Organization.' She also sent Miss Dix, the Superintendent of Nurses at Washington, her evidence before the Commission of 1857, and Harriet Martineau reported that Miss Nightingale's writings were 'quoted largely and incessantly in medical journals as a guide to military management in the Northern States'. No channel of communication with the Southern States was available, though Miss Nightingale wrote that she was 'horrified at the reports of the sufferings of their wounded'. It was being made evident, had she been willing to be encouraged, how far public opinion had been educated in the importance of health administration to an army in the field. The Secretary of War was petitioned to appoint a Sanitary Commission; plans were drawn up for the inspection of camps, for the introduction of female nurses into hospitals, and for the improvement and supervision of hospital diet and cooking. What she herself had done in the Crimea was reproduced. Though circumstances prevented much of this work being successful, the attempt showed a great change had taken place.

She became involved in a very large correspondence, advising charitable committees, organizations for sick and wounded relief, and religious bodies and women's associations who were working for army welfare. In 1865, when the war was over, the Secretary of the United States Christian Union wrote to her: 'Your influence and our indebtedness to you can never be known.'

To remain in Hampstead was impossible, and early in November 1861 she was persuaded to accept Sir Harry Verney's repeated offer of the loan of his London house and moved to 32 South Street. She was still very feeble, but she persisted in being alone.

She had only been in London a few days when she received another shattering blow: Arthur Hugh Clough died in Florence on November 12, 1861. Her grief for him was second only to her grief for Sidney Herbert. 'Oh Jonathan, my brother Jonathan, my love for thee was very great, passing the love of women,' she wrote to Sir John McNeill on November 18. Clough had united with intellect and wit an extraordinary ability to inspire affection. 'I do not know that I have ever cared so much for any man of whom I had seen so little as I did for Clough,' Sir John McNeill wrote on November 19. Miss Nightingale was blamed for misusing his talents, for hastening his end by driving him too hard, and Clough's family did not refrain from expressing their resentment. A fragment of a letter written to her by Jowett in December 1861 – the remainder of the letter has been cut away – advises her to 'disregard this attack arising from common misery at the death of our dear friend.'

Clough had complemented her as Sidney Herbert had complemented her; he gave her affection and sympathy, his brilliance and grace brought charm into her life. She gave him the energy, the conviction, the certainty which he had somehow fatally lost. His death inflicted a mortal wound. Coming only three months after the death of Sidney Herbert, when she was struggling to recover herself, the effect was crushing. She was totally unnerved. On November 18 she wrote to Douglas Galton, 'Now hardly a man remains (that I can call a man) of all those I have worked with these five years. I survive them all. I am sure I did not mean to.'

'Hardly a man (that I can call a man).' She had used that phrase before when Sidney Herbert died and she was speaking of the most faithful, the most devoted, possibly the most able of them all, of Dr Sutherland. He alone had done what she demanded, he alone had given up his whole life to the work, yet she had never loved him, never could love him. To love was essential to her; and with Clough lost, Alexander lost, Sidney Herbert lost, she felt herself horribly alone.

On November 19 Hilary Bonham Carter went to see her; she had collapsed and been very ill. 'She wept very much,' wrote Hilary to Clarkey on November 20. 'She thinks she may perhaps withdraw her hand from Government matters entirely.'

But she was not to be allowed to withdraw her hand. A fortnight after Clough's death she received an urgent appeal from the War Office. England seemed on the brink of war with the Northern States of America: two agents of the Southern States had been taken by force from the neutral British steamship *Trent* and carried prisoner to a Northern port. It was an outrage on the British flag, war seemed inevitable, and the Government decided to send reinforcements at once to Canada. On December 3 Lord de Grey wrote asking if he might call and be advised by her 'as to sanitary arrangements generally' for the expedition, including transport, hospitals, the clothing and feeding of the troops, and comforts for the sick.

Ill, shattered, sunk in grief though she was, she summoned energy to work night and day. She did far more than advise. She redrafted the proposed instructions to officers in charge of the expedition, and on December 10 Lord de Grey wrote to tell her that every one of her alterations had been adopted. She ascertained the average speed of transport by sledge and calculated the time required to transport the sick over the immense distances of Canada. She drew up schemes for relays of transport and for the setting up of depots containing necessary stores. She investigated the question of clothing and recommended that buffalo robes should be issued to the troops in place of blankets. Her astonishing capacity for detail was unimpaired. On December 19 she wrote to Douglas Galton: 'Your draft does not define with sufficient precision the manner in which the meat is to get from the Commissariat into the soldiers' kettle; and the clothing from the Q.M.G.'s store on to the soldiers' back. You must define all this. Otherwise you will have men, as you did in the Crimea, shirking responsibility.'

The Canadian Expedition was a successful piece of work. 'I have been working just as I did in the time of Sidney Herbert,' she wrote to Clarkey on December 13, 1861. 'We have been shipping off the Expedition to Canada as fast as we can. .... The War Office were so terrified at the idea of the national indignation if they lost another army, that they have consented to everything.'

Through the intervention of the Prince Consort war was avoided. Though mortally ill, he roused himself from his deathbed to insert, with his own hand, modifications in the British despatch which made it possible for the Northern States to withdraw without humiliation. It was his last public act, and a fortnight later he died. Miss Nightingale felt his death a great national disaster of which the nation was oblivious. 'He was,' she wrote to Clarkey, 'really a minister. This very few knew. He neither liked nor was liked, but what he has done for this country no one knows.'

The Canadian Expedition was a turning point. She was back in harness; work came rushing in; retirement was impossible. Weary, heartbroken, grief-stricken though she might be, her private feelings must be laid aside, and she must force herself to work once more. Jowett once said Miss Nightingale was the only person he had ever met in whom public feelings were stronger than private feelings. But that did not mean that her private feelings were weak; on the contrary, they were almost overwhelmingly strong, and in 1861, though her sense of duty forced her to dedicate herself to the *respublica*, she was unreconciled to her lot. She resented more than ever the sacrifice demanded of her, she fell even more frequently into frenzies of grief, rage, and disgust with the world.

She was horribly lonely. She had no friend; she had no helper. She was entering a period of great toil, relentless self-sacrifice, discouragement, and she had no single soul to give her support.

She looked round the world and what did she see? Women everywhere. The world was full of women, and not one of them would help. Rage seized her. What had she not endured from the pretensions, the foolishness, the frivolity, the selfishness of women! All her grief, her pain, concentrated itself into a passion of contempt and dislike for her own sex. On December 13 she began to pour out to Clarkey, who had just written a book on Madame Récamier, an enormous, disconnected diatribe on women: '... you say "women are more sympathetic than men".

Now if I were to write a book out of my experience, I should begin, *Women have no sympathy.* Yours is the tradition – mine is the conviction of experience. I have never found one woman who has altered her life by one iota for me or my opinions. Now look at my experience of men.

'A Statesman, past middle age, absorbed in politics for a quarter of a century, out of *sympathy* with me, remodels his whole life and policy – learns a science, the driest, the most technical, the most difficult, that of administration as far as it concerns the lives of men, – not, as I learned it, in the field from the living experience, but by writing dry regulations in a London room, by my sofa, with me.

'This is what I call real sympathy.

'Another (Alexander whom I made Director General) does very nearly the same thing. He is dead too.

'Clough, a poet born if there ever was one, takes to nursing administration in the same way, for me.

'I only mention three, whose whole lives were re-modelled by sympathy for me. But I could mention very many others – Farr, McNeill, Tulloch, Storks, Martin, who in a lesser degree have altered their work by my opinions. And, most wonderful of all – a man born without a soul, like Undine – Sutherland. All these elderly men.

'Now just look at the degree in which women have sympathy – as far as my experience is concerned. And my experience of women is almost as large as Europe. And it is so intimate too. I have lived and slept in the same bed with English Countesses and Prussian Bauerinnen, with a closeness of intimacy no one ever had before. No Roman Catholic Supérieure has ever had the charge of women of the most different creeds that I have had. No woman has excited "passions" among women more than I have.

'Yet I leave no school behind me. My doctrines have taken no hold among women. Not one of my Crimean following learnt anything from me – or gave herself for one moment, after she came home, to carry out the lesson of that war, or of those hospitals. I have lived with a sister 30 years, with an aunt four or five, with a cousin two or three. Not one has altered one hour of her existence for me. Not one has read one of my books so as to be able to save me the trouble of writing or telling it all over again.

'Hilary is the type of want of sympathy. Because she is the most unselfish, and because she has a "passion" for me. Yet have I not

influenced her by one inch. Nay rather all these women have influenced me, much more than I have them. Parthe always told
me, as a reproach, that I was "more like a man". Indeed I began
to think it was true.

'No woman that I know has ever *appris à apprendre*. And I
attribute this to want of sympathy. . . .

'It makes me mad the "Woman's Rights" talk about the "want
of a field" for them – when I know that I would gladly give £500
a year for a Woman Secretary. And two English Superintendents
have told me the same. And we can't get *one*.

'As for my own family, their want of the commonest knowledge
of contemporary history makes them quite useless as secretaries.
They don't know the names of the Cabinet Ministers. They don't
know the offices at the Horse Guards. They don't know who of
the men of today is dead and who is alive. They don't know which
of the Churches has Bishops and which not.

'Now I'm sure I did not know these things. When I went to the
Crimea I did not know a Colonel from a Corporal. But there are
such things as Army Lists and Almanacs. Yet I never knew a
woman who, out of sympathy, would consult one – for my work

'. . . A woman once told me my character would be more sym-
pathized with by men than by women. In one sense, I don't choose
to have that said. Sidney Herbert and I were together exactly like
two men – exactly like him and Gladstone. And as for Clough, oh
Jonathan, my brother Jonathan, my love for thee was very great.
PASSING THE LOVE OF WOMEN.

'In another sense I do not believe it is true. I do believe I am
"like a man", as Parthe says. But how? *In having sympathy*. I am
sure I have nothing else. I am sure I have no genius. I am sure
that my contemporaries, Parthe, Hilary, Marianne, Lady Dun-
sany, were all cleverer than I was, and several of them more un
selfish. But not one had a bit of sympathy.

'. . . Women crave *for being loved*, not for loving. They scream
at you for sympathy all day long, they are incapable of giving any
in return, for they cannot remember your affairs long enough to
do so. . . . They cannot state a fact accurately to another, nor can
that other attend to it accurately enough for it to become in
formation. Now is not all this the result of want of sympathy? If
you knew what it has been to me, having my aunt instead of S
Herbert, Hilary instead of Clough, etc. etc. etc. – not because the
man had power the women none. But simply from what I say of

want of attention. I'm sure I don't think what falls from my lips pearls and diamonds. Only, if they are not going to listen, I had so much rather not say it. I'm none too fond of talking. ...

'People often say to me, you don't know what a wife and mother feels. No, I say, I don't and I'm very glad I don't. And *they* don't know what *I feel*. Why, dear soul, Blanche went away and left her husband for a year! I am the only person who made any effort to save his life and gave him £500 to go abroad, my hard earned savings. And they are living on that now, his wife and sisters, at Florence. ...

'I am sick with indignation at what wives and mothers will do of the most egregious selfishness. And people call it all maternal or conjugal affection, and think it pretty to say so. No, no, let each person tell the truth from his own experience.

'Ezekiel went running about naked, "for a sign". I can't run about naked because it is not the custom of the country. But I would mount three widows' caps on my head, "for a sign". And I would cry. This is for Sidney Herbert, I am his real widow. This is for Arthur Clough, I am his true widow (and I don't find it a comfort that I had two legs to cut off, whereas other people have but one). And this, the biggest widow's cap of all, is for the loss of all sympathy on the part of my nearest and dearest. (For that my aunt was. We were like two lovers.)

'... This is the shortest day, would it were the last. Adieu dear friend. I am worse. I have had two consultations and they say that all this worry has brought on congestion of the spine, which leads straight to Paralysis. And they say I must not write letters. Whereupon I do it all the more.'

Her health was approaching a new crisis. She had already collapsed twice within the last six months, once when Sidney Herbert died in August and again when the news of Clough's death reached her in November. She was alone, devoured by grief, remorse, and resentment, unable to rise from her bed, unable to eat. In this condition she had forced herself to work day and night on the Canadian Expedition. On Christmas Eve 1861 she was dangerously ill, more dangerously ill than she had been since her collapse in the summer of 1857.

For some weeks she was expected to die. She longed to die, but her iron constitution triumphed, and by the middle of January she was able to sit up in bed. But a further stage had been reached in the decline of her health. After this last illness she became bedridden and did not leave her room for six years. She moved from house to house but could not walk – she had to be carried. She never saw the outside world, to exchange one set of four walls for another was the only variation in her outlook.

By the end of January 1862 she was convalescent, and she insisted that the Verneys should come back to their house in time for the spring season. Once more she had to find somewhere to live. In her present state of health a hotel was out of the question and W. E. N. offered to pay the expense of taking a furnished house. A suitable one was found in Chesterfield Street, Mayfair. She spent £100 on new carpets and curtains, and in this house, 'a fashionable old maid's house in a fashionable quarter', through the spring of 1862 she slowly struggled back to sufficient health to enable her to work.

She was gloomy and despairing now beyond description; hope had left her; she was like a man brought back to health so that he might be able to walk to the gallows. 'I have lost all,' she wrote to Fanny on March 7, 1862. 'All the others have children or some high and inspiring interest to live for – while I have lost husband and children and all. And am left to the dreary hopeless struggle. ... It is this desperate guerilla warfare ending in so little which makes me impatient of life. I, who could once do so much ... I think what I have felt most during my last 3 months of extreme weakness is the not having one single person to give one inspiring word, or even one correct fact. I am glad to end a day which never can come back, gladder to end a night, gladder still to end a month.'

The task which lay before her was indeed daunting. The reformers had been appalled at Sidney Herbert's death, but even so they had not fully realized his value. On June 6, 1862, Miss Nightingale admitted to Douglas Galton: 'One did not appreciate the power of Sidney Herbert's hand at the War Office while he was alive.' Her close intimacy with Sidney Herbert could not be repeated, but it was a malign stroke of fate which replaced Sidney Herbert by Sir George Lewis – no two people could have honestly found each other more difficult to understand than Sir George Lewis and Miss Nightingale. His virtues were of a kind which she was unable to appreciate. He was one of the best classical scholars in Europe, extremely industrious, and of unimpeachable integrity, but he lacked warmth. Greville said he was as cold-blooded as a fish. He had written several books in a restrained and polished style on classical and political subjects and had considerable wit. One of his sayings, 'the indiscretion of biographers adds a new terror to death', was often quoted by Miss Nightingale. In his position as Secretary of State for War he deserved sympathy; he had been unwilling to take office, and had accepted only out of public spirit. 'I can fancy no fish more out of water than Lewis amidst Armstrong guns and General Officers,' Sidney Herbert had written on July 16, 1861, adding that he was a gentleman and an honest man. But the nature of his breeding and integrity belonged to the eighteenth rather than the nineteenth century. He was no philanthropist, no reformer. He was able, as Richard Monckton Milnes said, to make up his mind to the 'damnabilities of the work'.

Miss Nightingale never would meet him. The spell which she had cast over Panmure she never would attempt to cast over him. On his side Sir George Lewis was friendly. In the spring of 1862 he suggested he should call, but she refused on the grounds of her recent illness. Knowing her to be a classical scholar, he sent her one of the classical *jeux d'esprit* in which he excelled, the nursery rhyme 'Hey Diddle Diddle' translated into Latin verse. She was not flattered, but enraged, and wrote to Douglas Galton that Sir George Lewis would do far better to keep his mind on the War Office. Meanwhile he followed up 'Hey Diddle Diddle' in Latin with 'Humpty Dumpty' in Greek.

However, owing to Sir George's lack of experience in army administration, Lord de Grey was becoming of increasing importance, and since the Canadian Expedition he had become her friend. Lord de Grey had been born to great position and great wealth. His father, Lord Ripon, was Prime Minister at the time of his birth, and his birthplace was 10 Downing Street. He united an instinctively aristocratic outlook with radical and even revolutionary views. When a young man he had been a member of the Christian socialist movement, and had written a pamphlet which was suppressed. His integrity was beyond question, his capacity for work great, and his sense of public duty very high.

When Miss Nightingale re-entered the War Office after the Canadian Expedition, she had Lord de Grey behind her and influence within the departments through Douglas Galton and Dr Sutherland. Outside the War Office she had a powerful friend and protector in Lord Palmerston, who had become Prime Minister again in 1859.

Almost immediately a crisis arose. On May 15, 1862, Sir Benjamin Hawes, Permanent Under-Secretary for War, unexpectedly died. Both the man and his office had been major obstacles in the path of reform. The Permanent Under-Secretary for War stood in an administrative bottle-neck. 'He was,' wrote Miss Nightingale, 'a dictator, an autocrat, irresponsible to Parliament, quite unassailable from any quarter, immovable in the middle of a (so-called) Constitutional Government and under a Secretary of State who *is* responsible to Parliament.' One of the fundamental principles of her scheme of War Office reorganization had been the abolition of the office of Permanent Under-Secretary.

The office had been attacked, and the attack had failed. Benjamin Hawes had beaten Sidney Herbert. Now Fate had intervened, and he was dead. Supported by Lord de Grey, she pressed urgently for reform. The office of Permanent Under-Secretary should be abolished, the work divided into two parts and performed by two Under-Secretaries, each directly responsible to the Secretary of State for War. The purely military work was to be done by the Military Under-Secretary, a post, already in existence which had been created by Sidney Herbert, and the health and sanitary administration of the army should be done by a civilian with the title of Assistant Under-Secretary. She made the bold suggestion that Douglas Galton should be allowed to resign his commission and be appointed to this post as a civilian.

A pitched battle ensued. Objections to splitting the office of the Permanent Under-Secretary into two were strengthened by the name of Sir Charles Trevelyan being put forward as Benjamin Hawes' successor. Sir Charles was an able man and an admirable administrator whom Miss Nightingale liked and respected. But she was not to be turned from her determination. The system was wrong; and 'there could be no more fatal mistake than to attempt to offset the evil of a system by introducing into it individuals of merit'. As Permanent Under-Secretary under the present system Sir Charles Trevelyan would still be 'an absolute despot though a wise one', she wrote to W. E. N. on May 24, 1862; and 'inasmuch as Trevelyan is a better and abler man than Hawes, it would have been *worse* for my reform of principle'.

She succeeded. It was agreed that the work of the office should be divided; its importance was halved, and it became no longer worthy of Sir Charles Trevelyan's consideration.

The next step was to secure the appointment of Douglas

Galton as Assistant Under-Secretary. Official opposition was determined, and Miss Nightingale appealed directly to the Prime Minister. Lord Palmerston spoke to the Commander-in-Chief, who told him that the appointment was 'simply impossible'. Lord Palmerston refused to be deterred. Miss Nightingale had convinced him, as six years ago she had convinced him in the matter of the plans for Netley Hospital. He had been thwarted then; he did not intend to be thwarted now. He ignored the Commander-in-Chief and directed Sir George Lewis to make the appointment, which he obediently did. Miss Nightingale, who despised Sir George Lewis's powers of administration, said he did not understand what he was doing. On May 24 Douglas Galton was appointed.

She had won a notable victory. With Douglas Galton in charge of the health and sanitary administration of the army, with Lord de Grey as Parliamentary Under-Secretary, with Sir George Lewis acquiescent, she saw War Office reorganization as a certainty in the near future; and she allowed herself to rejoice. On May 24, 1862, heading her letter 'The poor Queen's Birthday', she wrote to W. E. N.: 'I must tell you the first joy I have had since poor Sidney Herbert's death. Lord Palmerston has forced Sir G. Lewis to carry out Mr Herbert's and my plan for the reorganization of the War Office *in some measure*. And it may seem some compensation to you for the enormous expense I cause you, that, if I had not been here, it would not have been done. Would that Sidney Herbert could have lived to do it himself!'

Her jubilation was short-lived, for no radical change took place in the War Office. She had hoped too much both from the reform and the reformers. Douglas Galton and Lord de Grey, sincere, talented, and hard-working, did not possess the genius, the driving force which alone could have accomplished what she had called 'the fearful task of cleansing the Augean stable'. Where Sidney Herbert had failed, it was not surprising that Douglas Galton and Lord de Grey did not succeed. On August 8, 1862, she wrote to Sir John McNeill: 'Lord de Grey and Douglas Galton miscalculated their powers and their intelligence when they promised to re-organize the W.O. The administrative work they do well.'

It was final defeat, and she accepted it. She continued to press Douglas Galton and Lord de Grey in season and out of season for departmental efficiency, but the grand project of complete War Office reorganization was relinquished for ever.

It was not revived even when, on April 13, 1863, the situation was changed once more by the sudden death of Sir George Lewis. Three men were possible candidates for the office of Secretary of State for War – Lord Panmure, Mr Cardwell, and Lord de Grey. The prospect of having the Bison once more at the War Office was distasteful, but even more distasteful was the prospect of Mr Cardwell. Edward Cardwell, who ten years later carried through the most important army reforms of the century, was a man to whom Miss Nightingale failed to do justice. He had been a scholar of Winchester and Balliol; he was conscientious, industrious, eminently discreet, kind-hearted, and an excellent public servant. But he lacked charm. Nor did he commend himself to her by being the devoted disciple of Gladstone. They had already met in 1857, while she was working on the first Royal Sanitary Commission, and they had not been attracted to each other.

She plunged into a campaign to secure the appointment of Lord de Grey. Lord Palmerston had been the friend and admirer of Sidney Herbert, and she appealed to him, speaking not with her own voice but with the voice of Sidney Herbert. On April 15 Sir Harry Verney was sent down to read Lord Palmerston a letter she had written. On the afternoon of April 15 he wrote describing the interview. 'Lord Palmerston was so good as to admit me. I said I had seen you this morning, and that by your desire I requested him to allow me to read a letter to him from you. He said "Certainly"; and I read it to him rather slowly. Having read it, I said that you had mentioned this morning that within a fortnight of Lord Herbert's death, he had said to you more than once that he hoped Lord de Grey would be his successor. He took the letter and put it in his pocket. He then asked how you were and where, and I told him. There is a Cabinet at 5.30 this afternoon.' A copy of the letter was also sent to Mr Gladstone before the Cabinet meeting.

At the same time Miss Nightingale appealed urgently to Harriet Martineau for newspaper support. A draft of a telegram, written in her own hand and sent on April 16, 1863, runs: 'From Florence Nightingale to Harriet Martineau – Agitate, agitate, for Lord de Grey to succeed Sir George Lewis.' On the following day the *Daily News* published a leading article pressing for his appointment.

In fact the matter was already decided. Lord Palmerston had once more been convinced. On April 16, the day after the Cabinet meeting, he went down to Windsor and read Miss Nightingale's letter to the Queen, and the appointment of Lord de Grey was announced on April 22, 1863.

The appointment ensured that Miss Nightingale's influence in army affairs continued, but her position was not the position she had held when Sidney Herbert was alive. She had then been closely concerned in the internal affairs of the War Office. She had done the work of an administrator. In 1863, not only was her friendship with Lord de Grey very different from her close intimacy with Sidney Herbert, but there was no urgent task, such as the Royal Sanitary Commission of 1857 had been, to bring her directly into War Office affairs. She herself wrote in 1862 that she had done the work of a Secretary of State at the War Office for five years, but she was doing that work no longer. She had passed from being an administrator to being an adviser.

As an adviser her position was extraordinary. For the next four years every problem affecting the health and sanitary administration of the British Army was referred to Miss Nightingale, though she was not only a woman but an invalid who never left her house and for months on end did not leave her bed. She was one of the first experts in Europe on sanitary questions, her vast, detailed and personal knowledge was matchless, and she had infinite wisdom in the ways of Government departments. She had not been writing regulations for seven years in vain. The busy Minister, the harassed Under-Secretary who requested information from Miss Nightingale received it in a form he could immediately use. She was an adept at the carpentry of drafting: 'Send me the two versions and I will dove-tail them in,' she wrote in 1862. She had enormous knowledge of the history of the departments, she knew the course of every transaction for years past, she knew where to go for information, she knew where papers were to be found. Secretaries borrowed copies of documents from her which were inexplicably missing from War Office files. She saved trouble to busy men. Once there was work to be done she asked neither for credit nor consideration, only to be allowed to do it. Ministers, Under-Secretaries, Assistant Secretaries wrote to her daily asking her as an expert for expert assistance. It was as if she were indeed a retired Secretary of State with vast experience, willing to devote his life to anonymous and unpaid work. She drafted hundreds of minutes; she drew up warrants and regulations; she wrote official memoranda, letters, and summaries for the Ministers' use; she composed instructions.

Her genius for financial administration was extraordinary. She devised a cost-accounting system for the Army Medical Services, which was put into operation between 1860 and 1865 and, eighty years later, was still in use. In 1947 the Select Committee on Estimates reported favourably on it, commented that it worked admirably, though in other departments systems installed within the last twenty years had been discarded, and inquired with whom it had originated. They were told – Miss Nightingale.

War Office Abstracts list the questions on which she was engaged during one year as a new *Warrant for Apothecaries, Proposals for Equipment of Military Hospitals*, a scheme for the *Organization of Hospitals for Soldiers' Wives, Proposals for the revision of Army Rations, Warrant and Instructions for Staff Surgeons, Instructions for treatment of yellow fever, Proposals for revision of Purveying and Commissariat in the Colonies, Revised diet sheets for Troop-Ships, Proposals for appointments at Netley and Chatham, Instructions for Treatment of Cholera.* And these were sidelines; her main work for the Army was still concerned with the improvement of barrack and hospital accommodation and the reform and reconstruction of the Army Medical Department, the two contributions Sidney Herbert had succeeded in making which she carried forward as a sacred trust.

She exercised authority over plans for building and reconstructing barracks and hospitals. Douglas Galton had all construction works under his control and submitted almost all plans to her. He leaned, as Sidney Herbert had leaned, not only on her judgement but on her remarkable ability to tear the essentials out of great masses of detail, and her astonishing unwearying thoroughness. A specimen memorandum on the plans for the new general military hospital at Malta covers more than two dozen foolscap pages.

It was the toughest and driest work, only occasionally enlivened by a gleam of humour. In 1863 Douglas Galton sent her plans for model cavalry barracks; she returned them with the request that the horses should be provided with windows to look out of. 'I do not speak from hearsay but from actual personal acquaintance with horses of the most intimate kind,' she wrote on June 4, 1863. 'And I assure you they tell me it is of the utmost consequence to their health and spirits when in the loose box to have a window to look out of. A small bull's eye will do. I have told Dr Sutherland but he has no feeling.' On this letter Dr

Sutherland scribbled: 'We *have* provided such a window and every horse can see out if he chooses to stand on his hind legs with his fore feet against the wall. It is the least exertion he can put himself to.'

The task of reconstructing and reforming the Army Medical Service was the most difficult of all her tasks. Sidney Herbert had succeeded in forcing through new regulations, and as a result new duties and increased responsibilities had been given to medical officers, but almost nothing had been done to improve the conditions of the Service so that men of superior calibre might be attracted into it. 'You MUST do something for these doctors,' Miss Nightingale wrote to Douglas Galton on December 24, 1863, 'or they will do for you – simply by not coming to you.' She urged better training, increased pay, more privileges – but without success. Military expenditure was being cut to the bone and the pay of army doctors was an easy target. On April 6, 1864, she wrote to Sir James Clark: 'I have written threatening letters both to Lord de Grey and to Captain Galton. ... I have shown how, when we are expecting duties from the Medical Officer, such as sanitary recommendations to his Commanding Officer, which essentially require him to have the standing of a gentleman with his Commanding Officer – we are doing things such as dismounting him at parade, depriving him of presidency at Boards, etc., which, in military life, to a degree which we have no idea of in civil live, deprive him of the weight of a gentleman among gentlemen.' The cost of sickness and invaliding to the army was fantastically in excess of any paltry sums spent on the remuneration of doctors. Why would the officials ignore self-evident truths? She beat herself against their indifference and their stupidity. She became very angry. Week after week, month after month, year after year, through 1862, 1863, 1864, she kept pounding away, writing letters almost daily, in one of the most persistent of her campaigns. 'I wrote *for the tenth time* a statement of eight pages with permission to make any use of it they pleased, with my signature as to Lord Herbert's intentions,' she told Sir James Clark on April 9, 1864, pressing for an increase of pay for medical officers. 'You may think I am not wise in being so angry,' she wrote again on April 15. 'But I assure you, when I write civilly I have a civil answer – *and nothing is done.* When I write furiously, I have a rude letter – *and something is done* (not even then always, *but only then*).'

Constant and bitter opposition came from within the Medical Department itself: the senior members had scores to settle dating back to the Crimea. Opposition blazed up into fury when, in 1862, Miss Nightingale and the Medical Department found themselves on different sides in the controversy over the Contagious Diseases, i.e. Venereal Diseases, Act.

The proportion of men in the British Army invalided with syphilis was disconcertingly high, and since 1861 the War Office had been debating measures to reduce it. The Medical Department recommended that the continental system, by which prostitutes were licensed, inspected and, if necessary, forced to submit to medical treatment, should be adopted. To this system Miss Nightingale was passionately opposed. She considered the continental system morally disgusting, unworkable in practice, and unsuccessful in results.

In the hospitals and barracks of Scutari and Balaclava, in the crowded slums of London, she had come into close contact with prostitution. The quality of her mind, her common sense, her humanity freed her from contemporary prejudice. In 1862 she wrote to Edwin Chadwick that the causes of vice in the Army were not 'moral but physical'. 'They are,' she wrote, '1. Filthy crowded dwellings. 2. Drunkenness. 3. Ignorance and want of occupation.' The way to improve the soldiers' morals was to improve his living conditions. 'In civil life you don't expect that *every* workman who does not marry before he is 30 will become diseased,' she wrote to Douglas Galton in June 1861. 'In military life you do. Why? Because a workman may have occupation and amusement and consort with honest women. People always say a woman can't know anything about it. It is because I know more about the actual workings of the thing than most men that I cannot hold my tongue.' She advanced her views with such force that in 1862 she was officially invited to submit a paper, giving her first-hand experience in barracks and hospitals. Lord de Grey told Douglas Galton that he was shaken by her figures, but Mr Gladstone, though he said he should approach the question with circumspection, 'doubted the possibility of making a standing army a moral institution'.

She had made sufficient impression for the Government to proceed cautiously. In 1863 a committee was set up by the War Office to investigate the results of police inspection of prostitutes in what were termed 'protected' armies. The instructions for the committee were drawn up by Miss Nightingale, and she was invited to submit a list of suitable members.

Meanwhile the Army Medical Department was becoming more and more infuriated. Though Miss Nightingale's name did not appear, it was an open secret that she was the moving spirit. In July 1863 one of the few scurrilous attacks ever made on her came from a member of the Department, who wrote her anonymously she told Harriet Martineau, a letter of 'vulgar and indecent abuse'.

The suggestions put forward by the Army Medical Department were of incredible naïveté. One doctor told Dr Sutherland, 'quite gravely, that the only way would be to attach a certain number of these women to each regiment and place them under religious instruction'. Miss Nightingale remarked, 'the prostitutes who survive five years of this life should have good service pensions'. She had great faith in the effect of the inquiry made by the War Office Committee. She was wrong. Public opinion was not converted; the House of Commons was overwhelmingly in favour of police regulation and inspection; and in 1864 the Contagious Diseases Act became law. Miss Nightingale said that the War Office deserved the V.C. for their cool intrepidity in the face of facts.

The passage of the Act left her depressed. Was she succeeding to any degree, was she accomplishing anything at all? Ill, lonely, grief-stricken, toiling incessantly, living without the slightest relaxation, she was haunted by a settled conviction of failure. Any compromise was defeat. Everything was to be measured against perfection; if it fell short of perfection there was no good in it; and yet, if she would only see it, she was repeatedly presented with solid evidence of how much she had achieved.

In the spring of 1864 she was asked to supply information for a speech to be made in the House of Commons by Lord Hartington, the Under-Secretary for War, who was to defend the increase in the cost of the hospital and medical services for the Army, which had risen from £97,000 in 1853 to no less than £295,000 in 1864. Surely here was cause for congratulation; here was proof of extraordinary progress? She refused to be cheered. She sent Lord Hartington a detailed memorandum, setting out what the nation got for its money, which he used with success. She experienced no gratification and pronounced the speech to be very dull.

In the following year she had even more remarkable proof of

progress. Lord Panmure, now sitting in the House of Lords as Lord Dalhousie, made an attack on sanitary principles in general and sanitarians in particular; he attacked the Herbert Hospital as being 'all glass and glare' and providing only a fraction of the accommodation of his hospital at Netley; he attacked the wasteful system which Lord Herbert had inaugurated by paying attention to sanitarians; and, finally, he made a personal attack on Miss Nightingale herself. He 'could not help thinking that all these unnecessary knick knacks in hospitals were introduced partly from the habit, which prevailed at the War Office, of consulting hygienists not connected with the Army'.

She had prepared Lord de Grey with a reply, but defence was not necessary – the speech fell flat. Lord Dalhousie's attack on sanitary science was out of fashion. In eleven years so great a change had taken place that his voice seemed to come from a past age.

'Do not fear,' Lord Stanley wrote to Miss Nightingale on July 10, 1863, 'that Lord Herbert's work will be left unfinished: sanitary ideas have taken root in the public mind, and they cannot be treated as visionary. ... The ground that has been gained cannot be lost.'

If she felt any satisfaction, she did not record it. War Office affairs had become of secondary importance; memoranda, the drafting of minutes, warrants and instructions, the never-ending toil involved in the scrutiny of plans for barracks and hospitals had become sidelines. Another vast undertaking had come into her life which was overwhelming her and crushing her as no other previous labour had done – the Royal Sanitary Commission on the Health of the Army in India.

18

ONCE more, as Miss Nightingale approaches this enormous task, one is overcome by a feeling of hopelessness. It is too much to attempt; the labour is too gigantic, the questions involved too vast and too intricate; the difficulties of distance, language, communication must prove insuperable. She herself is now a bedridden invalid of over forty, shattered in health and overwhelmed with other work. Nevertheless, in 1862 and 1863 she reached the peak of her working life; her will conquered her physical disabilities; she drove herself to work as she had never worked before, and as, after this period, she was never able to work again.

The warrant setting up the Royal Sanitary Commission on the Health of the Army in India was issued on May 19, 1859. Sidney Herbert was then chairman, but he resigned when he became Secretary of State for War in June 1859, and was succeeded by Lord Stanley. However, he continued to be the moving spirit of the Commission and received daily reports from Miss Nightingale. The personnel of the Commission consisted of three sanitary experts, including Dr Alexander and Dr Sutherland, a statistician, Dr Farr and two members of the India Council. Unfortunately Dr Alexander died before the real work of the Commission began. After his death Miss Nightingale, although on friendly terms with all the members of the Commission, worked almost exclusively with Dr Sutherland and Dr Farr.

The Commission did not go to India but sat in London. The taking of evidence should have begun in November 1859, but owing to the great distances from which facts had to be collected the assembling of the preliminary data took more than a year, and the Commission did not begin to sit until the autumn of 1861. Sidney Herbert was then dead and the burden of the work fell on Miss Nightingale.

Sidney Herbert had left her a frightful legacy in the Indian Sanitary Commission. Without him, she repeatedly said, she would never have contemplated it. Lord Stanley, 'that noble and industrious lord,' was not altogether satisfactory to her. They were friends, but he was cool, 'singularly cool', one of his contemporaries called him, cautious and critical. His slowness to take action, refusal to be driven, threw Miss Nightingale into frenzies of irritation. She did not realize his value. The fact that the movement to improve Indian sanitation had behind it a man of Lord Stanley's known stability was in reality of the greatest assistance to her.

The whole undertaking bristled with difficulties. The situation in India was not favourable. In the daily life of India the passage of the Control of India Bill had made only a small upheaval. The power of the East India Company had long been a shadow; its Governor-General had become merely the agent through which the British Government found it convenient to rule. But in the administrative departments there was confusion. The final transfer of authority from the Great House, the headquarters of the East India Company, to the India Office was not accomplished without friction. The policy of the British Government of converting and absorbing the surviving administrative machinery of the East India Company resulted in the preservation of much that was inefficient and out of date. 'The state of the India Office is inconceivable,' wrote Miss Nightingale to Douglas Galton in 1862; 'just the routine of the old G.I.C. (General India Council) has survived the wreck – the worst part – they are just waddling on doing nothing.'

Difficulties were increased by the fact that the transfer was not yet complete. Many records were still at the Great House, and the representatives of the defunct Company were not always inclined to be co-operative. To compile satisfactory statistics as to the health and sickness of the troops it was necessary to obtain figures from two different sources – for the Queen's Troops and the Company's Troops; each source was resentful of the other.

Early in May 1859, before the warrant for the Commission was issued, Miss Nightingale had discovered that no satisfactory figures and records, on which she and Dr Farr could work, existed and that to obtain even ordinary documents relating to India in London was a hopeless task. She decided to obtain all her information at first-hand, and in consultation with Sir John McNeill and Sir Charles Trevelyan, at this time Governor of Madras, she drafted a *Circular of Enquiry* which was to be sent to every military station in India. She also wrote to 200 larger stations asking for copies of all regulations, including local

Regulations, relating to the health and sanitary administration of the army. Finally, she wrote individually to all military and medical officers of high rank in India with whom she was acquainted, or to whom she could obtain introductions, asking for their goodwill and co-operation.

As the reports returned from India, they were sent to Miss Nightingale, who analysed them, assisted by Dr Sutherland and Dr Farr. The task was colossal. Literally tons of paper were involved. When she took a house, the reports required a whole room to house them. When she moved, they filled two vans.

Eventually the Station Reports filled the second volume of the Indian Sanitary Commission's Report, a folio volume of nearly 1,000 pages in small type. 'The bulk of the Report,' wrote Henry Reeve, 'is truly appalling.' They are documentary evidence of immense value, providing a detailed picture, which is recorded nowhere else, of military life in India, both British and native, in the years immediately preceding and following the Mutiny. No official survey of India was undertaken until 1872.

As the analysis proceeded, it became clear that facts of such overwhelming importance were being disclosed that they must constitute the basis of the Commission's Report. Yet Miss Nightingale, who had originated the scheme and executed it, from the drafting of the questions to the analysis of the replies, was not a member of the Commission nor did she qualify as a witness. How was her work to be included? The solution proved the extraordinary place she had earned in official estimation. She was officially invited, in October 1861, to submit in due course Remarks' on the Station Reports, which should be signed by her and incorporated under her name into the report of the Commission. By August 1862 the analysis was complete, and she had written her remarks under the title *Observations by Miss Nightingale*.

After *Notes on Nursing*, *Observations* is the most readable of her writings. Ill, exhausted, and bereaved, she had never written anything with fiercer vitality. She intended *Observations* to be provocative. True, what had officially been requested – and was expected – was a convenient summary, but she perceived that the official request gave her a heaven-sent opportunity. The extreme difficulty of arousing public opinion on Indian sanitation was due first to blank ignorance, second to complete want of any source of information. *Observations* was designed first to startle and then

313

to inform. To make the text more interesting she introduced woodcuts, executed by Hilary Bonham Carter from sketches sent from India. Illustrations had never been used in a Blue Book before, and the Treasury was taken aback. An agreement was reached by which Miss Nightingale paid, not only for the drawings and the blocks, but for the extra expense involved in printing them.

She did not intend to allow *Observations* to be shelved until the publication of the Commission's report. She had a large number of copies privately printed, and during the autumn of 1862 sent them out, to the Queen – 'she may look at it because it has pictures' – to Harriet Martineau, to Cabinet Ministers and to the members of the India Council. The result was all she had hoped. 'Miss Nightingale's Paper is a masterpiece in her best style,' Dr Farr told Dr Sutherland on December 1, 1862, 'and will rile the enemy very considerable – all for his good, poor creature.' She told a frightful story with accuracy. 'The picture is terrible but it is all true,' wrote Sir John McNeill on August 9, 1862. 'There is no one statement from beginning to end which I feel disposed to question, and there are many which my own observation and experience enable me to confirm.'

Statistics showed that for years the death-rate of the British Army in India had been 69 per 1,000. 'It is at that expense,' wrote Miss Nightingale, 'that we have held dominion there for a century; a company out of every regiment has been sacrificed every twenty months.' This enormously high death-rate was not the inevitable result of the climate. The diseases from which troops died like flies were not specifically tropical diseases, they were the diseases which had produced the mortality in the Barrack Hospital, the result of overcrowding, lack of drainage, bad water, in fact, camp diseases rendered a hundred times more deadly in India by climate and the proximity of native populations living in conditions of appalling filth. At Agra, in the native city, the streets were impassable owing to heaps of refuse of the most disgusting nature, no latrines or privies existed – 'the population resort to the fields' – and corpses were buried within the compounds of houses. Native settlements had been allowed to spring up actually within British cantonments. At Bangalore there was a native population of over 100,000 within cantonments, at Kamptee of over 70,000, destitute of any means of sanitation, without latrines or privies, 'flinging all filth into the streets'. The bazaars were

314

'simply in the first savage state of social life'; they were allowed to spring up near barracks, and the stench from the bazaars affected the health of the troops.

Barracks were primitive in construction, built of lath and plaster with floors of earth varnished over with cow dung 'like Mahomet and the dung hill, if men won't go to the dung hill, the dung hill it appears comes to them'. The water supply was deplorable. Only two stations supplied a chemical analysis, one of which read 'like an intricate prescription', other stations contented themselves by describing their water as 'smells good' or 'smells bad'. At Hyderabad, in Scinde, the water 'visibly swarmed with animal life'. It was almost universal to use the same tank for drinking and bathing. At Bangalore the tank used for drinking was known to be the outlet for the whole drainage of a filthy bazaar of 125,000 inhabitants. The Commander-in-Chief wrote: 'The disgustingly filthy nature of the source from which the water used at Bangalore is taken has been brought to notice scores of times by me during the last $4\frac{1}{2}$ years but as usual nothing has been done.'

No drainage whatever existed 'in any sense in which we understand drainage. The reports speak of cess pits as if they were dressing rooms,' wrote Miss Nightingale. The Indian authorities were like the London woman who, when asked to point out the drains, said: 'No, thank God, Sir, we have none of them foul stinking things here.' Means of washing were practically nil, stations were either without lavatories or, if lavatories existed, they had no fittings. One station washed in 'earthenware pie dishes on a wooden form'. 'If the facilities for washing were as great as those for drink, our Indian army would be the cleanest body of men in the world,' commented Miss Nightingale. Drunkenness was universal and universally accepted. True, the sale of spirits to troops in the bazaars was strictly forbidden, but licences to sell spirits in bazaars were disposed of officially. 'The Govt sells licence to sell drink in the bazaar and orders men not to profit by it. The present law is like lighting a fire and then charging it not to burn anything.' A station was described as 'temperate' in which one man out of every three admitted to hospital was suffering directly from the effects of drink. No effort was made to prevent the soldier from drinking, the only question which appeared to interest the authorities was how he had better get drunk. He was encouraged to kill himself on spirits bought at

315

the canteen rather than on spirits bought in the bazaar. 'May there not be some middle course whereby the men may be killed by neither?'

The diet of the army was 'absurd'. The soldier received a standard diet at all stations irrespective of climate, altitude, or season. Means of cooking were inferior to the culinary amenities of Scutari, and the favourite situation for the kitchen was next door to the cess-pit.

Barracks were crammed with troops. One report stated that '300 men per room were generally accommodated without inconvenient overcrowding'. 'What is *convenient* overcrowding?' enquired Miss Nightingale. Married quarters were criminally inadequate. 'At Dum Dum 554 women and 770 children,' wrote Miss Nightingale, 'were crowded together while their husbands were fighting and as many died as in the Cawnpore massacre in the Mutiny.' The Principal of the Lawrence Asylums for children said that 'the children of British soldiers in the plains die so early that only one in five is found surviving the fifth year of residence there.'

Troops had no occupation, no means of recreation, and no opportunities for exercise. In the hot weather they were customarily confined to barracks from 8 a.m. to 5 p.m. At Cawnpore in the hot weather they were compulsorily confined to barracks for 10½ hours a day. They had nothing to do, nothing to read, nowhere to go; out of 24 hours they lay on their beds for 18.

Hospitals were inferior in construction and comfort even to barracks. Often they were merely sheds supported on poles. There was no system of hospital orderlies. Patients were washed and nursed by a ward coolie hired at 4 rupees a month who was not a soldier and, in a cholera epidemic, usually ran away. When a man was dangerously ill one of his comrades was sent for and then, in Miss Nightingale's words, 'The Regimental Comrade not knowing the language, nurses the Patient by beating the Coolie.'

The allowance of tubs and basins was 1 per 100 men. Washing was done by pouring water over the patients from a tin pot. There was no day-room for men well enough to get up. Convalescents spent 24 hours a day in bed. Privies were highly offensive and gangrene and erysipelas widespread. Troops would conceal their illness rather than go into hospital. 'You might say of them as of the Pasha's new fort on the Dardanelles. He would be much safer outside it than inside it,' wrote Miss Nightingale.

But bad as these conditions were they paled before what was

ndured by native troops. The native troops received no rations nd had no barracks, no lavatories, no baths, no kitchens, no anitary supervision of any kind. They lived in huts, used the round round them as privies without hindrance, and left cleaning to the rains. The squalor of their huts was indescribable, odies of animals and of human beings were left unburied for ays; the water they drank was stinking and they starved themelves of food to hoard their pay for their families. Consequently, hough temperate, they were decimated by disease.

To rectify these conditions involved a new problem. Of what se to improve and sanitate barracks and hospitals when next loor lay the bazaar and native city in all their filth? To improve onditions for the troops, the whole sanitary level of the country nust be raised. 'The salvation of the Indian Army must be rought about by sanitary measures everywhere,' Miss Nightinale wrote. The health of the army and the people of India went and in hand.

Another great mission had come to her. Work in military hositals had inevitably led her to civil hospitals; sanitary work for roops in India led her to work for the health of the peoples of ndia. By the time she had finished the report, she was as much oncerned with the one as the other. 'This is the dawn of a new ay for India in sanitary things, not only as regards our Army but s regards the native population,' she wrote to Harriet Martineau n May 1863.

Miss Nightingale proposed that a Sanitary Department should e established at the India Office to draw up and enforce a saniary code for all India. She did not succeed. The Sanitary Comnission on the Health of the Army in India was a military comnission, instructed to concern itself with military affairs. Between he civil and military administration in India there was friction; o suggestions for changes in civilian administration would be ntertained which emanated from a military Commission. She id, however, succeed in establishing some sanitary control at ome. Two Indian representatives were added to the Barrack and Iospital Commission, and its powers were extended to include ndia; as a result all sanitary works for the Army in India would ass through Dr Sutherland's hands. Miss Nightingale was easonably satisfied. The working Commission, she told Harriet Martineau, was 'not quite in the original form proposed, but in hat may prove to be a better working form because grafted on

317

what exists'. Lord de Grey was now Secretary of State for War and at the War Office the report would be received sympathetically. The sky seemed to be clearing, and Miss Nightingale relieved from the enormous burden under which she had staggered for two years, allowed herself a moment of elation. On May 19, 1863, she wrote to Harriet Martineau: 'I cannot help telling you in the joy of my heart that the final meeting of the India Sanitary Commission was held today – that the Report was signed and that, after a very tough fight, lasting over three days to convince these people that a report was not self executive, our working commission was carried.'

She was confident that the report would be read, not only by Members of Parliament, but by the public. So many copies of the Army Sanitary Report of 1857 had been sold that the Government had made a profit, and she anticipated that the sales of the Indian Report would be larger still. But disaster followed. Either a genuine mistake occurred, or, as she believed, she was the victim of a deliberate plot. It appeared that an attempt was made, if not to suppress the whole report, then at least to suppress the disclosures most unpalatable to the Government.

Unknown to her the Clerk to the Commission prepared a new and shorter edition: shorter because it left out the facts on which the report was based – the Station Reports, her abstract of the Station Reports and her Observations. The ground was covered by giving a *Précis of the Evidence*, executed with so little competence that reference was repeatedly made to passages in the sections which had been eliminated. This edition was to be the only one on sale to the public and was to be the edition presented to both Houses of Parliament. A thousand copies only had been printed of the original report, the type for the two enormous volumes having already been broken up. Even these copies were not obtainable. They were 'reserved' by the Government.

On July 20 Miss Nightingale wrote to Sir James Clark. 'There has been a perfect outcry (and I think a legitimate one) because the two folio book is not to be sold, not to be had, not to be published, not to be presented to Parliament, and that the 8vo makes references passim to a work which is not to be had.' She refused to believe that the affair was an unfortunate mistake. The Report contained too many inconvenient revelations. On August 11 she wrote to Harriet Martineau, 'the Government wished to suppress the sale'.

She could do nothing; it was impossible to set up and reprint the two enormous books. The position must be accepted, and she must set to work to see what could be salvaged from yet another wreck.

She discovered that the 1,000 'reserved' copies were to be had 'on application' by Members of Parliament. By an irony of Fate the address to which applicants were directed was the Burial Board, where the Clerk to the Commission happened to hold a post. She wrote to Members with whom she was acquainted begging them to apply and heard that for the first time in history there was a run on a Blue Book.

Next she published the *Observations*. Copies went out to India and were read by officials first with anger, then with conviction. Many years later Sir Bartle Frere was asked what had started the movement for sanitary reform in India. It was not the Blue Book, he said, which no one read, but a certain little red book 'which made some of us very savage at the time but did us all immense good.'

Finally, she persuaded Lord de Grey to allow her to rewrite the abridged edition. It was manifestly not only unjust but absurd that the only form of the report available should omit the evidence she had worked for years to obtain at the instruction of the Government. Might she not be allowed to prepare a new edition, with a *Précis of Evidence* accurately done, which could be circulated among army officers and officials; it was absolutely essential that these men should be correctly informed of the facts. Lord de Grey was doubtful – he was Secretary of State for War and did not wish to appear to be acting over the head of the India Office; but she persisted. 'Surely,' she wrote in August 1863, 'Sir Charles Wood [Secretary of State for India] will be very grateful to you for remedying his *mistake*.' The inevitable objection that the Treasury would not authorize the expense, she met by offering to pay the cost herself. With the issue of the revised edition all she could do on the report itself was finished. Its possibilities had been enormous, and it had been a crushing disappointment.

She began to work on setting up the administrative machinery which should put its recommendations into practice. For the moment all went well; an official despatch was sent to India recommending the formation of sanitary commissions in each Presidency; the two additional members to represent India were added to the Barrack and Hospital Improvement Commission;

and Miss Nightingale was asked to prepare a list of suggestions for sanitary improvements which might be sent to India and be the foundation of a sanitary code.

An outcry followed, furious opposition coming from official in the India Office, in the War Office, in India itself, and it became evident that sanitary reform would only be carried out in the teeth of resistance from every authority concerned. All through the summer Miss Nightingale stayed in London, ill, miserable and alone. Lord Stanley had gone to the country and would not be persuaded to return. The work of the Indian Sanitary Commission was at a standstill. All her gigantic labours seemed once more to be dissolving into thin air.

Suddenly, however, the scene changed. In October 1863 Lord Elgin, the Viceroy, was taken seriously ill. On November 20 he died, and Sir John Lawrence was appointed his successor.

The appointment of Sir John Lawrence as Viceroy of India opened a new period in Miss Nightingale's life. He was the first of a series of Indian officials of the highest rank who became her intimate friends, and through whose affection and admiration she gained an inside influence in Indian affairs approximating to some extent to the influence she had exercised at the War Office while Sidney Herbert was alive and in office.

Her position in Indian affairs was even more extraordinary than her position at the War Office. She had never been to India, she never did go to India, and yet she was considered an expert on India and consulted on its affairs by men who had lived there all their working lives. This knowledge was the reward of her enormous labours on the Station Reports. To her bedroom had come a return from almost every military station in India, not from one Presidency or one district but the entire Peninsula. Year after year she had toiled, examining, classifying, grouping. She possessed prodigious powers of absorbing, retaining, and marshalling masses of facts, and when she had completed her task the whole vast teeming country lay before her mind's eye like a map.

Sir John Lawrence had first called on her when she was in the midst of her work on the Station Reports in 1861. Both felt an instant attraction. He had striking personal beauty, immense height, curling golden hair, and flashing blue eyes. He was fearless, chivalrous, incorruptible, and deeply religious. Lord Stanley described his quality by saying that he had 'a certain Homeric simplicity'. His brother, Sir Henry Lawrence, also an Indian administrator of great ability, had conducted the famous defence of the Residency at Lucknow during the Indian Mutiny and been killed there on July 4, 1857. Miss Nightingale wrote that in the great mass of the Station Reports she could see improvement in the fearful state of cities and bazaars wherever the hand of the Lawrences could be traced.

The Lawrences came of an Ulster Presbyterian family, and John Lawrence was one of twelve children, all above the average in beauty and ability. His father had commanded the 19th Regiment of Foot, and he passionately desired to be a soldier. It was a crushing disappointment when he found he was to be taken into the East India Company's service with a civil and not a military appointment. 'A soldier I am and a soldier I will be,' he exclaimed, and he conducted his life with a soldier's fearlessness, directness, discipline, and austerity. Fate brought him the opportunity to display military genius. When the Mutiny broke out, he was Governor of the Punjab. His swift action and personal courage, his bold disposition of the troops in his province, and above all his popularity and influence with the native population, prevented upper India from rising.

In spite of a fiery temper – he had been seen hurrying out of church to belabour a disobedient servant – he was one of the band of Indian administrators who possessed deep sympathy with the native races. He was disgusted by the mass executions which followed the suppression of the Mutiny, and he repeatedly protested against the severities of certain military authorities. Miss Nightingale wrote to M. Mohl on January 1, 1864: 'His love of and trust in the native races – his fear and distrust of the British Military authorities are sad and remarkable because so true – at least I can vouch for cause for the latter. With great simplicity he implies that the natives are much more capable of civilization – even of sanitary civilization – than our Army authorities in India. He looks upon our occupation as a conquest and we as camped out all over India, having hitherto attempted little but Martial Law over the conquered country. You must not betray him. For I received a hint from Head Quarters to tell him to be more conciliatory with our Army.' When his appointment as Viceroy was announced, her delight knew no bounds. 'There is no more fervent joy, there are no stronger good wishes than those of one of the humblest of your servants,' she wrote.

It seemed that a golden age must be about to dawn, and even Lord Stanley allowed himself to be optimistic. 'Sir J. Lawrence's appointment is a great step gained,' he wrote to Miss Nightingale on December 1, 1863. 'I believe now there will be little difficulty in India.'

He went on to make a remarkable suggestion, the more remarkable coming from a man with his regard for official etiquette. The new Viceroy must be instructed in the Indian sanitary question; he wished him to learn, not from any official, but from Miss Nightingale. With a lifetime spent in India behind him, the Viceroy was to come to be taught by an invalid lady who had never been to India in her life. 'The plans are, in the main, yours,' wrote Lord Stanley. 'No one can explain them better; you have been in frequent correspondence with him. ... Your position in respect of this whole subject is so peculiar that advice from you will come with greater weight than from anyone else.'

Sir John Lawrence had only a week before he sailed, but he called on December 4, 1863. The interview, wrote Miss Nightingale, was one never to be forgotten. The Viceroy remained with her for several hours, the Indian Sanitary Report was discussed in detail, and he declared himself 'heart and soul for Sanitary Reform'. The machinery for putting the recommendations of the report into practice was to be set up at once: and *Sanitary Suggestions* were to be issued for the Government of India to work on immediately. The attack on the veracity and accuracy of the Indian Sanitary Report was dismissed as nonsense. 'Sir John Lawrence so far from considering our Report exaggerated, considers it under the mark,' she wrote triumphantly on December 10. At last it seemed that the mountain that was India was being moved.

In January 1864, assisted by Dr Sutherland, Dr Farr, and the celebrated civil engineer, Sir Robert Rawlinson, Miss Nightingale prepared her *Suggestions in regard to Sanitary Works required for the Improvement of Indian Stations*. It was the first sanitary code for India, the starting-point from which, she hoped, great new projects, bringing health and prosperity to millions, would be developed.

Once more hope faded. The *Suggestions* were sent to the War Office, but nothing happened. The Barrack and Hospital Improvement Commission inexplicably came to a standstill on Indian work. In April Miss Nightingale discovered the truth: the

War Office and the India Office had fallen out. The India Office did not intend to have action proposed to it by the War Office, and the *Suggestions* had been pigeon-holed.

At this stage Lord Stanley became annoyed. It was seven months since the Commission, of which he was chairman, had reported, and nothing had been done. He considered that a reflection had been cast on his work. He went to see Sir Charles Wood and promised to support him in case of any criticism in the House and Sir Charles Wood then accepted the *Suggestions*. The conflict in dignity between the War Office and the India Office was solved by a phrase on the title-page; though the Indian Sanitary Commission had been a War Office Commission, the title-page stated that the *Suggestions* had been 'prepared by the said Commission in accordance with letters from the Secretary of State for India in Council'.

As soon as the *Suggestions* were officially approved, Miss Nightingale had copies printed at her own expense, which she sent out to Sir John Lawrence. Delay continued in the official issue. The War Office and the India Office fell into an argument as to the number of copies to be printed; and a further two months passed before the official edition was ready. By this time Sir John Lawrence was having Miss Nightingale's advance text reprinted in India, and she wrote to Douglas Galton: 'It might be as well to hurry your copies for the India Office who will otherwise receive them first from India.'

Once more she was the mainspring of the work. No toil was so wearisome that she shrank from it, no detail too small to receive her attention. Across two continents her burning zeal infused Sir John Lawrence with new strength. Sir Bartle Frere, at this time Governor of Bombay, told her, 'Men used to say that they always knew when the Viceroy had received a letter from Florence Nightingale; it was like the ringing of a bell to call for Sanitary progress.' Her hand was everywhere. She had drawn up the instructions to the Presidency Commissions and widened their scope, so that they were not only to 'supervise the gradual introduction of sanitary improvements in Barracks, Hospitals and Stations' but also to improve the sanitary condition of 'Towns in proximity to Stations'. The *Suggestions*, the code by which the Commissions worked, were written by her; the report of the Commission itself was her work. Lying in her bed in London, she held the threads of a network which covered all India.

Again hope ran high. 'I sing for joy every day,' she wrote in June 1864, 'at Sir John Lawrence's Government.' In October 1864 in the full flood of optimism she wrote to him: 'I feel it a kind of presumption in me to write to you – and a kind of wonder at your permitting it. I always feel you are the greatest figure in history and yours the greatest work in history in modern times. ... You are conquering India anew by civilization, taking possession of the Empire for the first time by knowledge instead of the sword.'

It was a lyric rapture which, alas, bore little relation to reality. While the machinery of the Sanitary Commissions was being set up, all went well; President, Secretary, and Commissioners were appointed and coached in their duties, information was collected, and schemes formulated on paper. And then – nothing happened. The machinery was there – but it did not operate.

The truth was that Sir John Lawrence was not altogether successful in the office of Viceroy, which he had accepted solely from a sense of public duty. The personal courage and military genius which had saved the Punjab were not qualities which shone in the committee room. He had a violent temper, a rough manner, strong prejudices, a want of tact. The difficulties which faced him were appalling. Financially and administratively India was in fearful disorder. The year 1859–60 had shown a deficit of over forty million pounds sterling. The enormous increase in military expenditure, due to the Mutiny, seemed likely to increase and become an intolerable burden, since the extreme antagonism between Europeans and Indians prevented retrenchment.

The structure of government in India prevented the development of the country. The native merchant and trading classes, who alone could bring India prosperity, were depressed. The members of the Viceroy's Council and the Government officials in the Presidencies were almost without exception European officials. The tone of the Government was a military domination tempered by philanthropy. Interdepartmental jealousy was intense: the India Office hotly resented being given orders by the War Office, the Viceroy's Council in India resented being given orders by the India Office at home. Sir John Lawrence landed in January 1864; six months later Miss Nightingale told Sir John McNeill that he was 'discouraged by difficulties he could scarcely understand or anticipate'.

It became known that he was on bad terms with his Council.

The native population was distrustful. In June 1864 he described the feeling of hopelessness which overcame him when an attempt to improve the sanitation of Calcutta was construed by the Bengalis as an attack on the Hindoo religion. Miss Nightingale herself found the Indian mentality difficult to deal with. 'Nothing can give you any idea,' she wrote to Harriet Martineau in 1862, 'of the horrors of the disclosures as to the state of the stations which these Indians make themselves while declaring themselves to be 100 years before England.'

Delay succeeded delay; 1864 passed into 1865, and still nothing substantial had been done in India. Once more she was in despair. It was the old weary story of guerilla warfare, the odd point here, the odd point there, snatched from the India Office or the War Office at the cost of infinite toil. But where was the plan? The plan, the constructive campaign, had once more faded away.

19

Were the disappointments, the reverses she endured, the common lot of the reformer? She refused to admit it. She was convinced that she was singled out by a malignant fate. She herself, her work, everything she touched, was cursed. She beat herself against the callousness and indifference of the world; she exhausted herself in storms of resentment and despair. Such was the power of her extraordinary nature that in middle age the intensity of her emotion was unabated, and she continued to feel as violently, as blindly, as if she were a girl.

It was the old story, repeated now for more than twenty years. She must attain perfection, or she had failed. She must have everything, or she declared she had nothing. She refused to consider what had been done, only what had not.

Yet how much she had achieved was becoming plainer every day. Lord Stanley's coolness was notorious, but on July 25, 1864, he wrote to her: 'Every day convinces me more of two things; first, the vast influence on the public mind of the Sanitary Commissions of the last few years ... and next, that all this has been due to you and to you almost alone.' Jowett wrote on September 8, 1865: 'Considering what Ministers are, instead of wondering at their not doing all you want, I wonder at their listening to a word you say. A poor sick lady, sitting in a room by herself and they have only not to go near her and not to read her letters and there is an end of her. And yet you seem to draw them.'

She would not be comforted; she would allow no gleam of light to penetrate the darkness. Her physical condition was piteous — she could not move without assistance. 'I am so weak,' she wrote in a private note of 1864. 'Oh how weak I am!' Yet she drove herself mercilessly, and the demands of the work were becoming daily more enormous. She headed letter after letter, 'Written before it is light'. She wrote that she was like the washerwoman who said Heaven would be to have an hour a day with nothing to do. Among the Nightingale papers is a sheet on which she has drawn a large 'O!' and signed it 'F. Nightingale. Lonely and weak. March 10, 1866.'

More than health had been lost in these last years of incessant toil. Sometimes now she doubted. She had been able to be ruthless because she had been sustained by an unshakable conviction that what she did was right. Now she was not sure. A new note crept into her letters. Writing to Clarkey in May 1865 she said she felt like a vampire who had sucked Sidney Herbert's and Clough's blood. Writing to Jowett in July, she was humble and confused. 'You are quite right in what you say of me. ... I will try and take your advice. I have tried but it is too late. I lost my serenity some years ago, then I lost clearness of perception, so that sometimes I did not know whether I was doing right or wrong for two minutes together – the horrible loneliness – but I don't mean to waste your time.'

Once she had felt proud to be alone; now she dreaded the prospect of perpetual solitude. Her life was closing in. Every hour must be given up to work. Friends, affection, sympathy, all must be sacrificed. In the summer of 1865 Clarkey came over to London. She was Miss Nightingale's most intimate friend, and since Sidney Herbert's death they had drawn still closer. 'What I lost in Sidney Herbert you only (who lost M. Fauriel) can tell,' she wrote to her in 1862. Yet she could not spare time to see Clarkey. 'Clarkey Mohl darling,' she wrote on June 23, 1865, 'how I should like to be able to see you, but it is quite impossible. I am sure no one ever gave up so much to live who longed so much to die as I do. It is the only credit I claim. I will live if I can, I shall be so glad if I can't.'

The distinguished visitors who had called at the Burlington were no longer received. The Queen of Holland wished to call – 'I really feel it a great honour,' wrote Miss Nightingale in June 1865, 'she is a Queen of Queens. But it is quite impossible.' She would see no one who was not directly connected with her work. She made, however, one exception. In 1864 Garibaldi came to London; she had never forgotten her girlhood's passion for Italian freedom, and she had regularly sent money to his funds. Now she consented to receive him. On April 27 he came, using Sir Harry Verney's carriage to avoid notice. She was disillusioned. 'Alas, alas, what a pity that utter impracticability,' she wrote to Harriet Martineau on April 28. He was noble and heroic, but he was vague. He had, she wrote, no 'administrative capacity'.

He 'raved' for a Government like the English, but 'he knows no more what it is than his King Bomba did'. 'One year of such a life as I have led for ten years,' she told Harriet Martineau, 'would tell him more of how one has to give and take with a "representative Government" than all his "Utopias" and his "Ideal".'

Her consolation in her solitude was friendship with Jowett. By 1864 her intimacy with him was closer than her intimacy with any other human being, with the exception of Clarkey. They did not meet frequently, though he occasionally spent an afternoon with her when he was in London, but they exchanged many hundreds of letters. Nearly all Miss Nightingale's personal letters to Jowett, with the exception of a few rough drafts preserved among her papers, have been destroyed, but she kept most of his letters to her. Intense concern for her welfare, intense affection breathe from every line. He was devoted to her, and because he was devoted to her she would accept advice and even criticism from him. Jowett alone could tell her not to exaggerate, 'as you always do'; Jowett alone could tell her to be calmer, 'and don't try and move the world by main force'; Jowett alone could tell her not to despise people and, worse still, to let them see that she despised them; Jowett alone could scold her, and she would accept and almost enjoy the scolding. He had entered her life when she was crushed by the death of Sidney Herbert, and his remarkable powers of sympathy and understanding were offered her at a time when she was desperately in need of them. 'Mine has been such horrible loneliness,' she wrote to him on July 12, 1865. 'But how many women, maids of all work and poor governesses, have been lonelier than I – and have done much better than I. I think if I had had one friend – such a friend as you have been to me for the past six months. ...'

Friendship with Jowett took the only form in which friendship could have been fitted into her life. There was infinite solicitude, but there were no demands; there was constant and intimate communication, but the communication was by letter and brought no interruption to the routine of work. The friendship was eminently successful, as her friendship with Clarkey was eminently successful. Jowett was in Oxford; Clarkey was in Paris; she was in London. The problem of interference did not arise.

But even though Miss Nightingale had resigned herself to a 'desert-island' life she still had to find a place to live. Already her wanderings had been extensive. She had left 9 Chesterfield Street,

the 'fashionable old maid's house', in the autumn of 1862, faced
with a large claim for dilapidations which W. E. N. had eventually
to be asked to settle. She then went back to the Verneys' town
house, 32 South Street. In January 1863 she wrote to Sir Harry
Verney thanking him for 'a so comfortable three months in your
beautiful house'. The view down Park Street, where the houses
had stripes of yellow ochre and grass-green shutters, reminded
her of a French town, and even the public house at the corner
was, she said, 'a vestal for purity'. But the Verneys wanted to
come back to their house every year for the London season, and
she had to turn out, with all her papers and impedimenta – to
occupy the same house as Parthe was impossible. For a time she
had a floor in a hotel in Dover Street, but this proved expensive
and unsatisfactory, and in the spring of 1863 she took a furnished
house in Cleveland Row, St James's. In the summer she took
furnished rooms or furnished houses in Hampstead and Highgate
with varying success. She was an invalid, living alone without anyone to supervise her household, and servants were a difficulty.
Her life was unusual, her ideas of cleanliness and hygiene strange
to her contemporaries, and her standards exacting. 'Florence,'
wrote Fanny in 1864, 'has strange ideas about maids.'

Finding a house became a recurrent nightmare. Year after
year Fanny, Parthe, Beatrice Smith, Aunt Mai's daughter who,
since Parthe's marriage, had been living with Fanny, and the
devoted Mrs Sutherland 'panted' from agent to agent trying to
find a house that would satisfy her. She must be central, she must
be quiet, she had a morbid dread of being overlooked. In 1865
Mrs Sutherland sent a list of houses on which Miss Nightingale
scribbled comments. A house in Bolton Row which she knew
'looks as if it had been built for someone to do something wrong
in'. Park Street did not suit her – 'I stayed there once and was
perfectly sleepless.' Norfolk Street was 'too near Park Lane and
the grinding of luggage vans. Beautiful and noisy.' Every year the
problem arose, every year there was a house to be found, servants
to be engaged, arrangements to be made for moving in; every
year when she moved out there were dilapidations to be settled,
invariably with a dispute, because she spent her life upstairs and
did not know what her servants did downstairs. For some time
Fanny had been trying to persuade W. E. N. that the only solution was to buy a house, but he was unwilling. Fanny was extravagant; Embley and Lea Hurst were expensive to keep up; he was

irritated to find himself continually short of money. Yet Florence's accommodation cost more every year – the possibility that she could live on an annual income of £500 was never mentioned now. At last W. E. N. was forced to the conclusion that a permanent house might be an economy. On July 4 he wrote to Fanny: 'Saturday she goes to Ld Digby's big house at Hampstead £11 a week. 34 South Street, £500 the year, Hampstead if she stays there three months £120. Parthe urges me to offer £7,000 for 35 South Street, you are in the same vein. There is nothing for it but to say I will give £5000, you adding your money. Of course I shall consider the money sunk from the time I produce it and shall hope NEVER TO HEAR ANY MORE OF IT. Is it not of course too that the £7000 will not be accepted?'

The £7000 was accepted, and at the end of October, 'with everyone's united efforts,' wrote Fanny, Miss Nightingale was moved into No. 35 South Street, which henceforward became her home. The house suited her admirably; it was central, manageable, and backed on to the gardens of Dorchester House. She was able to enjoy fresh air, sunlight, and trees, and to observe birds, of which she was passionately fond.

A few years later the street was renumbered: No. 35 became No. 10, and at No. 10 South Street she continued to live until her death.

No further attempt was made to provide her with a companion. She moved into her new house to live alone. And, as she moved in, the spectre of desolate solitude, which had crept into her life as one by one her friends were taken from her, stalked nearer. The sword which seemed to hang over all those she loved fell again, and she sustained another great grief.

Of the friends of her youth she had loved none more than Hilary Bonham Carter. Soft, loving, a pleasure to look at and highly gifted, Hilary possessed a remarkable power of inspiring affection. 'I was so attached to her,' wrote M. Mohl on September 9, 1865, 'I have never known anyone so made up of kindness.' But Florence disapproved of Hilary, and in 1862 she had sent her away. Clarkey implored her to take Hilary back. 'My dearest,' she wrote on February 5, 1862, 'if she is as useful to you as a limb, why should you amputate her? ... the thing she likes best in the world is being with you and being useful to you ... I agree with you, she ought to do for herself, but I am not sure her nature can bear it. I give it to you as a problem, think on it.'

To make Hilary happy by giving way to her was criminal to Miss Nightingale. Clarkey could write, 'I can't alter Hilly, I can only give her a little enjoyment'; but Miss Nightingale despised such enjoyment. She refused to have her back, and Hilary continued to be maid-of-all-work to her family. In the late spring of 1862 her health took a sharp turn for the worse, and Clarkey, who saw her in the summer after an interval of nearly six months, noticed a great loss of power in her. She no longer made any attempt to paint or draw. She was in her sister-in-law's house. She had got up before six, to give her brother breakfast before he went to Scotland, though the house was full of servants; she came in worn out during the afternoon, and instead of resting insisted on going out to buy flowers and arrange them before dinner. 'Hilly is devoured by little black relations just like Fleas,' Clarkey wrote on August 19.

The break with Miss Nightingale was complete. 'I see nothing of Hilary,' Miss Nightingale wrote to Clarkey on February 14, 1863. 'I believe the fact is she *cannot* see me without an appointment and to her everything is possible but keeping an appointment.' In April 1865 Hilary broke down, was examined by a specialist, and said to have a tumour. In fact she was dying of cancer. On May 16, 1865, Jowett wrote to Miss Nightingale: 'She seemed to be at peace, but she had suffered greatly – poor thing – she said that she had mental trial in past times and that, she found, had been alleviated by trusting in God, but she did not find that physical suffering could be similarly alleviated.' On September 6 she died. 'Hilary was released this morning at half past eight ...,' Miss Nightingale told Clarkey. 'Oh dearest how she had suffered!'

Throughout the summer, while she was in London, solitary and overworked, Hilary's terrible illness had been preying on Miss Nightingale's mind. On Hilary's death she broke into frenzy. Rage seized her, made up of resentment, anger, and despair – despair at Hilary's wasted talents, at Hilary's wasted life, at the stupidity and indifference of Hilary's family, at the system of family life which permitted such things to be. On September 8 she wrote Clarkey an immense, furious letter: '... There is *not a single person*, except yourself, who does not think that Hilary's family were quite right in this most monstrous of slow murders – and all for what? There is something grand and touching in the Iphigenia sacrifice and Jephthah's daughter. But, if

331

Jephthah had made his vow to sacrifice his daughter to feed his pigs it would only be very dirty and disgusting. And I say, the Fetichism to which Hilary has been sacrificed is very dirty and disgusting. . . . I shall never cease to think as long as I live of you and M. Mohl as of Hilary's *only* friends. The golden bowl is broken – and it was the purest gold – *and the most unworked gold* – I have ever known. I shall never speak of her more, I have done. . . . How I hate well meaning people.'

Frantic and distraught, she refused to be comforted. Jowett tried to reason with her and in two letters written consecutively on September 7 and 8 besought her to be calm – 'don't look upon her life as wasted . . . we cannot quite tell what is lost or wasted in this world. Very few persons can ever be expected to carry on their lives on a systematic plan and persons who have great gifts least of all. (Poor thing, she must have passed through a real fire.) . . . I am very much grieved to hear you have been, during the last two months, much worse. . . . I am afraid you have been "agonized" by poor Miss Carter's state. . . . I think there are very few such people in the world, so mild and good and with so much intelligence. Do you think they really did suck the marrow out of her life? . . . I thought she had always a look of physical exhaustion.'

Hilary's death had opened old wounds. Miss Nightingale brooded over her relations with her family and the injuries she had received from them. Solitary, feverish, miserable, she revived old unhappiness and felt them as keenly, as implacably, as if nearly twenty years had not elapsed. She recalled to Clarkey how her family divided her asunder, 'soul from spirit, bones from marrow'. They had always been indifferent to her feelings; no member of her family had enquired after Sidney Herbert all the time he was ill, or after Clough all the time he was ill, though they knew she was distracted with anxiety. Even her cat, she told Clarkey on September 17, 1865, 'seemed to know something was wrong the day Hilary died and sat with his arms round my neck. Whereas Parthe, e.g. has never said one word but of hardness to me in all my trial years – my 4 years.' Clarkey remonstrated. 'Do not dearest dwell on the failings of your poor people,' she wrote in September; '. . . you let their want of comprehension corrode and eat into you. If you would but consider that it is not in the power of people to understand more than they can . . . do dearest be indulgent. I don't care so much about Parthe, tho' I am always charmed to be with her, but your father and mother are so fond of you, it is not their fault if they have not understood you.'

She would not be placated. To Clarkey and to Jowett she poured out letter after letter in which Hilary's death, her own past sufferings at the hands of her family, and Parthe's present behaviour united into one furious resentment. She complained that Parthe got people to write letters to her 'saying the hardest things,' that when she and Parthe met everything she said was 'distorted', that Parthe made trouble for her with Sir Harry Verney.

Hilary's death was succeeded by another blow, the death of Lord Palmerston on October 18. Miss Nightingale regretted him deeply. But personal regret was secondary to the serious loss she sustained in no longer having Palmerston to help her in her work. On October 18 she wrote to Dr Walker, the Secretary of the Bengal Sanitary Commission: 'He may be passing away even at this moment. He will be a great loss to us. Tho' he made a joke when asked to do the right thing, he always did it. No one else will be able to carry the things thro' the Cabinet as he did. I shall lose a powerful protector. ... He was so much more in earnest than he appeared. He did not do himself justice.'

Through the winter of 1865, her first winter in her own house, she was miserably ill and miserably unhappy. Death and failure, failure in India, her loss of power at the War Office combined to crush her. 'I just keep my place on at the War Office by doing all their dirty work for them, i.e. what they are too cowardly to do for themselves – *les lâches*,' she wrote to Clarkey in March 1865. She had been bed-ridden for four years, and she began to have severe pains in her back, described as 'rheumatism of the spine'. She was ordered to have a 'rubber', a masseuse, three times a week, but the state of her nerves made treatment difficult. The 'rubber' was forbidden to speak. She must come in noiselessly, do her work, and withdraw without speaking to Miss Nightingale on any pretext. The treatment had little effect, and she passed the winter and spring in constant pain. 'Nothing did me any good,' she wrote to M. Mohl in July 1866, 'but a curious little new fangled operation of putting opium under the skin which relieves one for twenty-four hours – but does not improve the vivacity or serenity of one's intellect.' Yet for the first time for many years her horizons were widening to include something beyond her work. During the long winter nights, lonely, sleepless, and in pain, she had once more begun to read and, encouraged by Jowett, she turned once more to Greek. She began to resign herself to the fact that the ideal companionship, the ideal sympathy, for which she so passionately longed, were never to be hers, and to make the best of such materials as life offered her.

Her emotions found an outlet in affection for her cats. Their soft silent movements never shocked her nerves, and she felt they gave her sympathy. 'Dumb beasts observe you so much more than human beings and know so much better what you are thinking of,' she had written to Fanny in 1862 when her grief for Sidney Herbert and for Clough was still fresh.

However busy and over-driven she might be, even during the gruelling hard work on the Indian Sanitary Report, she found time to write long letters, to Clarkey and M. Mohl, about her cats. One was a tom-cat named Thomas. 'I do not like to give Thomas away,' she wrote to Clarkey in 1861. 'He is stupid, ignorant, thievish and dirty. I don't think anyone would keep him and be kind to him as we are. He is so handsome that people in this country come from afar to see him. It was a shock to find he was not to be contented with one wife (*et soror et conjux*). Pussy had four kits more beautiful than herself, of which Thomas killed one and then hit his eldest daughter a tremendous whack on the side of the head, which she survives however. ... The only reason which makes the servants bear Thomas is that they think Pussy would pine without him.' Thomas was sent away and Miss Nightingale told Clarkey: 'He did nothing but disgrace himself. He got up the chimney, had to be put in the wash tub and when sent out into the garden at night was not favoured by the Phoenician Venus.' She had to have him back.

She mated her cats carefully and gave kittens away as a mark of special favour to carefully selected homes. In February 186 she wrote to M. Mohl: 'These cats are so capricious. The little cat wanted to get married. So I presented to her the two greatest partis in England, Messrs Bismarck and Benedek. She would have neither of them. Now she wants me to invite a hideous low cat out of the street – I won't. But I said she might go out if she liked. But she is too shy.'

She worked with a cat 'tied in a knot round her neck'. As many as six cats wandered at will about her room and made 'unseemly blurs' on her papers. On many of her letters and drafts is still to be seen the print of a cat's paw. She amused herself by playing with them and described to Clarkey her efforts to teach one of her kittens to wash itself; and the kitten saying 'what an awkward great cat that is'.

The winter of 1865 passed. It was two years since the Report of the Indian Sanitary Commission had been issued, and nothing had been accomplished. The work of pushing on without result, of enormous labours which perpetually came to nothing, was infinitely dreary, infinitely exacting; and the strain on her was increased by the irritation of her relations with Dr Sutherland.

As they worked together year after year, she became more and more dependent on him and found him more intolerably provoking. She admired the ability, but she disliked the man. 'I know he is your pet aversion as he is mine,' she wrote to Clarkey in 1862. 'Don't believe what Sutherland tells you he told you ... he only does it to annoy me,' she wrote to Douglas Galton in 1862, 'You know how queer he is.' As she grew more exacting, Dr Sutherland became more elusive. He lived at Finchley, which Miss Nightingale thought sufficiently distant, but in 1865 he moved even further away to Norwood. At Finchley he had been fond of gardening; at Norwood he had a garden of considerable size which absorbed a great deal of his time. 'I find – I don't know whether you find – it more and more difficult to rouse Dr Sutherland to do the work we *have* to do,' she wrote to Douglas Galton in 1866. 'He has always some *pond* to dig in his garden. Confound that Norwood.' Through the years of crushing labour on the Indian Sanitary Report she railed at him with increasing bitterness. She complained of his 'incredible looseness of thought and recklessness of action'. She scolded him for losing papers – 'it is as I thought, Sutherland took my copy of the Army Medical Schools Report and now he can't find it.' Sometimes he would not work, and she was infuriated. 'I could not get Sutherland to do a thing yesterday,' she wrote to Douglas Galton in 1866. 'He was just like one possessed.'

She had always been irritated by his deafness. He was becoming more deaf as he grew older and always found it totally impossible to hear anything if he were scolded. As her health deteriorated, speaking loud enough for Dr Sutherland to hear exhausted her. From 1864 onward she developed a system of communicating with him by scribbling on any piece of paper which happened to be handy; literally hundreds of such scribbles are preserved, written on odd scraps of paper of all descriptions, from the margins

335

of letters to pieces of blotting-paper. 'What *have* you done?' she wrote. 'You said you were going to lay it before your Committee, you had much better lay it before *me!*' 'Well I don't suppose the man will hurt you.' 'My dear soul! It's rather late for this.' 'There's fish for you at one.' 'You know they *could* only have let you out because you were incurable.' 'Write that DOWN.' 'Why did you tell me that tremendous BANGER.' 'Which means nothing but that you're too lazy to look at it.' 'You've looked at it? For five minutes on Wednesday.' 'What has become of the 8 copies of the Indian Report? Where is Barbadoes? Where are the three Registrar General papers.' On one occasion Dr Sutherland tried to be reconciled with her and she wrote, 'I *won't* shake hands until the Abstract is done.'

Irritated by him as she was, their intimacy was very close. No other person was part of her daily life as he was part. She saw him, however ill she might be, and he acted as host in her house. Many notes refer to visitors. 'These two people have come. Will you see them for me? I have explained who you are.' 'Was the luncheon good?' 'Did he eat?' 'Did he walk?' 'Then he's a liar, he told me he couldn't move.'

In the autumn of 1865 there was a serious quarrel. She was told by Douglas Galton that it was extremely probable that Dr Sutherland would be invited, as representative of the Barrack and Hospital Commission, to go to Algiers, Malta, and Gibraltar to investigate recent cholera epidemics. She became frantic. 'For God's sake,' she wrote to Douglas Galton on December 15, 1865, 'if you can, prevent Dr Sutherland going, he is so childish that if he heard of this Gibraltar and Malta business he would instantly declare there was nothing to keep him in England.' She had pledged her word to have the Indian reports and abstracts ready before Parliament met after the Christmas vacation – 'a thing I should never have done if I thought Dr Sutherland was to be sent abroad.' Dr Sutherland was offered the appointment, and in spite of all her entreaties he went. She was furious. On January 19, 1866, she wrote to Dr Farr: 'Dr Sutherland has been sent to Algiers, and I have all his business besides mine to do. If it *could be done* I should not mind. I had just as soon wear out in two months as in two years, so the work be done. But it can't. It is just like two men going into business with a million each. The one suddenly withdraws. The other may wear himself to the bone but he can't meet the engagements which he made with two.' Dr

Sutherland was aware he was in disgrace. 'I have been thinking,' he wrote to her from Algiers on January 28; 'will she be glad to hear from me? Or will she swear?' In spite of Miss Nightingale's anger he continued during 1866 to leave London for weeks at a time and owing to his absence in the Mediterranean Miss Nightingale was without him during an important crisis in Indian affairs. If Dr Sutherland had been in London, she was convinced the outcome would have been very different.

At the beginning of 1866 the Indian outlook brightened. Delay in carrying out the recommendations of the Commission was due to the fact that the Sanitary Commissions set up in the Presidencies were subordinate to their local Governments. Miss Nightingale had written to Sir John Lawrence urging that the Sanitary Commissions should be transformed into a public health service, standing on its own feet, responsible directly to the Viceroy and the Viceroy's Council, and kept active by a complementing department of experts at the India Office in London. On January 19 he replied that he agreed to the necessity for reconstructing the sanitary organization and had written a dispatch to the Secretary of State for India requesting a scheme. He did not send her a copy of the dispatch as he assumed she would see it. A week earlier Sir Charles Wood, Secretary of State for India, had been succeeded by Lord de Grey and with Lord de Grey at the India Office it seemed that the establishment of an efficient sanitary administration in India was certain. She asked Lord de Grey to send her a copy of the dispatch but he replied that he had received no such dispatch; it would come in by the next mail, no doubt. Several mails came in, but no dispatch appeared. Meanwhile it became evident that the Government was on its last legs. Within the next few months, even within the next few weeks, it would fall, and Lord de Grey would be Secretary of State for India no longer. She became frantic. March went by; April went by. The Government tottered nearer its fall, and still no dispatch appeared. Miss Nightingale scolded, implored, threatened, but without result. As a final resort Lord de Grey made a personal search, and the dispatch was discovered. 'At last the Sanitary Minute has been found,' he wrote on May 5, 'it was attached to some papers connected with the Finance Department and so escaped notice.'

She was ill and Dr Sutherland away, but by May 7 she had managed to submit a draft scheme to Lord de Grey. The Government was now on the verge of disaster and Lord de Grey harassed and distracted by party business. It was not until June 11 that she was able to extract instructions from him to proceed further. He then requested her to complete the scheme and add a survey of the sanitary question. It was a formidable task, but by dint of further desperate efforts she completed it and sent it to Lord de Grey on June 19.

She was twenty-four hours too late. On the previous day the Government had been defeated and had fallen.

The blow was crushing. 'I am furious to that degree,' she wrote to Douglas Galton on June 23, 'at having lost Lord de Grey's five months at the India Office that I am fit to blow you all to pieces with an infernal machine of my own invention.' In a letter to M. Mohl on July 12 she said she had come to the end of her endurance. She had lost the opportunity of establishing a public health service in India 'by twenty four hours!! I am well nigh done for. Life is too hard for me.'

The Tories were now in power, and Miss Nightingale was pushed further outside Government matters. On March 20 she wrote to Clarkey: 'While Sidney Herbert was alive I made most of the appointments. This is no bray. .... *Now* if you can fancy a position where a person can do *nothing* directly, nothing but by frightening, intriguing, "soaping" or going on all fours, that position is mine.' A letter she wrote to Douglas Galton on June 27, 1866, showed how conscious she was of being outside Government circles: '... now do write to an agitated female F. N. about WHO is to come WHERE. Does Gen. Peel come to the War Office? If so, will he annihilate our Civil Sanitary element? Is Sutherland to go all the same to Malta and Gibraltar this autumn? Will Genl Peel imperil the Army Sanitary Committee? I MUST know – ye infernal powers! Is Mr Lowe to come into the India Office? It is all unmitigated disaster to me. For, as Lord Stanley is to be Foreign Office (the only place where he can be of NO use to us), I shall not have a friend in the world. If I were to say more I should fall to swearing. I am so indignant – ever yours furiously. F. N.'

Meanwhile Parliament rose, and everyone of importance left London. Nothing could be done until the autumn, and again Miss Nightingale must resign herself to inactivity. Jowett begged her to visit her parents. She had not been home for nine years. Fanny was seventy-eight, it was feared that her eyesight was failing, and she recently had been in a carriage accident which had left her bruised and suffering from shock. In August, when the Nightingales invariably went north to Lea Hurst, she was not well enough to travel. It was essential that W. E. N. should go to Lea Hurst, and Miss Nightingale agreed to go home to be with her mother.

Fanny in her old age was to be pitied. Parthe had been her companion, but Parthe was now immersed in her own life. She was mistress of the historic mansion of Claydon and of a house in London. Sir Harry Verney, who was in Parliament, took an active part in public affairs, and Parthe had become 'very much the fine lady'. She was achieving success as a hostess, and she was writing novels, one of which was published in 1865 in the *Cornhill*. The admiring cousins and maiden aunts she had once patronized complained she had dropped them, and she did not often go to Embley. Fanny had Beatrice Smith, one of Aunt Mai's daughters, as a companion, Fanny's eyesight was now so bad that she needed constant help, and she was paid long visits by her sister, Julia Smith. Aunt Julia, like Hilary Bonham Carter, had never married and had spent her life looking after other people's families, but while Hilary was soft and soothing Julia was temperamental. Miss Nightingale as a girl had christened her 'the stormy Ju'. In her old age Aunt Julia found herself homeless and unwanted and was subject to fits of hysterical depression during which she wept for hours on end.

Elaborate arrangements were made to receive Miss Nightingale. She travelled in an invalid carriage, and at Embley six rooms were given up to her. Her rooms and her way of life were to be sacred. She worked incessantly, saw no one, and never left her room except to visit her mother.

The first reunion with Fanny was affectionate, but even now she would not give way to her mother. Fanny might be seventy-eight and almost blind, but she was not to be indulged. Miss Nightingale still implacably disapproved of the way in which Fanny frittered away her life. On August 21, 1866, immediately after her arrival, she wrote a long letter to Clarkey. She began tenderly enough: 'I don't think my dear mother was ever more touching and interesting to me than she is now in her state of dilapidation. She is much gentler, calmer, more thoughtful. ...' But as she proceeded, irritation and disapproval crept in, and at the end of the letter she wrote sharply, 'I can't think Mama much altered except her memory ... and her habits which have become worse, till now she is seldom up until 5 or 6 p.m. and then goes out in the carriage.'

Miss Nightingale was not an easy visitor. She required a great deal and was critical of the way in which her requirements were met. To argue with her was forbidden. Aunt Mai, writing to Parthe during the summer, repeated a letter from her daughter Beatrice who was acting as Fanny's companion: 'In *confidence*. Beatrice finds her concerns with Flo extremely difficult. Beatrice does her best but it is very difficult to explain anything to Flo because of her health. Her heart ... may *snap* at any extra effort or excitement. Her feeling for Beatrice partakes of the displeasure so often felt when something is done which she thinks might be better done.'

Only children were allowed to break into Miss Nightingale's solitude. Fanny kept up her custom of having children to stay through the summer months. During August Fanny wrote to Parthe: 'I thought our poor F. both excited and exhausted on Sunday, but to-day she has accepted our beautiful baby, who marched into her room, all alone with a flower in each hand, and played upon her bed for ¾ of an hour.'

When she returned to London, affairs in India could hardly have been more discouraging. Sir John Lawrence's dispatch on sanitary organization still lay unanswered at the India Office, and in India what amounted to the abolition of the Sanitary Commissions was being proposed. Their place was to be taken by a single 'Sanitary Officer' in each Presidency, who was also to be Inspector of Prisons; the Inspector-General of Prisons was to become 'Sanitary Commissioner to the Government of India'. It was difficult to see how, after professing to be convinced of the necessity for a public health service on the lines laid down by Miss Nightingale, Sir John Lawrence could have agreed to a scheme which made sanitary administration a sub-department of the prison department. She was bitterly disappointed – very far away were the days when she had sung for joy at Sir John Lawrence's government. Dr Sutherland frankly said Sir John Lawrence was hopeless. 'He is our worst enemy,' he wrote in 1866; and he advised her before she attempted to do anything further to wait until Sir John Lawrence's term of office ended in the following year. Douglas Galton urged her to approach Sir Bartle Frere, a well-known Indian administrator, who had just been appointed to a seat on the India Council. Before she had time to write, he had asked permission to call, and on June 16 she wrote to Douglas Galton: 'I have seen Sir Bartle Frere. He came on Friday by his own appointment. And we had a great talk. He impressed me wonderfully. I hope Sir B. Frere may be of use to us.' The friendship became one of the closest of her life and for the next two months Miss Nightingale and Sir Bartle Frere met almost daily. 'I need not tell you how entirely my services are at your disposal,' he wrote after their first interview.

Sir Bartle Frere, like John Lawrence, possessed deep sympathy with the Indian character. His outstanding achievement was his administration of the province of Scinde. In eight years the revenues were practically doubled, 6,000 miles of roads were built, the construction of railways was begun, the first postage stamps ever used in India were issued, and so loyal was the province that he was able, when the Mutiny broke out, to hold Scinde with only 178 European soldiers. In 1862 he was appointed Governor of Bombay. Here he demolished insanitary buildings, introduced scavenging services, established a town council, founded a school for the education of the daughters of Indian gentlemen, and opened Government House freely to Indian gentlemen and their wives as well as Europeans.

On July 24 Miss Nightingale wrote to Douglas Galton: 'If only we could get a Public Health Department in the India Office to ourselves with Sir B. Frere at the head of it, our fortunes would be made.' She did not go to Embley in the summer of 1867. There was fresh hope of progress, and she stayed in London. Meanwhile in the summer of 1867 Sir John Lawrence threw another project into confusion. Three years ago he had asked Miss Nightingale to draw up a scheme for the employment of nurses in hospitals by the Bengal Sanitary Commission. The difficulties were enormous, largely owing to the very poor level of the Indian Medical Service. Miss Nightingale made notes of an interview on the scheme with a doctor holding the high rank of Deputy Inspector-General in Madras. 'He came intent on proving to me that no matron was wanted. The Dr ought to be Matron and wretched coolie women under the Dr nurses. But luckily for me, he was drunk. And before he went (he was here 2½ hours) he had admitted everything. He described his lying-in hospital where "the pupils deliver all the ordinary cases, without a midwife, and without a doctor. The Dr comes in for the extraordinary cases." ... I said "They are by to see the extraordinary cases delivered?" Here he got so drunk that he spent at least half an hour explaining to me that there was nothing to be seen that "everything was under the bed clothes". The lying-in patients were all fed by friends from outside and were always naked in bed – except the Europeans. There were 80 leper beds at the General Hospital, always full. He says "It all answers very well!!"'

The attempt to employ female nurses must be made with the greatest caution, and she submitted a scheme for employing nurses in a single hospital and observing the result before embarking on any large undertaking. Sir John Lawrence turned the scheme over to the medical service, who blew it up into a grandiose plan for introducing female nursing on a large scale into seven hospitals simultaneously at great expense. This scheme, not Miss Nightingale's, was submitted to the Government of India, who, she said very properly, rejected it. She was angry. It seemed that Dr Sutherland was right and that Sir John Lawrence was her worst enemy. She decided to approach Sir Stafford Northcote, the new Secretary of State for India. He had indicated that he was prepared to hear from her; and on August 19, while she was debating whether the time was ripe to suggest a meeting, he wrote suggesting that he should call on her at South Street 'for a little conversation'.

She was nervous. 'Hope was green and the donkey ate it (that's me),' she wrote to Douglas Galton on July 16. Nevertheless she determined to be bold. If she made an impression on Sir Stafford Northcote, she would ask, then and there, for the establishment of a department at the India Office with Sir Bartle Frere at the head of it to control the sanitary administration of India.

On August 20 Jowett wrote her a cautionary letter. 'I am delighted to hear you are casting your toils about Sir Stafford Northcote. May I talk to you as I would to one of our undergraduates? Take care not to exaggerate to him.'

The interview had already taken place. Sir Stafford had called at South Street on the day Jowett wrote, and the meeting had been a triumphant success. When Sir Stafford had gone, Miss Nightingale scribbled a note for Dr Sutherland. 'Well – I've won this. We are to have a department in the I.O. for Sanitary business. I don't know if he saw how afraid I was of him. For he kept his eyes tight shut all the time. And I kept mine wide open. .... I *liked* Sir Stafford Northcote.'

342

Once more the establishment of a public health service for India seemed just round the corner. 'We will make 35 South Street the India Office till this is done,' wrote Sir Bartle Frere.

The phrase touched a chord. Ten years ago when Sidney Herbert was alive, the Burlington Hotel had been called the little War Office, and Miss Nightingale scribbled on Sir Bartle Frere's letter 'I miss him so'. The wound had never healed. She passed each anniversary of his death, 'that dreadful day', in meditation and prayer. On August 2 she dated her private letters '6 years ago', '7 years ago'. On this August 2, 1867, she wrote in a private note: 'I feel myself not only a shattered wreck of what I was but a phantom among phantoms. ... We do not know until the crash we are so many phantoms.' Her grief was tenacious; she refused to be resigned, as she refused to accept compromise. To be resigned, to compromise was to accept the second best. That she would never do, and she kept grief with her. Grief was waiting for her when she returned from her work, grief with grief's companions – frustration, resentment, and remorse. And yet – if she had ceased to feel, she would have slackened. As great waves of resentment and grief surged up in her, she worked harder, vowed more furiously that she would never give way, never succumb to the low standards, the inefficiency, and the indifference of the world.

In the autumn of 1867 she did some of the hardest work of her life. In October Sir Stafford Northcote wrote that he had 'cleared the decks' and was now ready to give full attention to sanitary matters. On October 23 he came to see her again. 'He comes,' she wrote in a note to Dr Sutherland, 'to be coached in the best way to announce the Sanitary Committee to the War Office. Now you're going to be good aren't you, for I *am* tired and feverish.' The second interview was even more successful than the first. The names of the members of the Indian Sanitary Committee were agreed, and Sir Stafford consented to establish the authority of the Sanitary Committee as supreme in India. Further, he asked her to prepare a digest of the progress of the whole Indian Sanitary question from the setting up of the Sanitary Commission in 1859 to 1867.

She plunged into work at once. By the beginning of December she had completed the instructions for the Sanitary Committee, the Digest, and had added a Memorandum of Suggestions and

Advice. In addition, on her own initiative, she had drafted an important dispatch requesting a report on sanitary progress with particular reference to the *Suggestions in regard to Sanitary Works required for the Improvement of Indian Stations* which had been sent to guide the Indian authorities as long ago as 1864. What was the present position, the Secretary of State for India wished to know? what results had been achieved? She submitted her draft with temerity – she was fully aware, she wrote, that Sir Stafford might disapprove the whole scheme. Sir Stafford accepted the suggestion almost in its entirety. The dispatch was sent; and the reports received from the Presidencies as a result were printed as a Blue Book in 1868 under the title of the *India Office Sanitary Annual*. In future reports were to be sent in by the Presidencies and published every year.

At last she had accomplished something. She had secured a Sanitary Department in the India Office with supreme authority in India; she had secured publication of annual reports which would prevent authorities in India from going to sleep. On February 16, 1868, she was able to write to M. Mohl in triumph: 'By dint of remaining here for 13 months to dog the Minister I have got a little (not tart) but Department all to myself, called "Of Public Health Civil and Military for India" with Sir B. Frere at the head of it. And I had the immense satisfaction 3 or 4 months ago of seeing "Printed Despatch No. 1" of said Department. (I never in all my life before, saw any Despatch, Paper or Minute under at least No. 77,981.)'

Again she had paid a heavy price for her success. The thirteen months during which she had shut herself up in South Street working day and night had further impaired her health. She was slowing down. 'I do the work in 3 hours I used to do in one,' she told Clarkey in July 1867. She would not spare herself on that account. If she were slower, the only consequence was that she must drive herself harder. Once more the limit of what she could inflict on herself was reached, and in December 1867, after the autumn of gruelling work for Sir Stafford Northcote, she collapsed completely. 'I broke up all at once,' she told M. Mohl in February 1868, 'and fled to Malvern on December 26 with a little cat.'

Urgent work called her back – not Indian work; for the moment there was a lull in Indian affairs. She had other calls on her as exacting as India. While she had been working on the Indian

344

Sanitary Commission, gigantic as her labours had been, they were not her only occupation. Her work for public health in England, for hospitals and the reorganization of nursing, had rapidly expanded and assumed enormous proportions.

20

In 1861 Miss Nightingale received a letter from a Mr William Rathbone of Liverpool, who was the eldest son of a dynasty of Liverpool merchants and shipowners, and the sixth William Rathbone in succession to be senior partner in the family firm. He inherited a tradition of philanthropy and liberalism. His grandfather had been a prominent Abolitionist, and his father, though the Rathbones were originally Quakers, had taken a leading part in the struggle for Catholic emancipation. As a young man he was an honorary visitor for the District Provident Society in one of the poorest quarters in Liverpool, and witnessed the miseries endured by the poor who were ill in their own homes. In 1859 he founded district nursing, starting in his own district with one trained nurse. Since one nurse proved ludicrously inadequate, he decided to establish, at his own expense, a body of trained nurses to nurse the sick poor in their own homes. Finding that trained nurses of the type he required – responsible, trustworthy, and experienced – did not exist, he wrote to Miss Nightingale asking her advice.

Her experiences as a girl in the cottages of Lea Hurst had taught her how urgent was the need for District Nursing and though she was overwhelmed with work on the Indian Sanitary Commission, she gave William Rathbone's scheme 'as much consideration', he wrote, 'as if she herself were going to be the Superintendent.' She came to the conclusion that the only satisfactory solution was to train nurses specially, and suggested that the Royal Liverpool Infirmary should be approached to co-operate in opening a training school with the guarantee that a fixed percentage of nurses trained should be reserved for the Royal Infirmary. In the following year, at William Rathbone's expense, a Training School and Home for Nurses was opened in connexion with the Royal Infirmary and proved an unqualified success.

William Rathbone, austere in spite of great wealth, unselfish, tender-hearted, devoid of sentimentality to the point of dryness, was a man Miss Nightingale could appreciate, and they became intimate friends. His admiration and affection for her was unbounded; he wrote that he was 'proud to be one of her journey men workers', and when she moved into South Street he presented her with a stand filled with flowering plants, which he kept renewed weekly until his death.

While following up cases from his district he visited workhouse infirmaries and found that, though the sick in the slums were miserable, the paupers in the workhouse infirmaries were more miserable still. The Liverpool Workhouse Infirmary was, on the whole, well administered, but the condition of the sick was wretched beyond description. Twelve hundred sick paupers were accommodated. As in all workhouse infirmaries, such nursing as existed was done by able-bodied female paupers, and owing to the fact that Liverpool was a seaport and the city had large harbour slums, many of the women were drunken prostitutes. These women, wrote Miss Eleanor Rathbone in a memoir of her father, 'were superintended by a very small number of paid, but untrained, parish officers, who were in the habit, it was said, of wearing kid gloves in the wards to protect their hands. All night a policeman patrolled some of the wards to keep order, while others, in which the inhabitants were too sick or infirm to make disturbance, were locked up and left unvisited all night.'

On January 31, 1864, William Rathbone made the first move to reform workhouse nursing. He wrote to Miss Nightingale and suggested that a staff of trained nurses and a matron should be sent to the Liverpool Workhouse Infirmary; it was not possible to send nurses from the training school, as the full number trained there was already bespoken by the Royal Infirmary and the District Nursing scheme. If she could find nurses and a matron, he offered to guarantee the cost for whatever term of years she thought advisable. It would not be easy to obtain permission to introduce the nurses, and he asked her to draft a letter which he could send to the Vestry who controlled the Workhouse Infirmary, asking for their co-operation.

A long battle ensued. 'There has been as much diplomacy and as many treaties and as much of people working against each other, as if we had been going to occupy a Kingdom instead of a Workhouse,' Miss Nightingale wrote to Rev. Mother Bermondsey in September 1864. Permission was not granted until March

1865, and by then she had also become involved in the reform of workhouse nursing in London.

In December 1864 a pauper named Timothy Daly died in the Holborn Workhouse. There was an inquest and it was found that death had resulted from filthiness caused by gross neglect; the newspapers took the case up, and there was a public scandal. Miss Nightingale seized the opportunity to write a tactful letter to Mr Charles Villiers, the President of the Poor Law Board. She ventured to write, she said, because Timothy Daly's case proved the overwhelming necessity for the improvement of nursing in workhouse infirmaries. He might perhaps be interested to hear what was going to be done in the Liverpool Workhouse Infirmary with a staff of trained nurses and a matron from the Nightingale Training School.

Mr Villiers replied immediately; he was a friend of Lord Palmerston; he was a friend also of Miss Nightingale's ally, Sir Robert Rawlinson the engineer. A champion of the people's rights, Charles Villiers had devoted many years of his life to the repeal of the Corn Laws to reduce the price of bread. His appearance was romantically handsome, his manner was charming, and his powers of conversation were considered unequalled.

At the end of January he called and by the end of the interview they were firm friends. Her powers inspired him with intense admiration. Sir Edward Cook describes him bursting out to a friend after he had received one of her memoranda: 'I delight to read the Nightingale's song about it all. If any one of them had a tenth part of her vigour of mind we might expect something.'

More than the reform of workhouse nursing was discussed. She had realized that it was virtually impossible to reform workhouse nursing without reforming workhouse administration, and she urged Mr Villiers to make use of the death of Timothy Daly to initiate an investigation into the whole question of the treatment of the sick poor.

'They are much more frightened by the death from the Holborn Union than they "let on",' she wrote to Sir John McNeill on February 7. 'I was so much obliged to that poor man for dying. It was want of cleanliness. Mr Villiers says that he shall never hear the last of it.' Early in February 1865 Mr Villiers sent his principal assistant, Mr H. B. Farnall, to see her. Mr Farnall was Poor Law Inspector for the Metropolitan District. He had been working for Poor Law Reform all his life and knew the almost

348

insuperable difficulties lying ahead, but she inspired him with an astonishing confidence. 'From the first,' he wrote in December 1866, 'I had a sort of fixed faith that Florence Nightingale could do anything.'

It was decided to start the investigation in London, and with Mr Farnall's assistance Miss Nightingale drew up a 'Form of Enquiry' to be circulated to every workhouse infirmary and workhouse sick ward in the Metropolitan district. Mr Villiers approved, and the forms were sent out in February 1865.

In March permission was at last given for the Nightingale nurses to enter the Liverpool Workhouse Infirmary, and on May 16 twelve nurses and a matron, Miss Agnes Jones, arrived. 'Mr Rathbone puts down £1,200,' wrote Miss Nightingale. The experiment was at first to be confined to the male wards only.

Once more she was desperately anxious. Once more everything hung on the endurance, the good behaviour, and the good sense of a band of young women. She had convinced Mr Villiers of the possibility of employing trained nurses in workhouses infirmaries, but what if the nurses failed? The task which lay before them was fearful enough to daunt the boldest. Fortunately the matron was a young woman of remarkable character and qualifications who had already been selected by Miss Nightingale as the only woman capable of becoming her successor. Agnes Jones was the daughter of Colonel Jones of Londonderry and the niece of Sir John Lawrence. She was 'pretty and young and rich and witty, ideal in her beauty as a Louis XIV shepherdess,' Miss Nightingale wrote to Clarkey. But beneath the prettiness was the soul of a martyr – she had inherited the passionate altruism and uncompromising moral rectitude of the Lawrence family as well as their good looks. Miss Nightingale's work in the Crimea had inspired her to become a nurse, and, following in her footsteps, she persuaded her family to allow her to go to Kaiserswerth in 1860. After two years there she wrote to Miss Nightingale, who advised her to complete her training by a year at the Nightingale School at St Thomas's. She entered in 1863 and was the best probationer Mrs Wardroper had ever had and Miss Nightingale's 'best and dearest pupil'. On completing her training, she went to the Great Northern Hospital and was working there as a sister when she was offered the post of Matron of the Liverpool Workhouse Infirmary in August 1864.

She knew how fearful would be the task, and she refused; but her conscience would not let her rest. Had she not received a call from God? After days spent in agony and prayer, she wrote again and accepted.

In 1868 Miss Nightingale wrote an account of Agnes Jones's work in *Good Words* under the title 'Una and the Lion'. The Lion symbolized the paupers Agnes Jones had to nurse, 'far more untameable than any lion'. She had nursed in great London hospitals, but, she said, until she came to Liverpool Workhouse Infirmary she did not know what sin and wickedness were. The wards were an inferno, the hordes of pauper patients more degraded than animals. Vicious habits, ignorance, idiocy, met her on every side. Drunkenness was universal – thirty-five of the pauper nurses had to be dismissed for drunkenness in the first month. Immorality was universal. Filth was universal. The patients wore the same shirts for seven weeks; bedding was only changed and washed once a month; food was at starvation level; spirits entered the infirmary freely. The number of patients was very large, 1,350, rising at times to 1,500. There were administrative difficulties; her position and powers were not properly defined; the supply of food to the Workhouse Infirmary was done by contract, and the doctors had no control over it; the task of training the pauper nurses was hopeless. 'It is like Scutari over again,' Miss Nightingale told her.

At first it seemed that the experiment was failing. The governor of the Infirmary had supported the scheme, but Agnes Jones quarrelled with him. She objected to his 'want of refinement', and he thought her too strict and unpractical. There were several serious disagreements, described by Dr Sutherland as 'Hibernian Rows'. Miss Nightingale intervened, smoothed things over, talked each side round. When affairs in Liverpool reached a deadlock, William Rathbone made a flying trip to London to consult her. She was adept at coaxing: Dr Sutherland, drafting a letter to the governor of the Infirmary, left a blank 'for you to fill in with soft sawder'.

Gradually the scene changed. Agnes Jones, under Miss Nightingale's influence, became less rigid, and her genius as a nurse and as an administrator made itself felt. Old women visiting their husbands reported 'wonderful changes in the House' since the London nurses came. Ladies who acted as charitable visitors began to sing her praises; the medical staff asked for more nurses. The results Miss Jones was achieving were so good that they intended to propose she should take over the female as well as the male wards. Most important of all, the cost of maintaining the sick in the Workhouse Infirmary under her régime was less, not more, than before. Agnes Jones, wrote Miss Nightingale, converted the Vestry to the conviction of the economy as well as the humanity of nursing pauper sick by trained nurses.

With this success behind her Miss Nightingale pressed for legislation. It was impossible to correct the abuses which existed in workhouses and workhouse infirmaries until they were put on a new financial and administrative basis. To make the changes required, there must be an Act of Parliament. She did not try to establish a position for herself within the Poor Law Board as she had done with the War Office and the India Office. Mr Villiers was over sixty, and, much as she liked him, she was forced to realize that he no longer possessed the energy to be her mouthpiece.

The Poor Law Board would never produce the necessary legislation – she must go over its head and approach Lord Palmerston. Once more she succeeded. He promised that if she would draft a Bill he would use his influence to get it through the Cabinet. Full of hope, she began work with Mr Farnall, but another blow fell. It was the moment when Lord Palmerston was taken ill, and on October 18 he died; her hopes of influence in the Cabinet were at an end.

She must depend again on Mr Villiers, but she was not left without hope. The answers to the 'Forms of Enquiry' sent out to London workhouses and workhouse infirmaries revealed facts so shameful that they could not be ignored.

Through the autumn of 1865 she worked on a scheme for the Metropolitan area which was intended to be extended later to other areas. The first necessity was to change the mental attitude which made the miseries of the hideous system possible. 'So long,' she wrote, 'as a sick man, woman or child is considered administratively to be a pauper to be repressed and not a fellow creature to be nursed into health, so long will these shameful disclosures have to be made. The sick, infirm or mad pauper ceases to be a pauper when so afflicted.' It was the conception on which she based her attitude toward the human race. Suffering lifts its victim above normal values. While suffering endures, there is neither good nor bad, valuable nor invaluable, enemy nor friend. The victim has passed to a region beyond human classification or moral judgements, and his suffering is a sufficient claim.

Administratively her scheme for reform was based on three essentials which she termed the A B C of workhouse reform.

(*A*) The sick, insane, incurable, and children must be dealt with separately in proper institutions and not mixed up together in infirmaries and sick wards as at present. 'The care and government of the *sick* poor is a thing totally different from the government of paupers. Once acknowledge this principle and you must have suitable establishments for the cure of the sick and infirm.'

(*B*) There must be a single central administration. 'The entire Medical Relief of London should be under one central management which would know where vacant beds were to be found, and be able so to distribute the Sick etc., as to use all the establishments in the most economical way.'

(*C*) 'For the purpose of providing suitable establishments for the care and treatment of the Sick, Insane etc., Consolidation and a General Rate are essential.' This point was vital; the future of workhouse nursing and administration turned on it. As long as the workhouses and infirmaries were paid for out of the parochial rates, and staff appointments were made by local authorities with absolute power, there would inevitably be jobbery. Moreover, 'to provide suitable treatment in each Workhouse would involve an expenditure which even London could not bear'. At the moment to be a sick or insane pauper in a poor parish was to be horribly penalized.

The memorandum reached Mr Villiers in December. Her case was unanswerable, and he agreed to press at once for a new London Poor Law Bill. From day to day the situation improved.The *Lancet* sent a special commissioner to inquire into the state of London workhouse infirmaries, and, as a result, the 'Association for the Improvement of the Infirmaries of London Workhouses' was formed. In April 1866 the Association sent a deputation to the Poor Law Board pressing for immediate improvement, and Mr Villiers gave an assurance that legislation might be expected almost immediately. Newspaper articles began to appear. Delane, editor of *The Times*, called to see Mr Villiers, and journalists, including Edwin Chadwick, applied to Miss Nightingale for facts. In April the Metropolitan Workhouse Infirmary Bill seemed almost a certainty.

Yet once more everything vanished into thin air. It was the

pring of 1866, and the Whig Government was tottering. Work-house infirmary reform was a controversial subject, and as Lord de Grey delayed on the Indian Public Health Service so Mr Villiers delayed on the Metropolitan Workhouse Infirmary Bill. 'He was afraid,' wrote Miss Nightingale to Harriet Martineau, on May 2, 'of losing the Government one vote.' On June 18 the Government fell, Mr Villiers went out of office, and the Metropolitan Workhouse Infirmary Bill was lost. 'It was a cruel disappointment to me,' wrote Miss Nightingale to M. Mohl on July 2, 1866, 'to see the Bill go just as I had it in my grasp. . . . Alas! There is a pathetic story of Balzac's in which a poor woman who had followed the Russian campaign was never able to articulate any word except "Adieu, Adieu, Adieu!" I am afraid of going mad like her and not being able to articulate any word but Alas! alas! alas!'

Though all the summer she was very ill, she would not accept defeat. Surely in some direction something could be done. 'I had hopes for a time from a Committee of the House of Commons (on which serves John Stuart Mill) on the special Local Government of the Metropolis,' she wrote to M. Mohl on July 12. She sent John Stuart Mill a copy of her scheme; she approached Edwin Chadwick, who had been asked to advise by John Stuart Mill, offering 'to express my conclusions more in detail in answer to written questions (as I have done to 2 Royal Commissions).' At first she seemed to be making an impression, and the Committee asked her for 'a long letter', then hope faded as the whole question was postponed. 'Because it is July and they are rather hot they give it up for this year,' she wrote.

Early in July she had written to Mr Gathorne Hardy, Mr Villiers' successor, and on July 25 he replied in a complimentary and discouraging letter. She must not apologize for writing to him; she had 'earned no common title to advise and suggest upon anything which affects the treatment of the sick'. Sufficient compliments having been paid, he made it clear that he had no intention of becoming involved with her. He had 'not advanced very far from want of time'; and he was 'necessarily very much occupied with other business'. In conclusion, he hastened to assure her he would 'bear in mind the offer you have made and in all probability avail myself of it to the full'.

He did not invite her to write to him again; he did not suggest calling on her. On the heels of the letter came further discouraging news. Mr Gathorne Hardy removed Mr Farnall from his post at Whitehall as Poor Law Inspector of the Metropolitan District and sent him to Yorkshire. There could be no plainer proof that Mr Gathorne Hardy did not intend to be on the side of reform. Miss Nightingale was forced to admit she could do nothing further and went to Embley.

But in October Mr Gathorne Hardy made Mr Villiers 'frantically angry'. On October 31 Miss Nightingale wrote to Douglas Galton that 'Mr Hardy told him, Mr Villiers, *twice*, in the Ho: of C, that had he only known how to use, with dexterity and wisdom, the weapon of the law, he would have found it a very sufficient weapon.' Fresh legislation, Mr Hardy asserted, was quite unnecessary, provided the existing Acts were properly understood and applied. Since Mr Villiers had devoted his term of office to framing new legislation, he took this as an attack on himself. He began to write to Miss Nightingale again, expressing his determination 'not to sit down under this kind of thing'; he intended to 'catch Mr Gathorne Hardy out' and show him that something more was needed to solve the problem of Poor Law Administration than 'a touch of Mr Gathorne Hardy's magic wand'. On October 28 Miss Nightingale wrote to Edwin Chadwick: 'I have had a great deal of clandestine correspondence with my old loves at the Poor Law Board this last two months. There is only one thing of which I am quite sure. And that is that Mr Villiers will lead Mr Gathorne Hardy no easy life next February.'

In October Mr Gathorne Hardy appointed a committee of sanitary and medical experts to report 'upon the requisite amount of space and other matters in relation to Workhouses and Workhouse Infirmaries'. Among the other matters was included nursing. He did not consult Miss Nightingale, but she put her pride in her pocket and asked to be allowed to contribute. The committee invited her to submit a paper on nursing; this opening gave her the chance to put forward her scheme of workhouse reform, basing her argument on the obvious truth that the organization, construction, and administration of workhouse and workhouse infirmaries were of vital importance to the nursing system. She had her paper printed and sent a copy to Mr Gathorne Hardy with a long, urgent letter. She was writing repeatedly to him, but with no effect. He neither consulted her nor informed her what he intended to do. It was a complete surprise when on February

8 he introduced a Bill which became law in the following month under the title of the Metropolitan Poor Act.

Miss Nightingale and her fellow workers felt they had been exploited. On February 11 William Rathbone wrote: 'I think Hardy's use of our experiment and of your name atrocious.' On a draft copy Miss Nightingale scribbled angrily 'Humbug', 'No principles', 'Beastly'. She was especially angry that no direct provision was made for the improvement of workhouse nursing. For this, however, she herself was to blame. She had written a letter to Mr Gathorne Hardy reporting the victory of the Liverpool Workhouse Infirmary nursing in which she gave a dramatic description of the opposition and difficulties encountered at the beginning. He took fright; either he was genuinely alarmed, or he was not sorry to find an excuse for postponing nursing reform; in any case, he publicly gave her letter as a reason for shelving the question for the present.

In fact, resentful though the Reformers might be, the Bill was a great advance. Miss Nightingale campaigned furiously in the hope of getting it amended, but Mr Gathorne Hardy was both conciliatory and skilful. He put forward the Bill modestly as 'only a beginning', freely admitting that criticism was justified but emphasizing he was doing everything that could be done at present. One by one her supporters became lukewarm. Mr Villiers, who had begun by calling the Bill a seven months' child born in the Whitehall Workhouse, admitted that it would 'set the ball rolling.' Miss Nightingale herself did not suffer her usual torments of despair. After the Bill was passed, in March 1867 she wrote to Rev. Mother Bermondsey almost with cheerfulness: 'We have obtained some things, the removal of 2000 lunatics, 80 fever and smallpox cases and all the remaining children out of the Workhouses – (and the providing for them out of a common fund in order to relieve the rates) the paying of all salaries of Medical Officers, Matrons, Nurses etc., out of a Metropolitan (not parochial) rate. ... Also; – the removing all other sick into separate buildings which are to be improved – and constituting fresh Boards of Guardians for these sick with nominees from the Poor Law Board. This is a beginning, we shall get more in time.'

Another battle was over, but the pause which succeeded it brought her no rest. During the summer of 1867 she was forced to stay in London working desperately on Indian sanitary affairs. She was conscious of driving herself too hard; if Sidney Herbert or Clough had been alive, she wrote to Clarkey, one of them would have made her stop. But now there was no one to stop her. Indeed, far from being able to spare herself, she was forced, in June 1867, to undertake a new and laborious work.

Part of the money raised by the Nightingale Fund had been devoted to establishing a training school for midwives in King's College Hospital. The school flourished and was considered by Miss Nightingale to be one of her most satisfactory achievements, when it was overtaken by disaster. Puerperal sepsis broke out in the lying-in wards following the delivery of a woman suffering from erysipelas and developed into an epidemic which closed the school. An investigation followed, in the course of which she discovered that no reliable statistics of mortality in childbirth existed. 'There appears,' wrote Miss Nightingale, 'to have been no uniform system of record of deaths, or of the causes of death, in many institutions, and no common agreement as to the period after delivery during which deaths should be counted as due to the puerperal condition ... the first step is to enquire, what is the real normal death rate of lying-in women? Compare the rate with the rates in establishments into which parturition cases are received in numbers. Clarify the causes of death and see if any particular cause predominates in lying-in institutions; and if so, why so? And, since the attendance on lying-in women is the widest practice in the world, and these attendants should be trained; ... decide this great question as to whether a training school for midwifery nurses can safely be conducted in any building receiving a number of parturition cases.' Ill, harassed, and overdriven as she was, she set to work with Dr Sutherland's assistance to collect facts and figures on which statistics could be based. It was a difficult task. Doctors were suspicious of interference and surprisingly ignorant – one doctor told her that ergot was a specific against puerperal sepsis. Institutions were unwilling to disclose their figures. However, working through hospitals and doctors to whom she was personally known, she collected preliminary facts which pointed to a startling conclusion. It seemed that in lying-in institutions and hospital wards the rate of mortality was much higher than when patients were delivered at home, however poor and unhygienic those homes might be. She determined that a great mass of further facts must be collected and began to correspond with doctors, matrons, sanitary experts, and engineers throughout the world.

The work involved was enormous. The analysing and tabulating was done by her, and the majority of letters were written in her own hand. The inquiry took three years to complete, years during which the pressure of other work was crushing. At the end of the inquiry she had accumulated a mass of information but had neither time nor energy to work it up. A book was planned, but Dr Sutherland had to put it into shape, and it was not until 1871 that a small volume was published under the rather apologetic title of *Introductory Notes on Lying-in Institutions*.

It is one of the most interesting of Miss Nightingale's works. She reached the conclusion that the use of small separate rooms was the answer to the high rate of mortality in maternity cases. The same conclusion was reached, independently, by Sir James Simpson, the pioneer of the use of chloroform in childbirth, with whom she corresponded. When she wrote, the great discoveries of bacteriology were still ten years ahead; Lister had only just made public his first experiments with carbolic acid, and the nature of infection was not understood. Miss Nightingale herself regarded 'the fear of entering a cab in which a case of fever or small pox has been for half an hour' as 'morbid' and wrote of 'the myth of scarlet fever being carried in a bedside carpet'. But she did establish, independently, the fact that, whenever a number of maternity cases were collected together, and whenever maternity cases were under the same roof as medical and surgical cases, the mortality rose. The lower mortality in cases delivered at home was due to the fact that 'however grand, or however humble, a home may be in which the birth of a child takes place, there is only one delivery in the home at one time'. In London workhouses the death-rate depended on the number of deliveries. Thirteen infirmaries which had no deaths at all in five years, had under sixteen deliveries per annum. 'When Waterford Institution had 8 beds in 1 room the mortality was 8 per 1000. When the wards were moved and the number of beds reduced to 4, the mortality fell to 3·4 per 1000.' In La Charité in Paris maternity cases were under the same roof as medical and surgical cases, and the mortality in 1861 was 193·7 per 1000, of which over 80 per cent were due to puerperal fever. In the Military Female Hospital at Shorncliffe, an old wooden hut, but close to the sea with sea breezes blowing through it, and in a makeshift wooden hut used as a hospital at Colchester there had never been a death from puerperal fever. 'In both these hospitals beds are seldom, or never, occupied all at one time. They have their own attendant and seldom is more than one bed occupied.' 'Not a single lying-in woman,' wrote Miss Nightingale, 'should ever pass the doors of a general hospital. Is not any risk which can be incurred infinitely smaller?'

She pressed for midwifery as a career for educated women. In 1871 there was an agitation for women to be admitted to the medical profession, an ambition with which she had little sympathy; the crying need was for nurses and midwives. She added a letter to her book: 'Dear Sisters,' she wrote, 'there is a better thing for women to be than "medical men", that is "medical women".' 'It is a good thing you are at Lea Hurst,' wrote Dr Sutherland in July 1871, 'or your "dear sisters" would infallibly break your head.'

The spring of 1867 brought final victory in Liverpool. District nursing was rapidly developing. The city had been divided into eighteen districts, each provided with trained nurses. In the Liverpool Workhouse Infirmary the cost of the scheme which William Rathbone had borne was officially assumed by the Vestry, and all the wards in the Infirmary, male and female, were placed under the authority of Agnes Jones. It was triumph, but triumph was short-lived. The winter of 1867 was a time of unemployment and distress, Agnes Jones was already cruelly overworked, and the work entailed by the additional wards was more than any human being could accomplish. 'All the winter,' wrote Miss Nightingale to M. Mohl in February 1868, 'she has had 1350 patients and to fight for every necessary of life for them. She has never been in bed until 1.30 a.m. and always up at 5.30 a.m.' An epidemic of typhus broke out, and Agnes Jones caught the infection. On February 19, 1868, she died. The last words she spoke to Miss Nightingale were: 'You have no idea how I am overworked.'

Her death was a catastrophe. There was no one to take her place. Personal loss Miss Nightingale could bear; on the third anniversary of Sidney Herbert's death she had written that the mere personal craving after a beloved presence she felt as nothing, a few years and it would be over – her bitter grief was for the work. After so much had been sacrificed, so much had been achieved, it seemed that the reform of nursing in workhouse infirmaries would collapse for want of someone to carry it on. She became frantic. Her bitterness against women, her distrust of women, her resentment of their pretensions flared up again. 'I don't think,' she wrote to Clarkey in April 1868, 'anything in the course of my long life ever struck me so much as the deadlock we have been placed in by the death of one pupil, combined you know, with the enormous Jaw, the infinite female ink which England pours forth on "Woman's Work".'

'The more chattering and noise there is about Woman's Mission the less of efficient women can we find,' she wrote to Sir John McNeill on February 7, 1865. 'It makes me mad to hear people talk about unemployed women. If they are unemployed it is because they won't work. The highest salaries given to women at all we can secure to women trained by us. But we can't find the women. They won't come.' Those who did come gave enormous trouble, and there were moments when it seemed that everything she had achieved by her work in the Crimea would be lost through their unreasonableness and stupidity. 'It is not money we want,' she wrote to Harriet Martineau in February 1865, 'it is workers. ... We don't aspire, altho' they are needed by the hundred and the thousand, to sending out nurses by the hundred and the thousand. What we want to do is to send a small staff of trained nurses and a *trained training* Matron, wherever we are asked. But the material, especially in the latter (the Matron) does not come to us. We have 23 nurses now in training at St Thomas's, our largest number. 18 is the number we can entirely support at St Thomas's but this is no difficulty at all, even at the moment some of our 23 are supported by others. We should never lack the money, but we want the workers. ... *Applications* have no superfluity at all from any description or class of persons rushing to be trained. We can scarcely make up our number of the right sort. ... We never have rejected one of the right sort for want of room. But really not many of *any* sort come to be rejected. ... We have always 10 times as many situations offered as trained persons to fill them. Indeed I am sorry to say that nurses of ours have been made superintendents who were totally unfit for it, and whom we earnestly remonstrated with, as well as with their employers, to prevent them being made Superintendents but in vain – such is the lack of proper persons.'

Many first-class nurses were prevented by lack of education from being able to attempt administrative work. Miss Nightingale considered that Mrs Roberts, who had been her head nurse in the Crimea, was the best nurse she had ever met, but in 1857 she wrote: 'She might be a first-class physician and surgeon, learnt by 23 years experience, but not by reading. She *can't* read! not literally.' Too often the woman with some education who took up nursing did so only because she had proved herself too unreliable for other work. In 1871 visitors to the Poplar Sick Asylum found the matron in a 'low dress with short sleeves, being merry'; she confided to them that she had successfully brought two actions for breach of promise. In military hospitals the difficulty was even greater. In July 1871 Miss Nightingale gave Henry Bonham Carter, Hilary's brother, an account of an interview with the superintendent of a military hospital who had been reported for improper conduct. 'She had nursed a Pole and he had left her all his money. Alternatively he had only given one small present to her little girl. Then she began raving about her social position, her poor husband, her character, her age. WHO might not take presents if she might not she screamed – she must have been heard all over Lord Lucan's next door – to frighten me. She began by declaring that she would not keep her superintendency a day, she would resign at once. Then it flashed, even over her, that her resignation might be accepted.

AND SHE SHOUTED –

that she didn't care for the W.O. – not a fig – she snapped her fingers at them. She would stay where she was and nobody should turn her out, that if the W.O. asked her to resign she would defy them and they should find her a match for them and she should not resign. (Do you know I have the strong impression that there is a great deal to know which we do not know.) She told me she was a saint, she screamed this at least 40 times. She repeated (I am sure 50 times) that she had made the nurses a cake 'with her own hands' as proof of her being 'a mother to them'. She screamed 'You have made me miserable, miserable, miserable' (30 or 40 times). But she was quite evidently trying to practise on me. (I have no doubt she has tried this practice on many, especially her husband.) And the cruelty of her eyes in saying this was frightful to see.

The obvious solution was for educated women of a good type to become nurses, but this path too was strewn with difficulties. The 'ladies' and 'nurses' controversy which had caused so much heart-burning in the Crimea was still raging. On September 13, 1866, Miss Nightingale wrote Dr Farr a long letter protesting against statements made by a Dr Stewart and a Miss Garret, who asserted that she had been 'compelled to give up employing lady nurses, had been forced to abandon the introduction of educated women into the profession of nursing' and had 'declared that educated women were unable to undergo the training necessary for the purpose'. It was just possible, said Dr Stewart and Miss Garret, that a middle-class woman might become a nurse but quite impossible for an upper-class lady.

The truth, wrote Miss Nightingale, was exactly opposite.

Be it known to Dr Stewart who draws a painfully invidious distinction between 'upper' and 'middle class,' that the *fact* is exactly the contrary from what he represents it. It is far more difficult to induce a middle-class' woman than an 'upper class' one to go through as Head Nurse the incidental drudgery which must fall to the province of the Head Nurse – or be neglected.

1. *No* nurses should do the work of 'scrubbers' – that therefore the Nurse whether she be upper, middle or lower class is equally able to go through the training of a nurse.

2. No Lady Superintendent – be she upper, middle or lower class – is qualified to govern or to train nurses, if she has not herself gone through the training of a nurse.

3. I don't exactly know what Dr Stewart and Miss Garret mean by the 'upper' class. ... Therefore I will wait to know before I mention many who have gone through the training of a nurse ... are equally qualified to be Nurses, Head Nurses, to attend an operation or to be Supt – and yet are of what is usually called the 'upper' class.

Dr Stewart and Miss Garret voiced views widely held by Miss Nightingale's contemporaries. The figure of the self-immolating sister of charity was fixed in the public mind, and few people could visualize the professional woman, trained, efficient, and highly paid whom she wished to call into existence. 'To make the power of serving without pay a qualification is, I think, absurd,' she wrote to Dr Farr on September 13, 1866. 'I WOULD FAR RATHER THAN ESTABLISH A RELIGIOUS ORDER, OPEN A CAREER HIGHLY PAID. My principle has always been – that we should give the best training we could to any woman, of any class, of any sect, 'paid" or unpaid, who had the requisite qualifications, moral, intellectual and physical, for the vocation of a Nurse. Unquestionably the educated will be more likely to rise to the post of Superintendent, but *not* because they are ladies but because they are *educated*.'

When the right type of woman had been secured and there were no religious objections, difficulties remained. The Nightingale nurses, carefully selected, trained under Miss Nightingale's own eye, were efficient, professional, educated, but they suffered from a feeling of their superiority. 'Intolerable conceit is one of our nurses' chief defects,' she wrote in 1871. One of the Nightingale probationers – 'a good girl,' wrote Miss Nightingale, 'but using the lowest kind of High Church slang' – refused to sit at meals with an old-style sister from St Thomas's on the ground that the sister was 'low'. When Nightingale nurses were sent to work under matrons who were not Nightingale-trained, they were patronizing. On one or two occasions Miss Nightingale was asked to recall her nurses because they were too difficult to manage. In December 1871 she described herself as 'transfixed with horror' to receive a complaint, forwarded by one of the medical staff, of the behaviour of a Nightingale nurse to the patients at the Highgate Infirmary. 'Sir. Is an assistant nurse to be allowed to incur the guilt of being insubordinate to the Patients? Is she to tell the Patients that the food is good enough for Paupers? Is it to be endured that she should say that they require a great deal of waiting on?' She commented in the margin, 'very fairly written and signed by two men patients'.

There were difficulties inherent in the very fervour which inspired the new nurse. The woman who was neither frivolous nor in financial difficulties, who was prepared to undergo a rigorous training and subsequently endure the conditions of work in the wards of a hospital or infirmary of the period, was likely to be animated with a fanatic's spirit. Miss Nightingale did not want fanatics; she did not want warfare, especially holy warfare. No one knew better that almost everything was wrong with the conditions and technique of contemporary nursing. But the way to improvement did not lie through rebellion; she had never been a rebel, and she did not mean to send out parties of rebel nurses. Authority must not be flouted but converted. Regulations must be observed, because regulations were essential to organization. If regulations were bad, they must certainly be changed, but until they were changed they must be observed. In a private note of 1866 she wrote: 'Women are unable to see that it requires wisdom as well as self denial to establish a new work.'

It was a difficult lesson to teach. Women who had trained as nurses inspired by a spirit of devotion found themselves sent to posts where their good intentions were frustrated and their skill wasted. Unhappy and rebellious, they appealed to Miss Nightingale, and when she preached patience, yielding, moderation, their disappointment was great. Many lost faith in her, and she frequently mentions having received letters of abuse from unhappy and disappointed nurses. But she would not change her policy. 'Do you think,' she wrote to a rebellious nurse on April 22, 1869, 'I should have succeeded in doing anything if I had kicked and resisted and resented? ... I have been shut out of hospitals into which I had been ordered by the Commander-in-Chief, obliged to stand outside the door in the snow until night, have been re-used rations for as much as 10 days at a time for the nurses I had brought by superior command. And I have been as good friends the day after with the officials who did these things – have resolutely ignored these things FOR THE SAKE OF THE WORK.'

The want of women to train as nurses, the difficulties perpetually arising from their unreasonableness and instability affected Miss Nightingale's attitude toward the feminist question. The movement for the higher education and emancipation of women was gathering strength, and between 1860 and 1870 the first organized efforts were made to enable women to enter the learned professions and to give them the vote. In September 1860 John Stuart Mill asked Miss Nightingale to support the movement to enable women to qualify as doctors on the same terms as men. But she was unsympathetic; her difficulties had left her with the conviction that women already had more opportunities than, at the moment, they were capable of using. She was impatient with 'female missionaries', with the 'enormous Jaw about Woman's Work'. She was convinced that her attitude was based on hard facts derived from her experience, but in fact the truth lay in her own nature.

Her outlook was aristocratic. Equality meant little to her, equality of the sexes, the goal of the early pioneers of feminism, least of all. She had never felt handicapped by her sex or wished to be a man. In all the long history of frustration recorded in her private notes, she never suggests she was frustrated by men because she was a woman. Stupidity frustrated her, not sex. She had been made aware that in the world of affairs suggestions from a woman were accepted less readily than suggestions from a man, but by using the right tactics she had been able to overcome that drawback. Statesmen, Cabinet ministers, public servants had willingly sat at her feet, and she assumed that any woman who chose to take the trouble could achieve the same position. In spite of the extraordinary power of her mind, she was a woman of intensely feminine nature whom men admired and spoiled. She preferred men to women, and sex antagonism, sex rivalry were foreign to her. Nothing exasperated her more than a desire on the part of women to imitate and emulate men. 'To do things just because men do them!' she wrote contemptuously. The exaggerated praise lavished on female achievement infuriated her – why should what was normal for a man be considered exceptional for a woman? Why should it be hailed as remarkable when a woman qualified as a doctor – not because she qualified brilliantly but because she succeeded merely in qualifying?

Dr Elizabeth Blackwell was the first woman to qualify as a doctor. She had studied in Paris and qualified in America, and was now a celebrity. On January 19, 1862, Miss Nightingale wrote that she would be 'inferior as a 3rd rate apothecary of 30 years ago'. 'Female M.D.s have taken up the worst part of a male M.D. ship of 50 years ago,' she wrote to John Stuart Mill on September 12, 1860. 'The women have made no improvement, they have only tried to be "men" and they have succeeded only in being third rate men. Let all women try ... these women have in my opinion failed ... but this is no prior conclusion against the reasoning.'

On September 23 John Stuart Mill replied, pointing out the limitations of her argument. 'When we consider how rare first rate minds are, was it to be expected on the doctrine of chances that the first two or three women who take up medicine should be more than you say these are – third rate. It is to be expected that they will be pupils at first, not masters. ... Neither does the moral right of women of admission into the profession depend at all upon the likelihood of their being the first to reform it.'

She remained unconvinced. The moral right did not interest her. The all-important consideration was the work waiting to be done in the world. Women made third-rate doctors and first-rate nurses; there were plenty of first-rate doctors; there was a shortage of first-rate nurses; what could be plainer than the conclusion that women ought to become nurses, not doctors? In July 1867 John Stuart Mill asked her to become a member of the first committee of the London National Society for Women's Suffrage. 'A Society has been formed for the purpose of obtaining the Suffrage for Women. The honour of your name as a member of the

General Committee is earnestly requested.' She refused. Again, the moral right of women to have a voice in the government of the country meant little to her. Her objections were practical – 'that women should have the suffrage,' she wrote on August 11, 1867, 'I think no one can be more deeply convinced than I. It is so important for a woman to be a "person" as you say. ... But it will be years before you obtain the suffrage for women. And in the meantime there are evils which press much more hardly on women than the want of the suffrage. ... Till a married woman can be in possession of her own property there can be no love or justice. But there are many other evils, as I need not tell you.' John Stuart Mill answered that he deplored on moral grounds the 'indirect influence' to which women were restricted owing to their want of political power. But Miss Nightingale, whose immense influence was all indirect, was devoid of moral qualms and influenced by only practical considerations. 'I have thought I could work better, even for other women, off the stage than on it,' she wrote simply.

She declined to believe in the vote as a universal panacea for the wrongs of women. 'If women were to get the vote immediately Mr Mill would be disappointed with the result,' she wrote in a rough draft of a letter to Jowett. The greater part of female misery was due to economics – not to the economic situation of women specifically but the economic situation of the whole nation. She instanced the 'frightful burden of pauperism, the overflowing workhouses. ... The wives and daughters of all these people are starving, does Mr Mill really believe that the giving of any woman a vote will lead to the removal of even the least of these evils?' In a sentence which she used again in writing to John Stuart Mill, she spoke of her own position. 'In the 11 years I have passed in Government offices I have never felt the want of a vote, because, if I had been a Borough returning two members to Parliament, I should have had less administrative influence.' Her arrogance was unconscious; her modesty was genuine. She insisted that the only difference between herself and other women was that she worked and they did not. She never could be brought to admit there was anything else.

Miss Nightingale never was to take any active part in the feminist movement. In 1867 she excused herself not only from becoming a member of the first committee but from joining the London National Society for Woman's Suffrage at all. 'I have no time,' she told John Stuart Mill at the end of her letter of August 11. 'It is 14 years to this very day that I entered upon work which has never left me ten minutes leisure, not even to be ill. And I am obliged not to give my name where I cannot give my work.' In 1868, however, she did become a member, and in 1871 her name was added to the general committee. In 1877 she retracted her opinion of women doctors sufficiently to sign a memorial urging the admission of women to medical degrees at the University of London, but she was never stirred by the cause of the emancipation of women, nor did she place it on the same level as the cause of public health in India and England, of reform of the health administration of the army or of Poor Law reform.

In later life she was conscious she had been unsympathetic, and in 1896 she wrote to Sir William Wedderburn asking him to tell her what the vote would do for the ordinary woman. 'I am afraid I have been too much enraged by vociferous ladies talking on things they know nothing at all about to think of the rank and file.'

Miss Nightingale was now forty-eight, and she thought of herself as old. Writing to Clarkey in 1868, she spoke of 'the course of my long life.' Jowett implored her to change her way of living. It was inconceivable that she could intend to spend the rest of her days shut up in one room in London. Early in 1868 fate intervened. Another change of Government destroyed much of her remaining influence in official departments. In March the Tories went out, the Liberals came in, and Mr Gladstone became Prime Minister.

21

THE accession of Mr Gladstone to power was a severe blow to Miss Nightingale. They had never been in sympathy; administration did not interest Mr Gladstone. He disliked soldiers, regarded an army as an undesirable and unchristian institution which, as the world progressed, every civilized nation would discard, and consistently opposed increased expenditure on the welfare of the British soldier. A standing army, he had said, can never be turned into a moral institution.

In Miss Nightingale's opinion the effect of Mr Gladstone was disastrous. 'The administrative state of things here is to me unimaginable,' she wrote to M. Mohl on June 10, 1869. 'The War Office is drifting back to what it was before the Crimean War. Pauperism which concerns hundreds of thousands is just left alone. ... One must be as *miserably* behind the scenes as I am to know how *miserably* our affairs go on.' 'What would Jesus have done,' she wrote in a private note, 'if He had had to work through Pontius Pilate?'

The tide had turned against her. Very far off were the days when she wrote 'Alexander whom I made Director General'. There were no longer men in the departments who spoke the words she put into their mouths and were instruments in her hand. In the War Office her influence was almost at an end; in 1869 Douglas Galton, her last remaining friend, resigned and took an appointment at the Office of Works, retaining out of his War Office appointments only his seat on the Army Sanitary Committee.

India alone was left. In Indian affairs, in spite of disappointments and unrealized hopes, she had achieved a personal position of very great authority, and it happened that the number of men holding important offices in India who were her intimate friends steadily increased.

The Secretaries of each Presidential Sanitary Commission corresponded with her, and officials of influence, Dr Hathaway, private secretary to the Viceroy, Dr Hewlett, sanitary officer for

Bombay, Dr Cunningham, sanitary adviser to the Government of India, 'quite prejudiced, very candid, brimming with information but – about 7 feet 2 in height', became her friends, and hundreds of letters passed between them. Mr Ellis, President of the Madras Sanitary Commission, Dr Walker, sent out to report on the sanitary condition of jails, Mr John Strachey, first President of the Bengal Sanitary Commission and later Lieutenant-Governor of the North-West Provinces – all three called on her and wrote to her. Miss Carpenter, the founder of schools for Indian girls, wrote to enlist her help in the cause of Indian female education. They corresponded, though Miss Nightingale was critical of work exclusively for women, and refused to take any interest in what she described, in a letter to Lord Napier of Magdala in 1869, as 'the curious good work of Mrs Johnson who took little brown girls, daughters of Sepoys, from dancing naked on the sea shore and taught them to illuminate'.

At the end of 1865 her old admirer, Lord Napier, was appointed Governor of Madras, and he wrote asking her to receive him, assuring her that he was 'at your orders for any day or hour', and reminding her that he had had 'the happiness and honour of having seen you at the greatest moment of your life, in the little parlour of the hospital at Scutari.' She was ill, but 'managed to scramble up to see him' on January 1, 1866, and made not only an enthusiast for the cause of sanitation in India but a personal friend. He signed his letters 'ever your faithful grateful and devoted servant'; he told her 'I think I am attached to you irrespective of sanitation'; he promised her '*You* shall have the little labour that is left in me'. In Madras his governorship was marked by solid achievements. Roads and schools were built, drainage and irrigation undertaken, wells sunk, jails remodelled, and hospitals reconstructed. Under the direction of Lady Napier the experimental scheme for gradually introducing female nurses into Indian hospitals, which had been mishandled and subsequently abandoned by Sir John Lawrence, was put into practice in Madras. Miss Nightingale wrote that Lady Napier was one of the most efficient women she had ever met.

In 1868 Sir John Lawrence's term of office as Viceroy ended. On November 23, a few days before he sailed, he wrote to Miss Nightingale: 'I think we have done all we can do at present in furtherance of Sanitary Improvement and that the best thing is to leave the Local Governments themselves to work out their own arrangements. If we take this course we shall keep them in a good humour.' Far from having done all that could be done, she considered that almost nothing had been accomplished; and as for the local governments being in a good humour, he forwarded by the same mail an official memorandum from Mr John Strachey sharply criticizing the Sanitary Department at the India Office. The instructions they sent out, he asserted, were written from an English point of view, old discussions were 'hashed up', no credit was given the Government for the fact that 2 million pounds sterling a year was being spent on sanitary works, and the department's latest scheme for 'a system of sanitary experts sending in reports' was 'politically foolish, and indeed absolutely dangerous ...' The memorandum was, Miss Nightingale scribbled to Dr Sutherland, 'the nastiest pill we have swallowed yet.' However, she restrained her irritation and wrote Sir John Lawrence a final letter 'to bless and not to curse'.

And yet when, in April 1869, he called on her with a present of 'a small shawl of the fine hair of the Thibetan goat' she succumbed once more to his personal beauty and charm. 'When I see that man again,' she wrote to M. Mohl on June 10, 1869, 'all the statesmen of the moment in England whether "in" or "out" seem to me like rats and weazels.' Nevertheless, her heart never quite ran away with her head, and she added, 'but when I see him I understand he will not do much in England.'

She had already established her influence over Sir John Lawrence's successor, Lord Mayo. On the morning of October 28, 1868, Dr Sutherland, arriving at South Street after a day's holiday, sent up a message that he hoped 'there was nothing much on to-day'. 'There is a "something" which most people would think a very big "Thing" indeed,' she replied. 'And that is seeing the Viceroy or Sacred Animal of India. I made him go to Shoeburyness yesterday and come to me this afternoon because I *could not* see him until you give me some kind of general idea of what to state.'

Lord Mayo came and stayed the whole afternoon. Though she liked him personally, his attitude toward the responsibility he had undertaken provoked her wrath. He said 'quite calmly' that he had not been able to free himself from his previous office as Irish Secretary until October 6, he was not going to be able to see Sir Stafford Northcote, the Secretary of State for India, at all because Sir Stafford would be busy first electioneering and then

369

staying with the Queen, and he was proposing to go out to India as Viceroy on November 6 'completely uninstructed'. 'He came to me to be coached and with Sir Bartle Frere I gave him his Indian education.' Jowett wrote that she had earned a new title, 'Governess of the Governor of India,' but Miss Nightingale replied that her correct title was 'Maid of all (Dirty) Work'.

Lord Mayo was, however, a willing pupil. 'He asked me,' she wrote to Dr Sutherland immediately after the interview, '(over and over again) that we should now, at once before he goes, write down something (he said) that would "guide me upon the Sanitary Administration as soon as I arrive".'

In Lord Mayo's guide to sanitary administration, Miss Nightingale showed a new purpose. Once more she took a step forward and passed from advocating engineering works to laying down an economic policy. Nothing could be done in India until India was fed; before sanitation must come irrigation – 'famine is the constant condition of the people'. Health was impossible, justice was impossible, organization was impossible, as long as the great mass of the people of India was vitiated and corrupted by being semi-starved from birth to death. Agricultural development, which implied irrigation, must come first of all. Before education, before any of the blessings of Western civilization were offered to them, the people of India must be fed, and henceforward irrigation works became Miss Nightingale's first aim for India.

A year later she added the Commander-in-Chief to her circle. In December 1869 she received a note from Lord Napier of Magdala asking if he might call. He came on the afternoon of December 14 and was instantly elected to a leading position in her gallery of heroes. 'Ah there *is* a man,' she wrote to M. Mohl on April 1, 1870. 'We were like a brace of lovers on our Indian objects.'

Robert Cornelis Napier had gained his title as a reward for his brilliant military feat of storming the supposedly impregnable fortress of Magdala in Abyssinia. He was tall, extremely handsome and was described by Sir Bartle Frere as 'one of the few men fit for the Round Table'. He had begun his career in the Bengal Engineers of the East India Company and risen to be head of the Public Works Department of the Punjab. One of his achievements was the construction of the Bari-Doab canal, 250 miles long, said to have turned a desert into a garden. He was humane

370

based his discipline on confidence not fear, and devoted himself to the welfare of his troops.

In March 1870 he came for a final conference. 'Make no ceremony with me, as an old Père de famille and do not think of getting up and thus fatiguing yourself,' he wrote to Miss Nightingale on March 18.

'He actually spent his last morning in England with me, starting from this house,' she told M. Mohl on April 1, 1870. 'And I sent away the C.I.C. to India without anything to eat! He said he had too much to talk about to waste his time in eating.' Between them they put on paper a complete scheme of Indian Army reform to be begun at once, ranging from barrack and hospital reorganization to the provision of education and physical training. Lord Napier asked her to write to Lord Mayo to prepare his mind for the proposals – 'a letter from you would have great weight as it was you who raised public opinion in England on these subjects,' he told her.

It was not Miss Nightingale's nature to console herself; yet some progress had been made in the ten years since the Indian Sanitary Commission began its work. Her enormous energy, her extraordinary history, her capacity to inspire boundless faith, were producing astonishing results. Mr John Strachey criticized her sharply more than once, but he told Sir Bartle Frere: 'Of the sanitary improvements in India three-fourths are due to Miss Nightingale.' Her fellow workers regarded her as exercising an almost supernatural influence. 'I have often known a scrap of paper on which you had written a few words – or even your words printed – work miraculously,' wrote Sir Bartle Frere in 1868.

Man after man who came to see her in the bright austere drawing-room in South Street fell under her spell. Many began with concealed hostility: she was a thorn in the side of bureaucracy, an interfering aggressive woman to be visited only because it was good policy. Few were not converted. Her sincerity, her disinterestedness, her astounding knowledge were irresistible. Honest men, able men, men who had the good of India and its peoples at heart, became her friends.

But her record was not one of uninterrupted success. She had never been to India, her knowledge was a paper knowledge, and her persistence had its drawbacks. She owed her success to her ability to persist in the face of opposition. Again and again she proved to be right. When she proved to be wrong, she paid the penalty. In 1867 she came to the conclusion that the barracks and hospitals of the army in India were not adequately ventilated. In the hot weather, when infection and disease were most rife, it was the invariable custom to keep the windows shut and only open the doors.* In the Crimea, in the hospitals of London, she had proved the supreme importance of fresh air. In the Crimea she had been told that if the windows of the men's huts were opened during the winter the huts would become so cold that the men would die of pneumonia. She had forced the windows to be opened, and the men had not died. Their health had improved, and they had contracted less pneumonia. Therefore, when she wrote a memorandum to the Government of India advising that the windows of barracks and hospitals should be kept open through the hot weather and was told the men would be made ill with heat, she would not be convinced. She persisted in pressing for a general order to open the windows, writing to doctors, to the secretaries of the Presidential Sanitary Commissions, to Commanding Officers, to the Viceroy himself, until Sir John Lawrence told her bluntly that nothing on earth, even a direct order from the Government at home, would induce him to issue instructions for windows to be kept open in hot weather.

A laugh went up throughout the length and breadth of India. Her supporters were forced to realize that she might be betrayed into ludicrous mistakes, and even to-day Miss Nightingale's attempt to open windows in the hot weather is not forgotten.

A more serious failure followed. The report of the Indian Sanitary Commission had urgently recommended improvement in barrack accommodation, and during 1864 an enormous amount of work was done by Miss Nightingale and Douglas Galton on a model barrack plan. She was aware from the beginning that no single standard plan could be laid down and applied throughout India. Her object was to define the essential features which every barrack must possess and to leave the local authorities to adapt them to local conditions. The scheme was approved by the Government of India and the War Office, and in 1865 she wrote

* It had long been established in India that it was only by keeping the windows and shutters closed as far as possible during the hours when the sun was up, that the lower temperature of the night could be partly retained to make the day endurable.

that she had got a grant of seven millions for 'my Indian barracks'. Work began at once, and the plans were passed to the Royal Engineers.

Presently disturbing news reached her. The Royal Engineers were acting 'in a high handed manner'. They were determined to erect barracks for the army without civilian interference or advice. By the end of 1869 it was evident that the scheme had gone fatally wrong. On December 4 Miss Nightingale wrote to Sir Bartle Frere: 'We begged and prayed to be allowed to put up in Poona and the Deccan where the winds are terrific and the ground rocky, one storied barracks – we were *ordered* to wait. Sir Robert Napier [Lord Napier of Magdala, the Commander-in-Chief] was *ordered* to wait until a 3rd class engineer colonel, an ordinary man such as you can find anywhere, sent us "standard plans", which we were to use and no other and which were extravagantly expensive.'

She was in despair. Not a single one of the new barracks, she wrote to Sir John McNeill in 1869, was erected in accordance with the recommendations. Everything had been sacrificed for the sake of an imposing façade of European design totally unsuited to the climate. Good water, drainage, shade, space, all had been neglected.

The troops moved in, and disaster followed. Cholera broke out at several stations, and early in 1870 there was an outcry in *The Times* against the senseless extravagance of erecting palatial buildings in which the troops died.

In February 1870 Dr Cunningham, Sanitary Commissioner to the Government of India, came home on leave and brought a melancholy account. Miss Nightingale scribbled a note of his remarks for Dr Sutherland. 'The immense block of masonry strikes one at once as quite incongruous and improper. Men cannot drag their cots downstairs out of doors. No pent house roof to keep the rooms shaded. A great palace without any kind of shelter from the sun. Government will provide punkahs and tatties only on one floor. Recreation rooms are there unused with nothing but tables and forms in them. The men lie on their beds for 18 hours a day and think of Cholera. Cholera has broken out in the new barracks at Allahabad, Lucknow, Morar, and typhoid as well as at Jullundur.'

After Dr Cunningham, Lord Napier of Magdala called. From reports he had received he feared the position with regard to the barracks was even worse than she supposed. 'Instead of the old sunburnt brick which was cool, they had to build the barracks of solid masonry and red brick in order to support the upper storey. The upper part is *never* cool. The slated roofs of the barracks made them intolerable, they scorched the tops of the men's heads. Commanding officers were abandoning the upper storey and even moving their men back into the old barracks, as at Jullundur. Water had not been laid on and there was no drainage. Land inside the cantonments had been sold to civilians at a high price because inside cantonments they were not subject to rates, but they also were not subject to Army Sanitary rules, and the vicinity of the barracks was filthy. Dirty coolies of the lowest class with their families had been allowed to live in the barracks before the troops moved in, and the barracks were consequently infected and filthy before the troops arrived.'

The cause of sanitary reform had received a serious set-back. The money had been forthcoming, the work had been promptly done, and the result was complete failure. That the reformers were in no way responsible, that what had been done was a complete contradiction of every essential laid down, was impossible to explain. Military secrecy, military etiquette veiled the issue in hopeless obscurity, and not only the public but officials within the War Office who had been in favour of sanitary improvements now associated them with sentimental extravagance.

She was losing ground on all sides. She was already shut out of the War Office and the Poor Law Board, and in India the barrack failure must weaken her hold on Lord Mayo. Yet she hardly rebelled. In the last two years she had changed. 'I assure you I don't let these things corrode into me now,' she told Clarkey in 1868. She had worn herself out. Her last collapse, in December 1867, had weakened her not only in body but in mind. Some of her energy, some of her power of fierce feeling had gone. 'I am becoming quite a tame beast – fit for a lady to ride or drive – as horse dealers say of their most vicious brutes,' she wrote to M. Mohl in September 1868.

In the summer of that year she went to Lea Hurst for three months. She had not been there since 1856, and the visit marked a change in her way of life. While she was at Lea Hurst, she read the novels of Jane Austen, who, she told Clarkey in September 1868, in her opinion, 'ranked second to Shakespeare in the English language for dramatic power,' and the plays of Shakespeare – 'I don't know whether Hamlet was mad, he would certainly have driven me mad.'

Jowett spent a week at Lea Hurst; Parthe stayed away: Miss Nightingale had refused to go there unless 'Parthe and her governessing are excluded'. She had long talks with W. E. N. on metaphysics. Her mother, she wrote, was 'more cheerful, more gentle than I *ever* remember her tho' of course she is much aged. Her memory is nearly gone but to me she is far dearer, far more *respectable* than ever before.' Fanny was now eighty, W. E. N. seventy-four.

By the end of September she was bored. She wrote on September 27 imploring M. Mohl to stay on at South Street 'until I come which will be, please God, on Friday or Saturday'. Family life, books, friends were not sufficient; she needed an outlet. For some time Jowett had been urging her to write and in the summer of 1868 she began a 'Treatise on the Reform of the Poor Law'. Her loss of power was at once apparent. Jowett had to send her a draft before she could start, and even then she found composition an intolerable strain. The treatise was shortened to an article, but even so she could not complete it.

She asked advice from Sir Harry Verney, who was an authority on the Poor Law, from Parthe, who had had two novels published, and from Dr Sutherland. In the autumn she wrote in despair to Dr Sutherland. 'I have adopted *all* your corrections and *all* Parthe's and *all* Sir Harry's: and they have taken out all my *bon mots* and left unfinished sentences on every page; and this *kind* of work really takes a year's strength out of me; and now you MUST help me.' The article was put into shape by Dr Sutherland and sent to Froude, who published it in *Fraser's Magazine* for March 1869 under the title 'A Note on Pauperism'. It attracted considerable attention, and Sir Robert Rawlinson told Miss Nightingale on March 11 that Carlyle had praised it. She used the arguments of her earlier papers and memoranda on Poor Law reform and added a suggested scheme for large-scale State-subsidized emigration 'to bring the landless man to the manless lands'.

In September 1869 she wrote a letter to Dr Sutherland enclosing a large mass of notes on the subject, asking him to expand them into a book after her death, and to do the same for *Notes on Lying-in Institutions*. But she herself, though she worked on for more than twenty years, never touched either again. The days of her great achievements, when she had written the huge volume of *Notes on Matters affecting the Health, Efficiency and Hospital Administration of the British Army* with her own hand in six months, were over. She was no longer capable of the sustained effort necessary to write a book.

The energy with which she had once sprung on opportunity like a tiger had also left her. Her life was growing calmer. It had been, she wrote to Jowett in 1865, 'a fever and not a fitful one. Neck or nothing has been all my public life. ... Could I help in the two Royal Commissions I have served, in the 9 years I have served in the W.O. [War Office] exclusive of the Crimea my whole life being in a hurry? If the thing were not done to the day, it were not done at all.'

Not only were the demands of the work less frenzied, but she herself was working in a less frenzied atmosphere. A new influence for reasonableness had come into her life. Clough had been succeeded as secretary of the Nightingale Fund by Hilary Bonham Carter's brother, Henry. Henry Bonham Carter was devoted to Miss Nightingale; he gave up more than forty years of his life to her service, but he would not become a slave. When it was getting late, says Sir Edward Cook, he used to say, 'Now I must go home to dinner'; he was an excellent man of business and invaluable to Miss Nightingale, but his soul remained his own.

In 1870 some of the urgency of old days returned. War was declared between Germany and France in June, and in July the 'National Society for Aid to the Sick and Wounded', subsequently called the 'British Red Cross Aid Society', was founded at a public meeting in London.

Miss Nightingale was pressed to give up all other work and take control. The need was very great and heartrending reports of the sufferings of the troops on both sides were being received. But she refused; one laborious memorandum on sanitation in India affected the lives of millions on whom, even if her health allowed, she could never turn her back to become the Lady with the Lamp once more.

Though she had declined to be in charge of the National Society, its activities were under her direction. Sir Harry Verney, Douglas Galton, and Miss Emily Verney, Sir Harry's daughter by his first wife, were on the executive committee. Henry Bonham Carter and Dr Sutherland were sent by the Society to visit both the French and German hospitals during the war, and after it the

port on the Society's work was written by Dr Sutherland under
r supervision.

She advised the executive committee on organization and
ministration. 'Those who undertake the work of aiding the
ck and wounded must not be sentimental enthusiasts but down-
ght lovers of hard work ... attending to and managing the
ousand and one hard practical details which never the less
ainly determine the question as to whether your sick and
ounded shall live or die,' she wrote on August 2, 1870. It was
ne years to the day, she noted, since Sidney Herbert died. She
vised on practical matters from the administration of field
nbulances to the pattern of hospital suits and cooking utensils.
e wrote to workers at the seat of war; she interviewed volun-
ers for service; she directed and supervised the purchase and
spatch of supplies. She collected money from her friends which
as used chiefly for the relief of prisoners of war and recorded
nding out £5,000 in one week. Once again in scenes of horror
d confusion it was found that the quickest way to get things
ne was to go to Miss Nightingale. The amount of correspond-
ce involved was a strain. 'Every man and woman in the world
ems to have come into it with the express purpose of writing to
e,' she told Madame Mohl in 1870. 'Would I could go to the
at of War instead of all this writing, writing, writing.'

At first her sympathies were with Germany. She considered
apoleon III a tyrant and disliked and despised the Empress
ugénie, whom she described as 'the Empress who was born to
a dressmaker.' Germany was the home of liberal thinking,
usic, and philosophy. M. Mohl, whom she loved dearly, was a
erman; she herself had been trained at Kaiserswerth on the
hine; Prince Albert, for whom she had a profound admiration,
d been a German. Germany stood for music, folk-songs, sim-
icity, and thought. She had to discover that since her girlhood
startling change had taken place, and that in place of Germany
d risen Prussia.

Through her influence a War Office ambulance had been equip-
d and sent out to the German Army with the double purpose
assisting the German wounded and of observing and noting
e treatment and requirements of wounded in a large-scale
odern war. The Germans would not allow it to be used. In
ovember 1870 she wrote to Douglas Galton that she had heard
at the War Office ambulance was 'cold shouldered' by the

Prussians. On December 12 he wrote: 'Every foreign ambulance has a Prussian N.C.O. in it. Ours is the only exception because . has only one patient in it – a casual.'

On November 4, 1870, she wrote to M. Mohl: 'Is it not qui' unknown in history that a philosophical, a deep thinking, th' most highly and widely educated and in some respects the mo' civilized nation in Europe – the Germans, should plunge hea' foremost into this abyss called Military Despotism? That the should not *see* that that (*soi-disant*) German Unity means onl' Prussian aggrandisement.' The alacrity with which Germa' philosophy and culture hastened to prostrate themselves befo' the Prussian war machine left her bewildered. 'Now if you tak' all the greatest names in science, in literature, or metaphysic' and religious philosophy, in art, of the last 70 or 80 years in a' Germany, will you tell me how many of these came out of Berlin' she wrote in a private note. 'Yet the higher civilization is to t' subjected to the lower.' 'The free translation of German nationa' ity is Prussian military supremacy.'

German behaviour after the defeat of the French finall' alienated her. 'After the fighting,' she wrote to M. Mohl i' February 1871, 'come the miseries of the poor people. Corr' spondents known and unknown write to me by every post.' Sh' deplored Bismarck's 'rapaciousness', his 'want of delicacy or o' any nobility'. She had loved the German language, the Germa' mind, and the German way of life. All that had perished. Ther' had been a death, but the death was of Germany, not Franc' Worst of all, she realized that a new age had dawned for Europ' 'Prussia,' she wrote in a private note of 1871, 'openly says she do' these things because the first Napoleon did them 64 years ag' And France will say, long before 64 years hence, she will do the' because the Corporal Emperor King did them so many years ag' Horrible as is the account of wounds and grief and starvin' people, it is as nothing compared with the principles which th' War has put forth and brought to life.'

In 1872 Jean Henri Dunant, a Swiss banker, paid a visit t' London. He had succeeded in turning what the world assured hi' was a Utopian dream into hard fact – he had brought about th' Geneva Convention and founded the International Committe' of the Red Cross. In 1872, after the Franco-Prussian war, Dunar' visited London and read a paper on the work of the Society. H' first words were these: 'Though I am known as the founder of th'

Red Cross and the originator of the Convention of Geneva, it is to an Englishwoman that all the honour of that Convention is due. What inspired me ... was the work of Miss Florence Nightingale in the Crimea.'

She had legendary prestige and enormous popular appeal, and while she had these she could not be without power. Throughout 1870 and 1871 she debated the possibility that she might 'seek office' again. If she pursued Ministers, she might make her way into the departments once more. But such a course was contrary to the policy of her life. She had succeeded because she had made herself an instrument in the hands of Ministers, because she had been sought out, not seeking. Mr Gladstone's Government had no place for her. She despised his Ministers; she called them contemptuously 'Gladstone's secretaries', and though they treated her with deference she felt they were antagonistic. 'Here is a note from Mr Cardwell,' she wrote to Sir Bartle Frere in 1870, 'which seems to me, I don't know why, a nasty one.' Her powers of working and concentrating had declined, and she had come to depend entirely on Dr Sutherland. 'The only way I can work now,' she wrote to him in 1870, 'is by receiving written notes from you, and working them up into my own language, then printing and showing you the work.' Among her papers are hundreds upon hundreds of drafts in Dr Sutherland's hand covering every subject from Indian sanitation to family letters. She had lost faith in herself and leant on his judgement. 'I have been through them all,' he wrote of the Indian Sanitary reports, 'and you may safely say they are very well done.'

Loss of influence in Indian affairs finally decided her. Her hopes of Lord Mayo were not being fulfilled. It was not that he wished to do too little, but that he wished to do too much. 'He got,' wrote Sir Bartle Frere in 1874, 'into the hands of men who were like the Fisherman's Wife who never would make the best of what the Enchanted Fish gave her but always wanted something better.' With Sir Bartle Frere's help Miss Nightingale had preached irrigation to him. To her dismay in 1871 he sent home what Sir Bartle Frere described as a 'wild and visionary project' for providing every *ryot* in India with water at once. When it was pointed out that the money to pay for the enormous works involved could not be raised by tax until after the water had been provided Lord Mayo 'sulked'.

Miss Nightingale was profoundly discouraged. She had lost

Lord Mayo and with him her influence in the Government of India; under the present Government her influence in the War Office and at the Poor Law Board was at an end; she was wretchedly ill and overwhelmed with other and vitally important work. She decided she would struggle no more with Government departments. As 1871 passed into 1872 she wrote on a sheet of paper: '1872. This year I go out of office.'

Almost immediately she had proof that her term of office had already ended. In February Lord Mayo, while inspecting a penal settlement, was assassinated by a convict. He was succeeded by Lord Northbrook. She knew Lord Northbrook personally; he had been a friend of Sidney Herbert, but he did not consult her or call to see her before he sailed.

She was deeply wounded. 'Why should you be troubled at the Governor General not coming to see you (as he most certainly ought to have done),' wrote Jowett on April 3, 1872. 'Put not your trust in Princes, or Princesses or in the War Office or in the India Office; all that kind of thing necessarily rests on a sandy foundation. I wonder that you have been able to carry on so long with them.'

It was sixteen years since she had returned from the Crimea to instigate the Royal Army Sanitary Commission of 1857. For sixteen years she had laboured in Government departments, sacrificing health, pleasure, friends. She had done the work of a Secretary of State, she had 'made the appointments'. Now all that was over. Henceforward she must lead a new life. What kind of life should it be?

22

HER first thought was to live again for hospitals. One of the most painful sacrifices of her life had been the renunciation of hospital work for administration after her return from the Crimea. She had declared repeatedly that of all people she was least suited to the writing of regulations, that pen-and-ink employment drove her mad, that she starved without human contacts. 'My life is as unlike my Hospital life when I was concerned with the souls and bodies of men as reading a cookery book is unlike a good dinner,' she wrote to Rev. Mother Bermondsey in 1864.

She had already arranged that, when she could work no more, she should be taken to a hospital. In January 1864 she wrote to Mrs Bracebridge: 'You know that I always believed it to be God's will for me that I should live and die in Hospitals. When this call He has made upon me for other work stops, and I am no longer able to work, I should wish to be taken to St Thomas's Hospital and to be placed *in a general ward* (which is what I should have desired had I come to my end as a Hospital Matron).'

She was not prepared for her present situation. She had assumed that only death would release her from the obligation laid on her by God to do administrative work. She had assumed that she would leave the work – instead the work had left her. She was 'out of office', but she was not dying. Indeed, though she was fifty-two and an invalid, her expectation of death was more remote than it had been for sixteen years; she determined, however, to apply to St Thomas's to enter a general ward as an ordinary patient; the cost of her establishment at 10 South Street was large and W. E. N.'s financial affairs were not prosperous. While she was 'in office', while she was toiling at the work God had called her to do, the expense had been justified; now she was out of office the expense should end.

She determined to leave the world. She, the most famous woman in two continents, the friend of queens, the adviser of governments and ministers, would end her days lying side by side with poor working women subject to the discipline and the rigours of a hospital general ward. It was a scheme which appealed to her sense of drama, but when she confided it to Jowett he was alarmed. 'Something which you said to me on Sunday has rather disquieted me,' he wrote on June 22, 1872, 'and I hope that you will allow me to remonstrate with you about it. You said that you were going to ask admission as a patient to St Thomas's Hospital. Do not do this. (1) Because it is eccentric and we cannot strengthen our lives by eccentricity. (2) Because you will not be a patient but a kind of Directress to the Institution, viewed with great alarm by the doctors. (3) When a person is engaged in a great work I do not think the expense of living is much to be considered; the only thing is that you should live in such a way that you can do your work best. (4) I would not oppose you living at less expense if you wish, though I think it a matter of no moment, but I would live independently. (5) Do you really mean to live as a patient? It will kill you. I do not add the annoyance to your father of a step which he can never be made to understand; I look at the matter solely from the point of view of your own work. I have cared about you for many years; and though I have little hope of prevailing with you, I would ask you not to set aside these reasons without consideration.'

She yielded. Her affection for Jowett triumphed, and she told him she would set aside duty and conscience for his sake and abandon her plan of entering St Thomas's. On July 11 he wrote that he was flattered he had prevailed and now would she not allow herself a little happiness? A new period in her life was beginning – 'will you try to hope and be at peace?' He urged her to be calm, to tranquillize herself, to achieve a philosophical attitude.

But it was impossible, for a great gap yawned in her life. She was out of office; the press of departmental work no longer made every day a fever – and how was she to occupy herself? In a private note of 1872 she wrote: 'Never has God let me feel weariness of active life, but only anxiety to get on. Now in old age I never wish to be relieved from new work, but only to have it to do.'

Though she had planned to live a life of austerity, poverty, and discipline in a general ward at St Thomas's, she had never intended to retire from participation in the affairs of the hospital, and a new chapter in its history had just opened. In 1871 it moved from its temporary quarters in Surrey Gardens to the new buildings in Lambeth with which she had been closely concerned. The plans embodied her ideas, and every detail of equipment had received her meticulous attention. On the subject of hospital floors alone she had exchanged almost a hundred letters with Dr Sutherland, Douglas Galton, manufacturers, matrons, doctors, architects.

The Nightingale Training School had also reached a crisis. Miss Nightingale's attention had been distracted from it by the demands of Poor Law reform and India. Now she found that it had fallen away from its original standards. In the spring of 1872 she began an investigation into the teaching and organization of the Nightingale School. In the new St Thomas's the school had larger quarters and would train more probationers. Finding an urgent need for reorganization and reform, she made a new plan for herself; she would live near St Thomas's and devote her life to the training school and the hospital. Mrs Sutherland was set to work to find suitable lodgings in the district.

But in the summer of 1872 a drastic change took place in her life, and she was forced to return home. For the past three or four years Fanny and W. E. N. had been an increasing anxiety. They were old and ailing; in 1872 W. E. N. was seventy-seven and Fanny eighty-three, and the management of their property and their two establishments at Embley and Lea Hurst had become unsatisfactory. The position was peculiarly difficult. Since W. E. N. had no son, by his uncle's will the properties of Embley and Lea Hurst passed on his death to Aunt Mai and next to her son, Shore. Uncle Sam had become an exacting invalid, and the close ties of blood (in addition to Aunt Mai being W. E. N.'s sister, Uncle Sam was Fanny's brother) made criticism easy and businesslike arrangements difficult.

As soon as Miss Nightingale visited her family again she found herself elected into being the man of business of the family though most unwilling to accept the position. 'People who have carriages and butlers and housekeepers and who drive out every day for their pleasure and dress and go out every day, ask me, who have none of these things and am always in bed – and am chained to the oar – ask ME to pay their bills and do their business,' she wrote to Clarkey in the summer of 1868: the gibes were at Parthe. Parthe was 'always, as she always had been, the spoilt child', Parthe fussed over her health and if she had an aching foot 'made a tohu-bohu and would not put it to the ground.' In fact

383

her sister was suffering from the first symptoms of the arthritis which in a few years turned her into a helpless cripple.

It was not possible for Miss Nightingale to see things go wrong without trying to put them right, and against her will she was forced into assuming responsibility. By the autumn of 1871 her parents' affairs were interfering with her work. On October 3 she wrote to Jowett from Lea Hurst: 'I was due in London to-day – but have been kept here for the last 6 weeks and shall be for a few days more by doing some most harassing and painful business (looking into things which had gone *very* wrong) for my father and mother (which has taken more out of me than two years of real Crimean work). Do not mention this please.'

In the summer of 1872 she was forced to leave London and spend eight months with her parents at Embley. In June Fanny's old housekeeper, Mrs Watson, died: she had been in the Nightingale's service for twenty-five years and had been the first person to welcome Miss Nightingale on her return from the Crimea. Her death was followed by confusion. The discipline of the household had become slack. The servants, wrote Miss Nightingale, did what they liked; not one of them was doing his or her proper work and to put the household of Embley into order was as difficult as organizing the Barrack Hospital at Scutari.

Month followed month, and it was impossible for Miss Nightingale to leave Embley. Life was filled with the misdeeds and complaints of housemaids, kitchen maids, footmen, cooks, and with unwelcome discoveries in household and estate accounts. 'I am so stifled by dirty anxious cares and sordid *defensive* business,' she wrote in a private note of August 1872. 'Like the maid of all work who has to wipe her dirty hands on her dirtier apron before she can touch clean people.'

She was imprisoned again. Her father and mother clung to her; they were old and helpless, and her heart forbade her to abandon them. Miss Nightingale was fifty-two, but she had lost none of her capacity to suffer. 'Oh to be turned back to this petty stagnant stifling life at Embley,' she wrote in a private note of the late summer 1872, 'I should hate myself (I *do* hate myself) but I should LOATHE myself, oh my God, if I could *like* it, find "rest" in it. Fortunately there is no rest in it, but ever increasing anxieties. Il faut que le victime soit mise en pièces. Oh my God!'

The change in her circumstances was startling enough to try the strongest nerves. In place of a life in which every hour was

384

filled with matters of vital interest and national importance, she found herself in an isolated country house alone with two aged semi-invalids, spending her days in satisfying their exactions, smoothing over their difficulties with their servants and straightening their finances.

Fanny and W. E. N. could not be left alone together. Fanny, childish, almost blind, her memory gone, was a responsibility that W. E. N., seventy-nine and failing, could not undertake.

The thought of work piling up in London, of the reform and reorganization of the Nightingale School crying out to be done, while she was held prisoner at Embley, was agony to Miss Nightingale. All through the winter of 1872 she chafed. In the spring of 1873 she could bear it no longer. She must be in London; Parthe was ill and could not help – Fanny must come to London. The drawing-room floor at South Street was fitted up as a bedroom and sitting-room, and Fanny came to London in the spring of 1873.

Once in London Miss Nightingale threw herself with desperate haste into the reconstruction of the Nightingale School. The first task was to tighten up the technical side of the training. Soon after she arrived back, her friend Mr Whitfield, the Resident Medical Officer of St Thomas's, who had supervised the medical training of the Nightingale probationers since the foundation of the school, gave up the post owing to the extra work involved in the new enlarged hospital. His place was taken by Mr Croft, one of the honorary surgeons. In April and May Miss Nightingale and Mr Croft drew up a new plan of instruction. The standard of examination was raised; probationers were required to undertake a course of reading planned by Mr Croft and Miss Nightingale and were also at intervals to submit their notebooks for her inspection. Within a few months he reported that the work was improving and 'the answers collectively are much better than they have been for years'.

*

Miss Nightingale held that the training and education of a nurse, or indeed any education or training, was made up of two aspects of equal importance. First, the acquisition of knowledge which was properly tested by the passing of an examination; second, the development of character which could not be tested by the passing of an examination.

To improve the development of character, she created a new

post. Mrs Wardroper, the matron of St Thomas's, who supervised the probationers, now found, like Mr Whitfield, that the new hospital made greater demands on her time. An Assistant Superintendent was appointed with the title of Home Sister; the Home Sister was to make herself the girls' friend; she was to encourage them to read poetry, to listen to music, to go regularly to church. She was to inculcate a standard which would keep the Nightingale nurses 'above the mere scramble for a remunerative place'.

All influences, however, were secondary to the influence of Miss Nightingale herself. She dominated the school. From 1872 onward she determined to make herself personally acquainted with every probationer, and as soon as a girl had completed a trial period she was interviewed by Miss Nightingale, who wrote a character sketch which formed the first item in a dossier composed of examination results, notes of further interviews, letters, and comments. Miss Nightingale invited the probationers' criticisms and comments on the treatment they received from the sisters and the value of their medical lectures; she invited comments from the sisters on the character and conduct of the probationers. When she received a complaint or a suggestion which seemed to her to be worthy of notice, she wrote a memorandum to the persons concerned.

It was work with human beings again, the work for which she had longed. After the long dry years of toiling at administration, her life was rich once more. 'I am over whelmed,' she wrote to M. Mohl on June 21, 1873, 'in a torrent of my Trained Matrons and Nurses, going and coming, to and fro, Edinburgh and Dublin, to and from Watering Places for their health, dining, tea-ing, sleeping – sleeping by day as well as by night.'

Her stay in London had to be cut short when Fanny's maid and her maid quarrelled; Fanny became unwell and had to be taken back to Embley. Just before they left Miss Nightingale was upset by the loss of her favourite cat Mr Muff, who had been left behind at Embley and allowed to wander in the woods, where he was presumably shot by a keeper; she was broken-hearted – 'I have no one now to say like Ruth "Intreat me not to leave thee",' she wrote to M. Mohl in June. 'Poor Mr Muff said it, if ever Ruth did.' By the end of June Miss Nightingale was back at Embley chafing at being separated from her work, miserable, frustrated. The solution, Jowett told her, was to resign herself to dropping active work and to concentrate on writing. He greatly admired the powers of her mind, he was convinced she had a message to give the world, and he believed she could spread her message more widely with more happiness to herself if she expressed herself by writing.

In October 1872 he suggested that she should write some essays for the reviews on the Idea of God. 'During the ten years and more that I have known you,' he wrote, 'you have repeated to me the expression "Character of God" about 1000 times, but I can't say I have any clear idea what you mean.' The suggestion attracted her – all her life, in periods of unhappiness, she had found relief in exercising her mind on philosophical ideas. She wrote three essays on the Laws of the Moral World which repeated the ideas she had earlier treated at length in *Suggestions for Thought*. Two of the essays were published by Froude in *Fraser's Magazine* for May and July 1873 under the titles 'A Note of Interrogation' and 'A Sub-Note of Interrogation: What will our Religion be in 1999?'

Jowett also invited her to help him in revising his translations of the Dialogues of Plato – she still had considerable facility in Greek – and he placed a high value on her interpretations. 'You are the best critic I ever had,' he told her in 1872. He used her suggestions in his introduction to the *Republic* and wrote, 'I am always stealing from you.' In July 1873 she sent him a letter on the *Phaedrus* and he told her that he had 'put in most of what you suggested.'

At the end of 1872 he asked her to make a selection of Bible stories for a Children's Bible. Enclosing her selection she wrote: 'The story of Achilles and his horses is far more fit for children than that of Balaam and his ass, which is only fit to be told to asses. The stories of Samson and of Jephthah are only fit to be told to bull dogs; and the story of Bathsheba to be told to Bathshebas. Yet we give all these stories to children as "Holy Writ".' She summarized the book of Samuel and the books of Kings as 'Witches. Harlots. Talking Asses. Asses Talking. Young Gentlemen caught by the Hair. Savage Tricks. Priests' Tales.' Jowett was delighted, and on February 10, 1873, told her that she would find her suggestions had been adopted almost entirely and that he blessed her every time he took up the book.

And now she discovered there was a message she wished to convey to the world. Its nature was surprising. It had nothing to do with sanitary reform. She had turned away from practical affairs to the life of the soul.

Miss Nightingale was a mystic. She was not a contemplative. Like St Elizabeth of Hungary, she was an administrator. The union of a busy and active life with the practice of mysticism was normal. Yet mysticism had come to be regarded as apart from ordinary life, a practice confined to saints enclosed in convents and hermits in their cells. In the autumn of 1872 Jowett suggested she should compile a book of extracts from the medieval mystics translated by herself showing the application of mysticism to present-day life. 'You will do a good work,' he wrote on October 3, 'if you point out the kind of mysticism which is needed at the present day.' She began to work on a book and drew up a title-page, 'Notes from Devotional Authors of the Middle Ages. Collected, Chosen and freely translated by Florence Nightingale'.

As she worked she sent her extracts to Jowett, who became interested in her comments. On April 18, 1873, he wrote suggesting that she should add a preface to the book, formulating her conception of mysticism and giving guidance as to how mystical books should be used. 'I think it is clear,' he wrote, 'that this mystic state ought to be an occasional and not a permanent feeling – a taste of heaven in daily life. Do you think it would be possible to write a mystical book which would also be the essence of Common Sense?'

Through the summer and autumn of 1873 work on the book and the preface was her chief solace. She needed a 'taste of heaven in daily life'; her own had become a round of coaxing servants, humouring her parents, struggling to persuade them to allow her to straighten their neglected affairs. By December she wrote to M. Mohl that she was 'completely broken'.

Worse, however, was to follow. In January 1874 she escaped for a few weeks: Parthe was a little better, and as she and Sir Harry Verney were able to come to Embley Miss Nightingale hurried to London. On January 10 she heard that her father was dead. He had gone upstairs before breakfast to fetch his watch, slipped on the stairs, and died instantly.

The affection between them had been very deep. 'His reverent love for you,' wrote Richard Monckton Milnes on January 13, 1874, 'was inexpressibly touching.' But grief only too soon took second place as she became overwhelmed by the painful anxieties, the innumerable difficulties which arose out of his death.

Embley and Lea Hurst were now the property of Aunt Mai. It was not easy to turn out, it was melancholy to see Fanny, eighty-six and childish, weeping because she could not understand why she was being sent away from her home. Unhappiness and friction were inevitable, and the family temperament intensified them. 'We Smiths,' Fanny had once said, 'all exaggerate'; they were tenacious, voluble, determined, and easily offended. 'We are a great many too strong characters,' wrote Miss Nightingale to Clarkey in 1874, 'and very different; all pulling different ways.'

She went down to Embley to be with her mother. All the painful wearing business, she wrote, was left to her, and in a few weeks she was reduced to misery. She asked Aunt Mai to offer Fanny a home at Embley, but Aunt Mai refused – she was seventy-six, was crippled with arthritis, and had the responsibility of an invalid husband. The only solution was for Miss Nightingale to give up her work and take her mother to Lea Hurst, which the Smiths did not require and after which Fanny 'craved and longed'. 'I am utterly exhausted,' Miss Nightingale wrote. 'Not a day passes without the most acute anxiety and care. Oh the cruel waste of time, of all real work. If our family could neither read nor write, if they had only a limited number of Serb words at their disposal, and if postage were 5/– an ounce, how happy life would be. How happy it was in the Crimea on account of these things; that *was* living in spite of misery.' Her sole comfort was the consideration and good sense of Aunt Mai's son, the heir to the property, 'my boy Shore'. The affection she had lavished on him was, she wrote, a thousandfold repaid.

Once more there was no escape. Old, feeble, and unwanted, Fanny had a claim which to Miss Nightingale it was impossible to reject. The weary business of clearing up at Embley dragged on. 'Everything has gone from my life except pain,' she told Clarkey on June 8, 1874. In July Embley was given up, and she took Fanny to Lea Hurst.

Fanny's mind had almost failed, and she was blind. Surrounded by familiar objects, though she could not see them, hearing familiar voices, though she did not recognize them, she was at peace. But when she was taken to strange places, heard strange voices, she became agitated and unhappy, and wept. In her lucid moments she returned to the past. 'Where is Flo?' she asked one day. 'Is she still in her hospital?' Then she gave a sigh, 'I suppose she will never marry now,' she said. So very dim were her apprehensions that Parthe and Clarkey thought Miss Nightingale was making an unnecessary sacrifice: but she could not leave her mother to strangers. The tenderness which helplessness and suffering evoked in her were on Fanny's side now.

Many times in her life she was desperately unhappy, but never unhappier than during the summer and autumn of 1874 at Lea Hurst. It was, she wrote, 'utter ship wreck'. Writing had been a solace; now writing was impossible, and her book on the mystics was laid aside never to be resumed. Every minute which could be snatched from struggling with domestic problems was devoted to trying to preserve some part of her work. She habitually rose before dawn – her letters are headed '5 a.m.', '6 a.m.', '4–8 a.m.', 'Before it is light'. An enormous amount of writing was required. 'Because I am not in London I have to write 100 letters to get one thing done,' she wrote. Weariness grew on her. She implored Dr Sutherland not to crease her drafts 'because I have no strength to re-write.' She doubted herself – 'I am afraid I am dreadfully prolix.' 'I have put in far too much detail,' she told him in 1874.

One night, she recorded in a private note, the shadow cast by the night-light on the wall reminded her of Scutari. 'Am I she who once stood on that Crimean height? "The Lady with a Lamp shall stand." The lamp shows me only my utter ship wreck.'

Despair alternated with passionate self-reproach. She reminded herself that if failure were God's will then to rebel was the worst failure of all. She must force herself to believe that her present sufferings were not useless but part of God's scheme for the world. 'I *must* believe in the plan of Almighty Perfection to make us all perfect.' She must not snatch the management of the world out of God's hands. In practical details she found herself apt to give the Divine Will directions. 'I MUST remember God is not my private secretary,' she wrote on an odd scrap of paper.

It had never been her habit to live in the past, but now circumstances forced her thoughts backward. A succession of deaths removed figure after figure who had played an important part in her early life, and as she pined at Lea Hurst not only her present but her past seemed slowly dying. 'My friends drop off one by one,' she wrote to M. Mohl in May 1873. 'Every individual who formed my committee in 1857, many of them hardly older than myself, is dead. And I hang on.' In August 1872 Mr Bracebridge died. A host of memories rushed in on her. Her stay in Rome – 'I never enjoyed any time in my life so much as my time at Rome' – his sympathy in her early struggles, his work for her at Scutari. When he died, she was at Embley and miserably unhappy. Mrs Bracebridge's overwhelming grief provoked her to bitterness. In a private note she wrote: 'Sometimes I think that I am glad that when I go there will be no such heart rending grief felt for me as when two are parted, who have lived for nearly half a century with each other and for each other – or as I felt when Sidney Herbert died and feel every day more and more. On Friday he will have been dead 11 years.' She added, 'There are things worse than death.'

In May 1873 John Stuart Mill died suddenly at Avignon. His death, she told M. Mohl, was a great shock to her. On January 31, 1874, a week or two after W. E. N., Mrs Bracebridge (Σ), died after a long and painful illness. 'A dreary end for her who had been all warmth and radiance,' Miss Nightingale wrote. Mrs Bracebridge had been Sidney Herbert's close friend and for years had spent the anniversary of his death with Miss Nightingale. 'This is to me like the last parting with my past,' she wrote in a private note.

In July 1874 another familiar figure vanished – Lord Dalhousie, formerly Lord Panmure. 'I felt the death of Panmure, my old *enemy*, tho' I was always *friends* with him,' wrote Miss Nightingale to M. Mohl on August 11, 1874, '. . . it was the last breaking up of old associations, of strife and struggle for noble aims and objects; the last ghost disappearing of my Sidney Herbert life. . . . He used to call me "a turbulent fellow".'

Chapter after chapter was closing; figure after figure left the stage. She alone survived – but for what?

Shore and his wife Louisa tried to help her, Shore suggesting that he and his wife should have Fanny at their house in London for some months between October and July when it was most urgent for Miss Nightingale to be in London for her work. Parthe could do nothing – her health was worse, and it was evident that she was seriously ill. In addition to Shore Miss Nightingale had a new helper, Miss Paulina Irby.

Paulina Irby had cherished a passionate admiration for Miss Nightingale since girlhood, and, inspired by her example, had gone to Kaiserswerth to be trained as a nurse. She was a Greek scholar and a woman of great nobility of character who had devoted her life to the relief of the sufferings of the Christian populations of Bosnia and Herzegovina, struggling to emancipate themselves from Turkish rule. She had stayed constantly at Embley, and now looked after Fanny and took some of the domestic burden off Miss Nightingale's shoulders.

It seemed that here were arrangements by which she might have been relieved, but they did not work smoothly. In the summer of 1875 Fanny went to stay with Shore and his wife, but became so ill and unhappy that she had to be taken away. South Street was deserted at a critical juncture, and Miss Nightingale found herself in a villa at Norwood. On June 18 she wrote to Clarkey: 'I am "out of humanity's reach" in a red villa, like a Monster Lobster in charge of my mother by doctor's orders, as her only chance of recovering strength enough to see her old home (Lea Hurst) after which she cruelly craved. ... It is the only time for 22 years that my work has not been the first reason for deciding where I should live and how I should live. Here it is the *last*. It is the caricature of a life.'

Miss Nightingale's conscience deprived her of Paulina Irby's help. Paulina was becoming absorbed by the Nightingale family troubles, but she was in England to collect funds for her work in Bosnia, not to look after Fanny. In a private note written in the summer of 1875, Miss Nightingale sternly reminded herself that, however great the temptation to keep Paulina, the decision to send her away must be right. The world, as she had so often told Hilary Bonham Carter, was divided into devourers and the devoured. Was she, F. N., now to become a devourer and allow Paulina to sacrifice herself and her work to 'my poor mother's state and the family affairs'? At the beginning of 1876 Paulina was sent back to Bosnia.

In January Miss Nightingale sustained another great bereavement – M. Mohl died of 'a peritonitis'. She was heartbroken: she had cared deeply for M. Mohl, and he had been devoutly attached to her. No shadow had fallen on their friendship for nearly forty years. 'It seems,' she wrote, 'as if a great light had gone out of the world.'

By the autumn she was on the edge of breaking down. On November 28 she wrote to Clarkey from Lea Hurst: 'The good Shores have taken my mother back. But I am so worn out – not having had one day's nor one hour's rest since my Father's death 3 years ago come January, that I am staying on here for a few days silence. An eternity of silence seems too short to rest me.' Difficult as 1876 had been, 1877 was more difficult still. 'O my darling,' she wrote to Clarkey in June 1877, 'how impatient you are when your sister does but propose to you a companion – think of me – not proposed but obliged – and this is the fourth year and such companions – obliged to take charge of poor Mother, companion and a pack of new and strange servants.' In July she told Douglas Galton: 'I don't know when I shall be able to work again.' The summer of 1877 was disastrous. One of the servants developed smallpox; there was a scandal, and defamatory articles appeared in the local paper. This unpleasant business forced her to stay on in the country, and in October she wrote to Douglas Galton that there was no chance whatever of her being able to do anything in London during the autumn.

In her opinion the property at Lea Hurst was not well administered, and she set herself to do some of the work which she considered it was the Smiths' duty to do. Family relations were strained further, and she added the innumerable small unpleasantnesses and disappointments inseparable from local affairs to her burden. In June 1878 she wrote to Lord Napier of Magdala that she was 'ground to powder in the country by family affairs. I have not had one day's rest since my father's death 4½ years ago. ... There is no one to do anything for the place or the people.'

And yet she managed to keep control of the Nightingale School. When she could get to London, she saw her nurses and probationers constantly; every girl who trained at the school was still personally known to her; and, above all, she wrote to probationers, to nurses, to matrons, to those who were still in the school and those who had left it. Once a girl had become a Nightingale nurse, she did not slip out of Miss Nightingale's hands when her training was completed. Miss Nightingale did not approve of her nurses taking posts which had not been arranged by her. A nurse who had trained under her close supervision went to a post arranged and approved by her and continued to receive letters of advice. When Miss Torrance was appointed matron of the Highgate Infirmary Miss Nightingale sent her more than 100 letters in the first year, and had about the same number of replies. 'It takes a great deal out of me,' Miss Nightingale wrote to Clarkey in 1875. 'I have never been used to influence people except by leading in WORK; and to have to influence them by talking and writing is hard. A more dreadful thing than being cut short by death is being cut short by life in a paralysed state.'

Miss Nightingale insisted that her school should perform the dual function which was her conception of education: it must not only teach the mind, but it must form the character. 'It must be,' she wrote in August 1875, 'a Home – a place of moral religious and practical training – a place of training of character, habits and intelligence, as well as of acquiring knowledge.' In this conception she held the place, if not of a mother, then certainly of a favourite aunt. She was Fanny's daughter in hospitality and generosity, in the pleasure she took in seeing people comfortable and well fed. Fruit, game, jellies, creams, country eggs, and butter flowed from her to the Nightingale Home; she had sheaves of flowers sent up from Claydon for the Home and the hospital wards. When a nurse went to a new post, Miss Nightingale sent flowers to welcome her. Nurses who were ill had special dishes cooked for them by her cook. Nurses who were travelling found her manservant waiting at the train with a luncheon basket. Nurses who were run down were fed up at her expense. 'Get the things out of my money,' she wrote enclosing a detailed diet sheet. If a nurse were prescribed a change or a rest, she came forward. Sometimes the nurse would be sent to the seaside at her expense, sometimes asked to stay at Lea Hurst or Claydon. When Miss Nightingale was in London, she invited hard-worked nurses for what she called 'a Saturday to Monday in bed' at South Street. Her girls were encouraged to feel that she was always behind them. 'Should there be anything in which I can be of the least use, here I am,' was a favourite ending to her letters. To be of use included the practical and the spiritual. She never ceased in countless letters, in numberless interviews, to hold up before her nurses' eyes the spiritual nature of their vocation, to instil into them not only the high standard of efficiency on which she was adamant but a sense of the presence of God.

She was repaid. Her nurses constantly sought her advice. From all over the world they wrote to her, addressing her as 'Dear Mistress', 'Beloved Chief', 'Dearest Friend'. In spite of her exile she made the Nightingale School as much an expression of her own personality as if she had presided over it in the flesh.

She found great pleasure in the company of young women. She liked young people, and young people liked her. The spectre of a solitary old age, of 'horrible loneliness', haunted her, and in the torrent of nurses', dining, sleeping, tea-ing, coming to her for advice, confiding their difficulties to her, young, enthusiastic, affectionate, she enjoyed the human warmth of which she had been starved. All her life she had expressed personal feelings in terms of hyperbole; exaggeration was the custom of the age and set in which she had been brought up. Her old friend Lady Ashburton wrote of 'the deep joy of communion with my beloved' after spending a day with Miss Nightingale when she was nearly sixty, and repeatedly addressed her as 'Guiding Star of my life'.

Miss Nightingale wrote and talked to the young women to whom she became attached in these terms, and her attachments, like all her emotions, verged on the inordinate. Two young women in particular won her affection. Miss Pringle, whom she christened 'the Pearl', and Miss Rachel Williams, called the 'Goddess Baby'. Both were excellent nurses, both became matrons of important hospitals, and both were extremely good-looking. Rachel Williams in particular was strikingly beautiful. '. . . It was quite a pleasure to my bodily eyes to look at her,' Miss Nightingale wrote when Rachel paid her first visit as a Nightingale probationer. 'She is like a queen; and all her postures are so beautiful without being in the least theatrical.' The letters she wrote to Rachel Williams and to Miss Pringle were highly coloured. The ups and downs of hospital life, the minor crises inseparable from taking a new post or deciding to go for a holiday, became dramas. In January 1874 Miss Nightingale wrote: '. . . I am well aware that my dear Goddess-baby has – well a baby side, I shall not be surprised at any outburst, though I know full well that in the dear Pearl's terrible distress you will do everything and more than everything possible to drag her through. . . . Only don't break yourself down dear child.' In December 1874 she urged Rachel to come and see her in London. '. . . Telegraph to me any day and come up by the next express. . . . And I will turn out India, my Mother and all the Queen's horses and all the Queen's men, together with one sixth of the human race and lay my energies (not many left) at the Goddess's feet.'

Miss Pringle, the Pearl, was addressed as 'Dearest little Sister', 'Extraordinary little Villainess', 'Dearest ever Dearest'. Miss Nightingale reproached her tenderly for not having eaten her dinner, and sent it after her in a cab. She implored her to take a much needed holiday – 'Dearest very dearest. Very precious to me is your note. Make up your mind to a long holiday; that's what you have to do now. God bless you. We shall have time to talk.' When Rachel Williams or Miss Pringle lunched or dined, she took pains to tempt them. 'Dishes for Miss Williams,' runs a note to her cook in 1879: 'Rissoles, or fillets of sole à la Maître d'hôtel, or oyster patties, or omelette aux fines herbes, or chicken à la mayonnaise with aspic jelly, or cutlets à la Béchamelle.' She delighted in beauty and charm, and the friendship of these lovely and intelligent girls filled an important place in her life. The friendship had also its practical side. The work was furthered; they were the ablest young women she had ever trained.

With enormous effort the Nightingale School could be controlled by correspondence, but there was other nursing work which could not, and in 1874 Miss Nightingale had to turn her back on an important opportunity. She was fully alive to the importance of district nursing, but when in 1874 William Rathbone asked her to help him in organizing a district-nursing scheme for London she had to refuse – family difficulties prevented her from undertaking anything which required her to be in London. She could not personally organize, but she did everything that could be done from a distance. In 1874 she wrote a pamphlet, *Suggestions for Improving the Nursing Service for the Sick Poor*, and, in accordance with her suggestions, William Rathbone founded the Metropolitan Nursing Association. In April 1876 she wrote and signed a letter to *The Times* which was reprinted as a pamphlet under the title *Metropolitan and National Association for providing Nurses for the Sick Poor. On trained nursing for the Sick Poor, by Florence Nightingale*. It went into two editions. Finally, the first Superintendent of the District Nursing Scheme for London, appointed in 1876, was Miss Florence Lees, one of her ablest nurses, who had served with distinction during the War of 1870.

It was frightful to Miss Nightingale to turn her back on work, frightful not to attempt to do something which ought to be done – and she did not turn her back often enough. In her nursing work she had the Nightingale School to give her direction, but in her Indian work she became confused.

She had fully intended to go out of office in India as well as in England, but she received appeals she could not ignore. The state of India was sufficient to produce frenzy. There was so much to be done; the problems were so enormous, so urgent, so innumerable. As soon as investigation was made in any direction, fresh abuses emerged. In her mental state she was incapable of crying halt. At this period a letter from her was described at the India Office as 'another shriek from Miss Nightingale'. Irrigation led her to the land question, to rights of tenure, to usury, to taxation, to education, to communication. She toiled not at one issue but at twenty.

In 1874 she met Sir Arthur Cotton, the great master of irrigation. His record was impressive. He had irrigated Trichinopoly and South Arcot in Southern India by building, with immense success, two dams across the river Coleroon. There was not an individual in the province, it was said, who did not consider the damming of the Coleroon the greatest blessing that had ever been conferred on it. The financial returns were, respectively, 69 and 100 per cent. He had dammed the Godavery river and irrigated the Godavery district. The Godavery district was in a desperate state after a severe famine, and the district was almost depopulated. After the irrigation works were completed, the district became one of the most prosperous in India, and the population doubled. Was it not clear that the answer to the problem of India was irrigation? But Sir Arthur Cotton failed to persuade the Government to undertake large-scale irrigation works, and it was a favourite catchphrase to say that he had water on the brain.

Miss Nightingale sent Lord Salisbury schemes prepared by Sir Arthur Cotton, she demanded a commission on irrigation, she asked that the Government should collect statistics on the cost of irrigation works and their return.

In 1877 famine ravaged the Presidencies of Bombay and Madras, four million people perished, and irrigation became a burning issue. In 1878 a committee of the House of Commons was appointed to inquire into the possibility and desirability of preventing such famines in future by constructing public works, especially irrigation works, with money raised on loan. Sir Arthur Cotton was summoned to give evidence. Losses in the famine had been enormous, but the districts irrigated under Sir Arthur Cotton's schemes had not suffered. Nevertheless, the committee was hostile to Sir Arthur Cotton, and the recommendations contained in its report were in contradiction of his views.

It was a major defeat, and it brought Miss Nightingale into conflict with the India Office, which did not wish to admit the seriousness of the famine. When early in 1878 she applied for figures relating to the famine, she received a snub. On February

7 an official minute was addressed to her: 'The Revenue Committee is of opinion that an intimation should be made to Miss Nightingale to the following effect. The various objects of high interest to which she refers are engaging the earnest attention of the Govt of India ... in addition to this a special enquiry is about to be made by a carefully selected Commission on the subject of Famines. ... While then the Secretary of State would on public grounds deprecate the researches which Miss Nightingale wishes to make, as possibly interfering with and embarrassing the comprehensive enquiry of a Commission appointed by the Govt of India under the orders of H.M. Government, he would as a matter of official propriety, point out to Miss Nightingale, whose active and intelligent philanthropy is universally recognized, that to open the Records of a Public Office to the free inspection of a private individual, however distinguished for character and ability, would constitute a very inconvenient precedent.'

She was in disgrace. And while she became unpopular at the India Office, at the same time she lost more of her influence in India itself. Lord Northbrook was succeeded by Lord Lytton. She did not sympathize with Lord Lytton, he did not call on her and they never corresponded.

Seldom had Miss Nightingale sunk lower in misery than now. Her life with Fanny continued to present difficulty after difficulty, and Clarkey begged her to come back to London. Fanny was completely childish; it was very doubtful if she realized where she was. Why did Flo persist in burying herself in 'that absurd place Lea Hurst'? 'Why do you abuse me for being here?' wrote Miss Nightingale on September 13, 1879. 'Do you think I am here for my own pleasure? Do you think any part of my life is *as I please?* Do you know what have been the hardest years of my life? Not the Crimean War. Not the 5 years with Sidney Herbert at the War Office when I sometimes worked 22 hours a day. But the last 5 years and three quarters since my father's death.' The autumn dragged on. She had never in her life done anything she did not feel was morally justified, and she did not feel morally justified in leaving her mother now. But release was near. On February 2, 1880, Fanny died peacefully at the age of ninety-two, after regaining consciousness for a few hours, during which she listened to her favourite hymns.

23

THE conflict which had embittered Miss Nightingale's life for more than forty years was over. At sixty years of age she was free. Not because Fanny was dead, but because she had become reconciled with Fanny and with Parthe as well. All her life resentment against Fanny and Parthe had been a poison working within her. During these last difficult years, before Fanny's childishness, helplessness, and blindness, before Parthe's suffering, resentment had melted away.

A change came over her, and the *bonté*, the pervading benevolence, which had been her chief characteristic as a young woman, returned. She became gentler, calmer, even tolerant. In 1881 Uncle Sam died; she became reconciled to Aunt Mai, and they began to correspond affectionately again. With Parthe, for the first time since their childhood, she became intimate. She began to visit Claydon, where a room was set aside for her and called 'Miss Nightingale's room'. As Parthe's illness increased, Sir Harry leaned on her. 'You are our Family Solicitor,' Sir Harry wrote to her in January 1881, 'to whom we all turn when we get into a scrape.'

Failure began to weigh less heavily on her. Had she achieved nothing, need she reproach herself quite so desperately? On New Year's Eve, 1879, Jowett had written to her: 'There was a great deal of romantic feeling about you 23 years ago when you came home from the Crimea. (I really believe that you might have been a Duchess if you had played your cards better!) And now you work on in silence, and nobody knows how many lives are saved by your nurses in hospitals (you have introduced a new era in nursing): how many thousand soldiers who would have fallen victims to bad air, bad drainage and ventilation, are now alive owing to your forethought and diligence; how many natives of India (they might be counted probably by hundreds of thousands) in this generation and in generations to come have been preserved from famine, oppression and the load of debt by the energy of a sick lady who can scarcely rise from her bed. The world does not know all this, or think about it. But I know it and often think about it, and I want you to, so that in the later years of your course you may see (with a side of sorrow) what a blessed life yours is and has been. ... I think that the romance too ... did a great deal of good. Like Dr Pusey you are a Myth in your own lifetime. Do you know that there are thousands of girls about the ages of 18 to 23 named after you? Everyone has heard of you and has a sweet association with your name.'

Could Jowett be right? Ordinary happiness she had never wanted – 'miserable as I am,' she had written in 1867, 'I had rather be as I am than as I see the mass of London Ladies.' In 1872 when on her way to Embley, she had caught a glimpse of Lord Stanley, now happily married and absorbed in his country estates, his wife and his library. 'I saw them both at the station,' she wrote to Clarkey, 'they did not see me. (They were going to see the Queen.) I did not want to speak to him. I wanted to observe him. I saw it all at a glance. I should not have known him, so complacent, so obese, so happy – so bustling. All the great visions dropt away. (I was glad I had not to speak to him.) All quite forgotten, what once he was, or might have been. O happiness – like the Bread Fruit Tree, what a corrupter of human nature thou art!'

She could look back without regret, and now she found she could do more – she could look forward. On June 30, 1881, she wrote to Clarkey: 'I cannot remember the time when I have not longed for death. After Sidney Herbert's death and Clough's death in 1861, 20 years ago, for years and years I used to watch for death as no sick man ever watched for the morning. It is strange that now I am bereft of all, I crave for it less. I want to do a little work, a little better, before I die.'

Opportunity was on its way. At the moment she became free, opportunities for work for India, for nursing, even for the army, presented themselves once more. The political scene had just been transformed by the unexpected triumph of the Liberals in the General Election of April 1880. When Lord Lytton's term of office ended in May, her old friend and close ally, Lord de Grey, now Lord Ripon, was appointed Viceroy of India. Once more official doors were thrown open to her, and as Lord Ripon's Indian policy unfolded she was enthusiastic.

Two main measures of reform were proposed by Lord Ripon,

both highly controversial. The storm centre was the Ilbert Bill, introduced by Sir Courtenay Ilbert, which gave Indian magistrates, under certain conditions, power to try and sentence Europeans. An absurd situation had arisen. Since 1858, Indians had been allowed to enter the Civil Service, and, in spite of the fact that promotion was by no means made easy, certain of them reached the rank of District Magistrate; yet because Englishmen could be tried only by English magistrates, an Indian District Magistrate could find himself without authority to try cases which were within the authority of his subordinates.

Hostility to the Ilbert Bill, in fact a carefully guarded and by no means revolutionary measure, was based on racial grounds. Hysteria swept the country. Englishwomen wrote that it was an insult to subject English womanhood to native judges; atrocities committed during the Mutiny were recalled; Indian papers joined in with violence; insults and recriminations were freely exchanged, and India blazed from end to end with hatred.

Almost equally detestable, not only to Europeans but also to a large number of commercially successful Indians, were the proposals for land reform in Bengal and Oudh which endeavoured to protect the *ryot*, the Indian peasant, from oppression and exploitation by placing authority and responsibility in the hands of the head man of each village, thus laying the first foundations of a degree of local government.

Behind the frantic opposition to Lord Ripon's reforms lay the grim shadow of the Mutiny. The Mutiny had done irreparable damage. The atrocities committed by Indians on Europeans on lonely stations, the equal atrocities committed by Europeans on Indians as, against the advice of such men as John Lawrence and Lord Napier of Magdala, the European victors avenged themselves in rivers of blood, left a wound which has never yet healed. In 1882 the wound was fresh.

Miss Nightingale had received her Indian education in a different school. The great Indian administrators who taught her – John Lawrence, Bartle Frere, Lord Napier of Magdala – were men to whom racial hatred was unknown. It had been their creed that the future of India must lie in giving ever-increasing authority to Indians. In this spirit the Queen's Proclamation of 1858 had been drawn up, in which the Crown, assuming the government of India, declared it to be the Sovereign's intention that '... our subjects of whatever race or creed be impartially admitted to our service, the duties of which they may be qualified by their education, ability and integrity, duly to discharge'.

Through Lord Ripon, Miss Nightingale believed, light was coming to India at last. After the interminable delays, the endless disappointments – it was twenty years since she had written the 'Sanitary Suggestions' for John Lawrence – a new age was dawning. 'At last,' she wrote to Lord Ripon in June 1883, 'we have a government of India which steadfastly sets its face to carry out ... the spirit of the Queen's Proclamation.' In private notes she called Lord Ripon 'the saviour of India'. She described his term of office as the beginning of a golden age. 'It is the Millennium!' She acted as a reference library for Lord Ripon, and he used her encyclopaedic knowledge of Indian administration, reaching back for over twenty-five years, to guide him through the tangled jungle of Indian affairs. Her familiarity not only with facts but with persons was of immense assistance to him in a situation where opposition came not only from without but within. 'How well you know the subject of Indian sanitation,' he wrote in July 1883. 'If only there were a modicum of your intelligent sympathy to be found in the India Council!'

For the next four years she was absorbed in crusading for Lord Ripon's reforms. 'We shall want all the help you can give at home,' Lord Ripon wrote in 1882. She interviewed every official with whom she could get into touch, she saw native gentlemen, English Members of Parliament, journalists, missionaries. She no longer regularly wrote accounts of interviews, as she had done twenty years ago, but among her papers are notes of visits from Lord Roberts, who called to see her before going out as Commander-in-Chief Madras in 1881; from Sir Courtenay Ilbert, the introducer of the Ilbert Bill; Sir Mountstuart Grant Duff, Governor of Madras; Mrs Scharlieb, the celebrated doctor, who practised medicine among the women of India; Lord Reay, Governor of Bombay; Mr W. R. Robertson, Principal of the Agricultural College in Madras. Their visits were not visits of compliment. Again and again, as in the case of Lord Roberts, Mrs Scharlieb, Sir Mountstuart, Mr Robertson, she put against the name 'he desired me to write notes for him'. She wrote an explanation of the Ilbert Bill for Queen Victoria, she described Lord Ripon's land-reform proposals in a paper entitled *The Ryot, the Zemindar and the Government*, which was read aloud by Mr

Frederick Verney, one of Sir Harry's sons, at a meeting of the East India Association in June 1883.

She was drawn back into army work when in April 1880 she received a letter from General Gordon asking her to help his cousin Mrs Hawthorn, wife of a colonel in the Royal Engineers, in putting before the War Office facts concerning the neglect and ill-treatment of patients in military hospitals by orderlies. Miss Nightingale wrote a memorandum which was submitted to the Secretary for War. She was not successful. In his reply the Secretary for War stated that he 'failed to be convinced'. 'I have seen such answers in the Crimean War time,' she wrote to Douglas Galton in August 1880. 'The patient died of neglect and want of proper attendance; but by regulations should not have died, therefore the allegation that he is dead is disposed of.' Out of the failure, however, came friendship with General Gordon.

She was instinctively in sympathy with him. His intense Evangelical religiousness did not grate on her; she cared only for saintliness, nothing for the form in which it was expressed, and his attitude toward his soldiers and the people of India exactly corresponded with her own. 'I gained the hearts of my soldiers (who would do anything for me) not by my justice, etc., but by looking after them when sick and continually visiting the Hospitals,' he wrote to her on April 22, 1880. He came to see her repeatedly, and they discussed religious experiences. Both were aiming at the same end – a life of union with God producing practical good works. The bond became closer when in May he was appointed private secretary to Lord Ripon, an appointment greeted by universal astonishment. Before he sailed he presented Miss Nightingale with one of the religious writings which he described as 'little books of comfort'. On his way out he wrote to her, on May 30: 'On board this vessel nothing but discontent with their lot from Indian officers. ... The element of all government is absent, i.e. the putting of the governors into the skin of the governed. The old Indian was obliged to do so, he was bound in some way to consider the sympathies of the native.'

Gordon's understanding of Oriental races was indisputable. He had already a brilliant record of success both in India and China, but a large section of the official world detested him, and he was evidently unsuited to the post of private secretary to the Viceroy, involving official contacts and requiring the tact and social dexterity he lacked. As soon as he reached India he resigned, and after successfully executing a short mission in China he came home.

A period of vacillation followed, but his want of direction did not irritate Miss Nightingale. His difficulty was to find an employment which would satisfy his conscience. The fact that he felt he must exercise his own moral judgement, that he could not undertake to carry out any order he felt to be unethical, closed almost all avenues of official employment to him. As an additional difficulty he had no money on which to live. He spoke of going to work among the sick poor in Syria because Syria was cheap. She entreated him to go to India where there was so much to be done. But India was closed to him. 'I would have gone to the Cape, I would have gone to India as you suggest,' he wrote in January 1881, 'but I would never do so if I had to accept the shibboleth of the Indian or Colonial middle classes. To me they are utterly wrong in the government of the subject races, they know nothing of the hearts of these people, and oil and water would as soon mix as the two races. Men may argue as they like, our tenure of India is very little greater than it was 100 years ago. The people's interests not having been involved or interested in our prosperity or disasters are equally indifferent to either, in fact they hope more from our disaster than from our prosperity.' The Government would not allow him to enter India. 'I consider my life's work done, that I can never aspire to do, or seek employment where one's voice must be stilled to one particular note – therefore I say *it is done*. . . . I cannot visit the sick in London; it is too expensive. I can do so in Syria and where the sick are there is our Lord. My dear Miss Nightingale what am I to do? My life truly is to me a straw, but I must live. I would do anything I could for India but I am sure my advent there would not be allowed. The door is shut.'

Eventually, to assist a friend, he accepted an appointment in Mauritius, and from Mauritius he was called to Basutoland to negotiate with a rebellious chief. In spite of another success the Cape Government refused to renew his appointment, and he found himself back in England in November 1882 once more unemployed. After wandering in Palestine for a year, he accepted a mission to the Belgian Congo at the request of the King of the Belgians, and went to Brussels, but the Belgian Government refused to sanction his employment. While in Brussels he received a telegram on January 15, 1884, from the British War Office. The victories of the Mahdi in the Sudan made instant action necessary, and Gordon was asked to go out as Governor-General. On January 18 he went to the War Office, and so great was the urgency that he left for Egypt the same night. He was not able to see Miss Nightingale, but he wrote to Sir Harry Verney, on January 17, 1884, 'I daily come and see you in spirit, you and Miss Nightingale.' A year later, on Monday, January 26, 1885, Khartoum, which Gordon had defended brilliantly against overwhelming odds for 317 days, fell and he was murdered.

A tremendous outburst of indignation against the British Government followed. Miss Nightingale did not share it. Whether Gordon had succeeded in his mission or not, whether he had been betrayed by the British Government or not, was unimportant. On February 7, in a letter to Mrs Hawthorn, she spoke of the creed which she and Gordon shared. Suffering, disappointment, lack of success are the tribute which it is the soul's greatest privilege to present to God. In Gordon's death he had shown 'the triumph of failure, the triumph of the Cross'. 'With him,' she wrote, 'all is well'.

She took an active interest in the Gordon Home for Destitute Boys, founded in his memory. In 1887, sending the yearly report to a friend, she scribbled: 'Ask them to tea. The *roughest* boys first.'

However, Mrs Hawthorn's allegations were substantiated by independent evidence. Miss Nightingale persisted, and a Committee of Enquiry was set up in January 1882 under the chairmanship of her old Crimean acquaintance, Sir Evelyn Wood. The first results were disappointing, the committee merely reporting that 'improvements in the system of nursing are both practical and desirable'. A member of the committee commented to Miss Nightingale: 'This seems rather a mild opinion considering that all the independent evidence went to show that the orderlies were often drunk and riotous, that they ate the rations of the sick and left the nursing of the patients to the convalescents.' Before the report could be issued, the Egyptian campaign of 1882 had begun under the command of Lord Wolseley, and much more serious defects became apparent. Miss Nightingale was asked for nurses, and a party of twenty-four, under the charge of a Nightingale-trained matron, went out. Reading their reports, she exclaimed: 'It is the Crimea over again.' The proportion of sick was unduly high, and only the small number of troops involved and the short duration of the campaign prevented disaster. In October 1882 the Committee of Inquiry was reconstituted under the chairmanship of Earl Morley with instructions to inquire into the organization of the Army Hospital Corps and army hospital supply, organization, and efficiency in the field generally, including nursing.

She played a leading part in the second Committee of Inquiry, suggested witnesses, sent briefs for their examination, and outlined the facts to be elicited. As a result of this work she regained some influence at the War Office and became close friends with the Director-General of the Medical Department, Dr Crawford. 'We have not had a man of such unflagging energy since Alexander,' she wrote to Douglas Galton on November 1883.

She was working in administration again; she had influence at the War Office again, but how strange was the road by which she had returned! Nursing had brought her back to the War Office. The sacrifice of her personal life, the long bitter years of administrative toil, the thankless labour, the perpetual struggle with exhaustion had come to nothing. 'How little is left of all the good work of 1856 and that five years until 1861 for the Army,' she wrote to Sir John McNeill in February 1881. But out of the forty unsatisfactory tiresome creatures she had landed at Scutari, out of the drunkenness, the scandals, the back-biting had grown an immense work.

In 1884, when the Gordon Relief Expedition was sent to Egypt, female nurses were officially requested by the Government. Miss Nightingale selected and engaged the party. Some were sent up the Nile to Wady Halfa. In 1850, during her Egyptian travels, Miss Nightingale had been at Wady Halfa, miserably unhappy. 'How little could I ever have thought there would be trained nurses there now!' she wrote to Miss Pringle, the Pearl, on October 11, 1884. The nurses proved unquestionably successful. There were difficulties with orderlies; there was a shortage of medical supplies; there was a shortage of experienced sisters, but there was good-will on the part of the authorities. 'Government are now doing all they can,' Miss Nightingale wrote to Rachel Williams in the autumn of 1884. 'In my day they were hopeless.'

Her health improved. She visited Claydon; she stayed in a hotel at Seaford during the spring of 1881; she made a habit in fine weather of taking drives in the London parks with Sir Harry Verney. In 1882 she made her first personal visit to the Nightingale Training School; in November she went with Sir Harry

Verney to Victoria station to see the return of the Guards from the first Egyptian campaign; a few days later she attended a review, sitting on the platform next Mrs Gladstone; and on December 4 she was present at the opening of the Law Courts, where Queen Victoria spoke to her and expressed herself pleased to note that Miss Nightingale was looking well.

But the structure of her life was still rigorously laid out for work, and she still refused to see anyone without an appointment. She still wrote into the small hours, still sent letters dated 'Before it is light', still attempted more than any human being could accomplish, still continued to speak of herself as being on the verge of the grave; yet one by one the figures who had filled her life were steadily disappearing, and she remained, helpless, almost bedridden, but still alive.

Clarkey began to fail, and her indomitable gaiety – at the age of eighty-six she had been seen dancing to a German band – faded. Through the winter of 1882 she became feebler, and in May 1883 she died. The enormous series of letters in which Miss Nightingale had poured out her inmost thoughts and feelings for more than forty years ceased, and a curtain fell on her private life.

In 1882 her very old friend Dr Farr, the statistician, died. In 1883 Sir John McNeill died, her constant friend and counsellor; 'always so kind and fatherly', Aunt Mai had written in 1858. In 1884 Sir Bartle Frere died, and in 1885 Richard Monckton Milnes, now Lord Houghton, the man she had once adored.

But in her old age she no longer raged; she no longer resented what had not been accomplished; now she looked forward. On Christmas Day, 1885, when she was sixty-five, she wrote: 'Today, Oh Lord, let me dedicate this crumbling old woman to Thee.' *Nihil actum si quid agendum* was no longer her motto. How much she had changed was proved when, during the next few years, the tide turned against her again.

Unexpectedly Lord Ripon resigned. So great was the personal animosity against him that he considered his best course was to secure a suitable successor and go home. Lord Dufferin was appointed and on November 6, called on Miss Nightingale, the fifth Viceroy of India to receive his Indian education at her hands. Unfortunately a series of what Miss Nightingale described as 'political earthquakes' followed. Lord Salisbury's Government was defeated in the general election of December 1885. Mr Gladstone came into power, only to be defeated on the Home Rule

Bill. Another general election took place in 1886, and Lord Salisbury returned to power once more. In the excitements of these changes it was hopeless to expect any general interest in Indian reform. She wrote that it was 'excruciating', but she resigned herself.

In 1886 she was introduced by Lord Salisbury to Mr W. H. Smith, Secretary of State for War. He wished to begin a programme of welfare work for the troops and asked for her assistance. A scheme was drawn up, and its accomplishment seemed certain when once more Fate stepped in.

Lord Randolph Churchill, who had been Chancellor of the Exchequer, unexpectedly resigned. The Government was reconstructed, Cabinet offices were redistributed, Mr W. H. Smith left the War Office and went to the Treasury, and the Army Welfare scheme was shelved. 'We *are* unlucky,' she wrote to Douglas Galton on December 23, 1886.

In 1887 Queen Victoria celebrated her Jubilee, and Miss Nightingale, too, considered 1887 her Jubilee year: her 'voices' had called her first in February 1837 and she had now completed fifty years of service. Retrospection was universal, and Miss Nightingale retraced her own long and eventful past.

On August 5, 1887, she wrote to Aunt Mai, now completely crippled with arthritis: 'Dearest Aunt Mai – thinking of you always, grieved for your sufferings, hoping you have still to enjoy. In this month 34 years ago you lodged me in Harley Street (Aug. 12) and in this Month 31 years ago you returned me to England from Scutari (Aug. 7th). And in this month 30 years ago the first Royal Commission was finished (Aug: 7). And since then 30 years of work often cut to pieces but never destroyed. God bless you! In this month 26 years ago Sidney Herbert died, after five years of work for us (Aug. 2). And in this month 24 years ago the work of the second Royal Commission (India) was finished. And in this month, this year, my powers seem all to have failed and old age set in.'

Old age had come, and she accepted it. The storms had passed, and tolerance had replaced the uncompromising desire for perfection. She was entering on her last period of active work, and she enjoyed an Indian summer. Her health had improved, her mind was at rest, and her work in all directions bore a late harvest.

In India Lord Dufferin succeeded in passing with some

amendments Lord Ripon's Land Tenure Bills and pressed for irrigation. In 1888 the Government of India set up a Sanitary Board in every province which possessed independent and executive authority. It was partial fulfilment at last of the scheme for an independent public health service which she had so urgently pressed on John Lawrence twenty-four years before.

Scheme after scheme came, if not to perfection, at least to partial fulfilment. The drainage of the great Indian cities, especially Madras, progressed at last. The drainage of Black Town, the worst quarter of Madras, was begun in 1882 and the work extended in 1887. For twenty years she had been preaching the importance of the Indian village, with its traditional community life, as the unit through which any educative scheme must be developed. In 1889 her efforts were to some extent rewarded by the Bombay Village Sanitation Act, which aimed at educating each village as a self-contained community, the channel of communication being the head man.

In 1891 she managed to focus attention on the progress of Indian Sanitation by arranging that the International Congress of Hygiene and Demography, to be held in London, should include an Indian section. Indian gentlemen were sent as delegates and were entertained at Claydon. 'Sir Harry Verney renews his invitations to Claydon to the native Indian delegates ...', she wrote to Douglas Galton on August 1, 1891. 'Do you remember it is thirty years tomorrow since Sidney Herbert died?'

Lord Dufferin's term of office came to an end. On board ship on his way home he wrote to Miss Nightingale: 'Among the first persons whose hands I hope to come and kiss will be yours.' He was succeeded by Lord Lansdowne, a close friend of Jowett. Lord Lansdowne came to see her to receive his Indian education before he sailed, and he corresponded regularly with her. 'He did much for us in every way,' she wrote.

She was over seventy-one when she embarked on a complicated undertaking which proved her final crusade for the people of India, a scheme to make urgent sanitation a first charge on taxation. The method of taxation was complicated. Very broadly, a certain amount of taxation was fixed, while another amount, known as 'cesses', varied from time to time and was devoted to various purposes. She proposed that when a village lacked a pure water supply, lacked drainage, lacked any means of disposing of its refuse, and when it was suffering from cholera or typhoid, its

cesses should be applied to remedying these conditions before being applied to any other purpose. She prepared a memorandum setting out the scheme in detail; it was signed by Douglas Galton and other sanitary experts and forwarded to the Viceroy in April 1892.

The familiar history of delay followed. There was a party which thought the cesses should first of all be applied to the making of roads: sanitary works were important, but increased means of communication were the right way to create the prosperity which would enable sanitary works to be paid for. Another party, while agreeing that a pure water supply and 'simple latrine arrangements' were more than desirable, considered they should be made a charge on the revenues of the provincial government. Years passed by. Miss Nightingale argued, urged, reminded, interviewed. An enormous quantity of correspondence accumulated. Not until 1894 did she receive an official answer. The Government of India could not see its way to accept her suggestion, but would press the claim of sanitation upon local governments and administrations as opportunity offered.

It was her last campaign. She was still to do an immense amount of work for India, but in an advisory capacity; her vast knowledge, her long experience, and the weight of her prestige were called on again and again, but controversy was at an end.

*

In 1887, the year of Miss Nightingale's jubilee, the following hospitals, institutions, and organizations had matrons or superintendents who had been trained at the Nightingale School: the Westminster Hospital, St Mary's, Paddington, the Marylebone Infirmary, the Highgate Infirmary, the Metropolitan and National Nursing Association, the North London District Association, the Cumberland Infirmary, the Edinburgh Royal Infirmary, the Huntingdon County Hospital, the Leeds Infirmary, the Lincoln County Hospital, the Royal Infirmary, Liverpool, the Workhouse Infirmary, Liverpool, and the Southern Infirmary, Liverpool, the Royal Victoria Hospital at Netley, the Royal Hospital for Incurables, Putney, and the Salisbury Infirmary. Parties of nurses under a Nightingale-trained superintendent had also gone to the United States of America, Sydney, Montreal, India, Ceylon, Germany, and Sweden. Training schools modelled on the Nightingale Training School and supervised and directed by

Nightingale superintendents had been established at Edinburgh, at the Westminster Hospital, at the Marylebone Infirmary, at St Mary's, Paddington. The school in connexion with the Edinburgh Royal Infirmary was under the direction of Miss Pringle, and that at St Mary's, Paddington, of Rachel Williams.

Though she had expressed regret that she could not give herself to district nursing, Miss Nightingale held the threads of the movement in her hands. William Rathbone consulted her on every point. 'In any matter of nursing Miss Nightingale is my Pope and I believe in her infallibility,' he wrote. During the years following 1880, she formed the movement. Somehow, 'by hook or by crook', she managed to meet nearly all the trained nurses who took up district nursing and to keep up a correspondence with them. The function of the district nurse was established and defined by her. The district nurse must be a sanitary missionary, not an almsgiver, and to be a sanitary missionary she must be trained.

Finance was a constant difficulty. In 1884 Miss Nightingale wrote to Sir Harry's daughter-in-law, Margaret Verney, that ladies were ready enough to give money to pauperize the patients but not so ready to give money to train and pay nurses. The nurses were kept paupers in order that the patients might be pauperized. In 1887 Queen Victoria decided to devote the major part of the money which had been presented by the women of England as the 'Women's Jubilee Gift' to the cause of 'nursing the sick poor in their own homes by means of trained nurses', and the Jubilee Institute for Nurses was founded.

Results were beginning to exceed her highest hopes. In private notes during 1887 and 1888 she recalled the first beginnings of the work, her attempt in 1845 to train at Salisbury when her parents behaved as if she had wished to be a kitchenmaid, the difficulty of finding nurses to go to the Crimea, and Agnes Jones's experiences in the old infirmaries, when the police were regularly called in to establish order in the wards.

Yet though Miss Nightingale's influence in nursing was dominant, there was opposition to her. It was never contested that her results were not superior, but it was held that the form of training she demanded, the close supervision, and the exactions of her school, could not produce nurses in the numbers which were now necessary.

In 1886 a proposal was made which aimed at giving the trained

411

nurse official recognition and at placing her qualifications on a standard basis. A committee of the Hospitals' Association proposed than an independent body of examiners, not connected with the training schools, should be created. This body would set an examination, and when a nurse had passed it she would be entitled to have her name placed on a register of nurses. Thus a standard of technical excellence in nursing would be established, and the public would be protected against employing nurses who were incompetent or disreputable.

It was the beginning of a battle which split the nursing world in two. Miss Nightingale opposed the proposal for two reasons. First, she did not think the time was ripe for the step. In forty years' time, she wrote, the nursing profession might be ready, but at the moment nursing was still too young, still too unorganized, and contained divergences too great for a single standard to be applied.

The second ground on which Miss Nightingale opposed the scheme was of greater importance. The scheme as put forward was a contradiction of what she believed the training of a nurse should be. She was not necessarily against registration, but she was passionately opposed to the kind of registration proposed. The qualifying of a nurse by examination only took no account of the character training which she held to be as important as the acquisition of technical skill. A nurse, she said repeatedly, could not be tested by public examination as if she were an engineer. Nursing was a vocation as well as a profession, and the two must be united. When a nurse received a certificate from her training school, the matron was able to guarantee by personal knowledge that her pupil possessed the qualities of character as well as the degree of technical skill which were essential to the calling of a nurse. Devotion, gentleness, sympathy, qualities of overwhelming importance in a nurse, could never be ascertained by public examination. 'Nursing has to nurse living bodies and spirits. It cannot be tested by public examination, though it may be tested by current supervision,' she wrote.

In thirty years she had, she said, 'raised nursing from the sink' by training character. She had caused training schools for nurses to be called 'Homes' to emphasize the fact that they were places in which character was to be developed, general culture acquired, and a moral standard learned. Now the object of the training school was to be made not the training of character but the granting of a certificate. 'You cannot select the good from the inferior by any test or system of examination,' she wrote in 1890. '... Most of all and first of all must the moral qualifications be made to stand pre-eminent in estimation.'

The British Nurses' Association continued to agitate, and in 1888 a new committee was set up to conduct an inquiry among the training schools and the medical profession on their opinion of a nurses' register. It at once became apparent that opinion was divided and that feeling was running high. Miss Nightingale feared disaster. The nursing world would be divided into two camps. Political differences would become all-important, and the work would take second place. She recalled the division of opinion over the Reform Bill when she was a child: those who were for the Bill would refuse to sit at dinner with those who were against it.

In 1889 the situation crystallized. The British Nurses' Association published its policy. Its main object was to provide for the registration of qualified British nurses, and to accomplish this it intended to apply for a Royal Charter incorporating the Association and authorizing the formation of a register. Nurses would be deemed to be qualified who were certified by an outside Board as having attained a certain standard of proficiency, and it was suggested that a preliminary qualification should be three years' training in a hospital. The manifesto made a considerable impression, and Princess Christian, Queen Victoria's daughter, accepted the Presidency of the Association.

Miss Nightingale had received a set-back. When the time came to petition for the Royal Charter, Princess Christian would approach Queen Victoria. Moreover, Princess Christian had great influence and was much beloved for philanthropic work. 'This makes things awkward for us,' wrote William Rathbone. Through Sir Harry Verney, Miss Nightingale made overtures to Princess Christian. It was not possible, she found, to compromise, because Princess Christian felt that she was in honour bound to support the policy of the Association of which she was President, and Miss Nightingale settled down to fight.

For the next four years the battle absorbed her. She was seventy, and the work was exhausting. Her friends regretted her absorption, for the point at issue seemed unimportant. 'It is a comparative trifle,' wrote Jowett in May 1892, 'among all the work you have done.'

413

Miss Nightingale, however, was convinced she faced a major crisis: the principle which had governed her work for nursing was at stake. Nor in her opinion would the register as proposed protect the public. The fact that a nurse's name was on it would only mean that at a certain date she had satisfied the examiners in certain tests; it would tell nothing of her subsequent record. If a register were to be useful, it should be kept up to date, and include a description of each nurse's character and a recent recommendation from a surgeon or physician.

In 1889 the British Nurses' Association announced that it was applying for a Charter. Miss Nightingale, supported by the matron of the London Hospital and the matron of St Thomas's and most of the training schools, declared her intention of opposing the application. Two years of controversy followed, and a large number of pamphlets were issued on both sides. Then in 1891 the British Nurses' Association applied to the Board of Trade to be registered as a public company without the addition of the word 'Limited', the object of the company being to form a register of nurses and to lay down what should be the qualifications necessary for registration. Miss Nightingale presented a case opposing the application, and registration was refused. In the same year a committee of the House of Lords reported on the condition and organization of London hospitals. William Rathbone was called to give evidence as to the desirability of the proposed register of nurses; he gave evidence against the proposal, and the committee in its report did not recommend the formation of a register of nurses.

But the British Nurses' Association was not yet defeated. Later in the year it obtained permission from the Queen, through Princess Christian, to use the title 'Royal', and they then petitioned the Queen herself for a Royal Charter. The petition, in accordance with precedent, was referred to a special committee of the Privy Council and was heard in November 1892.

On both sides this was felt to be the decisive moment, and Miss Nightingale rallied all her forces. A campaign fund was raised and two counter-petitions opposing the grant of a Royal Charter presented, one signed by the Council of the Nightingale Fund, admittedly the pioneers of the training of nurses, and the other by many thousands of matrons, lady superintendents and principals assistant, doctors, and nursing sisters, as well as by superintendents and principals of training schools. The list was headed by the signature of Miss Nightingale. In addition, a letter from her was read to the Committee of the Privy Council by William Rathbone. Eminent barristers appeared on both sides, and two Law Lords sat on the committee.

The hearing took a week and was completed by the end of November, but the decision was not announced until six months later – in May 1893. The result was victory for no one. True, the Royal British Nurses' Association was granted a Royal Charter, but not in the terms it had sought. The word 'register' was removed, and the Charter conferred only the right to the 'maintenance of a list of persons who may have applied to have their names entered thereon as nurses'.

The battle was over, and Miss Nightingale put it behind her. In 1894 she talked and corresponded with Princess Christian regarding a scheme for the formation of a war reserve of nurses by the Royal British Nurses' Association. 'We should, I think,' she wrote, 'be earnestly anxious to do what we can for Princess Christian as she holds out the flag of truce, in order to put an end, as far as we can, to all this bickering which does such harm to the cause.' In 1893 she dedicated a lecture on Sick Nursing and Health Nursing, which was read at the Chicago Exhibition of Women's Work, to Princess Christian.

It was a tranquil end to her last great battle. She was an old lady now, and though her mind was still keen and her energy still remarkable, another change was taking place. Her horizons were narrowing; the world was receding; for the first time personal relationships were becoming of paramount importance in her life.

24

IF life had used her hardly, she was compensated now. Few human beings have enjoyed a fuller, happier old age. She was treated with an almost religious deference – ministers, kings, princesses, statesmen waited at her door, and her utterances were paid the respect due to an oracle. To millions of women all over the world she was the symbol of a new hope, the sign of a new age. Nor was she separated from the common joys of life. Though she had never married, she enjoyed the pleasures of matriarchy. In the lives of a large circle of young people, Shore's two daughters and his sons, Clough's son and daughters, and Parthe's step children, she held the place of a powerful, generous, and respected grandmother.

In old age an extraordinary atmosphere of peace flowed from her. She was formidable still; she preserved her rule of seeing only one person at a time and bent her whole attention on her visitor, making you feel, it was said, like a sucked orange, but she was animated now by the purest benevolence. To confide in her was irresistible. She delighted to concern herself with the small crises of daily life. Clough's son brought her his love affairs, Shore's daughters their examination papers. No detail was too small to command her interest – the character of a servant, the quality of a joint of meat, the treatment of a cold. She delighted to write birthday letters, to send gifts of jellies, fruit, creams, special soups to invalids, to make presents. Sir Harry Verney, suffering from eye trouble, was sent a special lamp-shade; a girl cousin working too hard received concert tickets; Margaret Verney, going on a night journey by train, was sent sandwiches, coffee, and a special cushion for her head.

Her sympathy extended itself beyond her family. Her butcher, the policemen on duty at the Park gates near her house, everyone who served her, came within the circle of her benevolence. Their family affairs received her earnest consideration; their health was the object of her solicitude.

To enter her house was to receive an instant impression of

416

whiteness, order and light. 'You have such a beautifully tidy house,' wrote a schoolgirl cousin. Her bedroom at the back of the house had French windows opening on to a balcony; there were no curtains only blinds, the walls were painted white, and the room was bathed in light.

A stand of flowering plants stood in the window, kept filled throughout the year by William Rathbone, and more flower-boxes stood on the balcony. The house backed on to the gardens of Dorchester House, and outside the windows were trees, flowers, and lawns. Birds twittered, and in summer the sunlight filtered through green leaves. Miss Nightingale's bed stood with the windows on her right; behind it was a shelf of books. She had a table beside her bed on which stood a reading-lamp with a green silk shade and a vase of fresh flowers – a large box of cut flowers was sent weekly by Lady Ashburton from Melchett Court. The furniture was unpretentious. There were an armchair, a bureau, a bookcase, another larger table. On the walls were a photograph of John Lawrence's portrait, a lithograph of the ground about Sebastopol, and a few water-colours. The room conveyed an exquisite and fastidious freshness. Flowers were never faded; vases sparkled like crystal; the pillows and sheets of Miss Nightingale's bed were spotless and without a crease.

On her good days she got up after luncheon and received visitors in the drawing-room below, lying on a couch wearing a black silk dress with a shawl over her feet, and a scarf either of delicate white net or fine quality lace round her head. 'No gentle-woman ever wears anything but real lace,' she told one of Shore's daughters; she was fond of the Buckinghamshire lace which Sir Harry Verney had made for her. The decoration of the drawing-room was severe, relieved by a profusion of flowers. The windows were curtained in plain blue serge; the walls were white. Round the room were hung the engravings of Michelangelo's ceiling in the Sistine chapel which she had bought in Rome, and there were several bookcases full of books. When young relatives waited in the drawing-room before going up to see Aunt Florence, they found the books consisted solely of Blue Books, with one exception, a copy of *The Ring and the Book*.

Visitors, even when staying in the house, never saw Miss Nightingale except by appointment. She never took a meal with any-one, but she did her own housekeeping and took immense pains over her household. 'Florence's maids and little dinners perfect,'

W. E. N. had written in 1867. Her staff consisted of five maids, her own personal maid, and a man known as 'Miss Nightingale's messenger', who was an old soldier and a member of the Corps of Commissionaires.

The household was highly organized. The proper duties to be performed in the house and in the kitchen at every hour through the day were marked on a chart. The food was ordered by Miss Nightingale, and she was particular as to quality. In March 1889 she wrote to a new butcher for 'a fore quarter of your best small mutton. I prefer four year old mutton.' The following week she wrote, 'the neck "ate" better than the shoulder tho' off the same piece,' and ordered 'a neck of mutton well hung and a leg well hung. Please tell your man to wait, as I always pay weekly.' The neck proved 'very good but the leg not so good', and she ordered '13 to 14 pounds of good sirloin of beef to try'. Meanwhile the butcher's wife had fallen ill. Miss Nightingale was all sympathy, gave advice, sent lemon jelly and when, unhappily, the woman died, wrote: 'God be with you and with your children is the earnest prayer of Florence Nightingale.'

Her taste in food was fastidious. Each day's menu was submitted to her, and she made suggestions and criticized the previous day's dishes. 'Remember I am a small but *delicate* eater,' she wrote. A couple of oyster patties, or a little broth, and a fried sole was a favourite order. 'Why was the glue pot used?' she wrote against 'stewed cutlets', and against minced veal, 'the meat hard and remember mincing only makes hard meat harder'. 'Minced beef for my dinner' was the heading of another note. 'The beef must be from the under-cut of the sirloin; mince the beef over a plate which must catch the juices which fall. The meat must be uncooked.' 'Sauces and gravies are not to be thickened with flour. The bones of the meat are simmered down with vegetables to make the stock, which is then reduced to make the sauces. Use plenty of herbs for flavouring.' Turnips were to be served by 'squeezing out all water, putting through a hair sieve and adding a gill of cream.' 'Brisket of beef must be cooked with herbs, onions, carrots, celery in a light broth on the hot plates from 10 a.m. to 9 p.m. *Never* too fast.' 'Roast pheasant must be hung not too near a good fire and basted every minute or two with good butter for an hour. Roast chicken must be larded *all over*.' 'Tell Miss Nightingale the luncheon was a work of art,' said the Crown Princess of Prussia.

Miss Nightingale's account of an interview in 1886 demonstrates how little she now inspired awe, how readily she felt sympathy. The interview was with a girl who wished to be a nurse. 'She showed,' wrote Miss Nightingale, 'a natural, unconscious, unrestrained interest in interesting things which I liked very much.' Three remarks struck her favourably. 'Oh I do so want to go inside the House of Commons some day just to hear Mr Gladstone speak once.' 'May I just look round the books to see if there is a Tennyson?' 'Oh, I'm not a bit tired now.'

It was even possible to disagree with her without disturbing her good humour. In 1895 she received a letter upbraiding her for opposing the registration of nurses. She scribbled a note on the margin for Henry Bonham Carter: 'Shall I royally disregard it – or shall I give them a BUSTER.'

As her character blossomed into benevolence, her physical appearance changed. The slight, tall, willowy girl whose elegance had struck everyone who saw her, whose small head had been set on her neck with the grace of a stag, who had loved to dance and been light as thistledown on her feet, the thin, emaciated, mature woman with lines of suffering deeply engraved on her face, underwent a surprising metamorphosis. She became a dignified stout old lady with rather a large good-humoured face. The shape of her head seemed to change; the face became wider, the neck shorter, the brow much more prominent. Surgeon-Major Evatt, who knew her in her old age, said she resembled Mr Gladstone, and a relative, introduced to her as a boy, retained as his recollection that she looked 'so jolly'.

Much of her life centred upon Shore and his wife and daughters and the children of Blanche and Clough; she followed them through their various stages of development, sent eggs and Egyptian lentils when one became a vegetarian, read pamphlets when another became an ardent advocate of co-operation, helped on several occasions with cheques for foreign tours. But her closest association was with the Verney family – to the Verneys she was indispensable.

Each year Parthe became more crippled with arthritis, and in 1883 she had a serious illness. She suffered a great deal, and no nurse could control her. Her household fell into confusion, and Sir Harry, now eighty-two, was distracted; so Miss Nightingale went down to Claydon and took command.

To deal with Parthe required endless patience. She was witty and gay, she could write with charm, she was responsible for rescuing and editing the famous Verney Papers which for centuries had been lying neglected in the attics at Claydon, but her good qualities became obscured by her physical condition. She created difficulties about money, she became jealous of her eldest stepson, Sir Harry's heir. She made undesirable favourites among the servants, she wrote painful letters. No one but Florence could soothe her. After 1883 Parthe was completely crippled, and Miss Nightingale became an essential part of the Verney family life. In addition to her old and deep affection for Sir Harry she was greatly attached to Sir Harry's son, Frederick Verney, who had been ordained a deacon and did social work in London. She corresponded with him, and on several occasions he read her papers to scientific and political meetings.

She also became intimate with the wife of Sir Harry Verney's eldest son – Margaret Verney. In 1869, after their first meeting, Miss Nightingale described her as 'a sort of heavenly young woman. I do not know that I ever saw anyone exactly like her. Only that she is witty and makes jokes she would be exactly like the Virgins and Saints of Fra Angelico.' Margaret Verney – Miss Nightingale's name for her was 'Blessed Margaret' – in addition to saintliness and beauty had capability. She had, wrote Miss Nightingale to Frederick Verney in 1896, 'administrative power, that power of detail which makes works succeed and is called capacity for business'.

The burden of Parthe's illness had fallen on Margaret Verney, and Miss Nightingale alone could help her. When Parthe wrote Margaret a letter 'so outrageously discourteous' that she 'destroyed it as if it were a viper ... I have no wish in the world but to be a daughter but there are some things Mama *must* not say to me,' Miss Nightingale persuaded Parthe to apologize; when she had been at Claydon, Parthe was much easier to manage. 'I write with a very thankful heart to-night for Mama has been so kind and gentle,' wrote Margaret on September 8, 1887, 'and I feel as if the echoes of your loving words and thoughts and prayers still linger here and have an influence for peace.'

The intimacy grew swiftly. 'Dearest Miss Nightingale' became 'Dearest Aunt Florence', and innumerable letters passed between them breathing affection and solicitude. In 1888 she called Miss Nightingale, 'the presence which to all of us brings such balm of sympathy and peace'. In 1889 she wrote, 'I long so much to see you. Thank you so much for all you have been to us.' In 1894: 'Have you been able to sleep? You cannot think how I *long* to be able to do something for you. .... If you could invent some wood to hew or water to carry, you would make me so very happy.' In 1892, when Miss Nightingale wrote to ask if a certain date would be convenient for her to come to Claydon, Margaret replied 'there never could be found in any almanac any day when it was not convenient and delightful that you should come here.'

In May 1890 Parthe died. It had become her custom to spend Sunday at 10 South Street when she was in London. On Sunday, May 4, she was carried into the drawing-room, evidently very ill; she spent that day with her sister, and the next day went down to Claydon. A week later, on May 12, she died. Their reconciliation had been complete. For seven years Parthe had been a difficult invalid, but Miss Nightingale's patience had never failed. 'You contributed more than anyone to what enjoyment of life was hers,' wrote Sir Harry on May 15, 1890. 'It was delightful to me to hear her speak of you and to see her face, perhaps distorted with pain, look happy when she thought of you.'

Parthe's death brought Miss Nightingale even closer to the Verney family. She went down to Claydon at once and stayed with Sir Harry until the autumn. He became the principal object of her life. He visited her every day, and if she went to London she wrote to him daily; when they were both in London, he called on her every morning. He was now nearly ninety, still mentally alert and still magnificently handsome, and she was seventy. One of the few photographs she ever allowed to be taken shows them sitting together on a garden seat at Claydon, smiling at each other. Her health had so far improved that occasionally she was able to take a short stroll leaning on his arm.

It was inevitable that she should interest herself in the management of the estate and inevitable that, having investigated accounts, condition of cottages, health of neighbouring villages, water supply and sanitation, she should find much that needed improvement. It was uphill work. Sir Harry was old; Parthe had been extravagant and careless. Even the treasures in the house itself had been neglected – Margaret found one of the historic family portraits used as a partition to separate stored apples. An immense amount of work was done by Miss Nightingale and Margaret to straighten out the confusion. In the house a degree of order was established, and the drains attended to. 'You know,' wrote Margaret in January 1892, 'how one goes through phases of discouragement at Claydon. You have established two definite steps forward which we never could have done without you.'

In the villages Miss Nightingale embarked on a new scheme. She wished to support the work of the District Nurse with Lady Health Missioners, women who were to be trained to teach village mothers the elementary principles of health in the home. Miss Nightingale was convinced that the best way to develop sanitary education, in England as in India, was to use the village as a unit. And, she insisted, 'the work *must* be personal'; the Health Missioners were 'not to lecture the village women but to work *with* them'.

It was a curious reproduction of the work she had done in her best days in India, a reproduction in miniature with Buckinghamshire in place of India, the Aylesbury district in place of Bengal. Even the conclusion repeated itself. Progress was impossible without water. Village sanitation in England, as in India, turned on water supply. 'Prizes to cottagers for cleanliness are not desirable,' she wrote to the Medical Officer of Health in November 1891. 'The prizes ought to be for handy water supply – to the authorities. . . . It is very pretty in a picture the group at the well of mother and children. It is not pretty in practice. The first possibility of rural cleanliness lies in *water supply*.'

Year succeeded year, and it seemed that Time had decided to pass Miss Nightingale by, that her Indian summer would last for ever while round her familiar faces were disappearing. In 1889 Aunt Mai died at the age of ninety-one. In July 1891 Dr Sutherland died. His last articulate words were for her – 'give her my love and blessing', he told his wife.

In 1893 a great grief awaited her: she lost Jowett. During the past few years they had drawn even closer. 'The truer, the safer, the better years of life are the later ones,' he had written to her in 1887. 'We must find new ways of using them, doing not so much but in a better way.' In October 1890 he had a heart attack and was expected to die. 'I am always thankful for having known you,' he wrote in a farewell letter on October 16. He recovered and in November 1890 she went over to Balliol from Claydon to see him, and stayed the night. In May 1892 he had another attack which greatly weakened him, but he still managed to visit her. 'I want to hold fast to you dear friend as I go down the hill,' he wrote. In August 1893 he was seen to be sinking – he became too weak to hold a pen; and on September 18 he dictated his last letter to her: 'Fare you well ... How large a part has your life been of my life.' On October 1 he died.

Four months later she had to bear another great grief: in February 1894 Sir Harry Verney died at the age of ninety-three. Six months later, in August 1894, Mr Shore Nightingale, 'My boy Shore', died, whose kindness, she was never tired of saying, had been one of the great recompenses of her life. 'I have lost the three nearest to me in twelve months,' she wrote. But there was no bitterness, none of the resentful anguish which had torn her apart thirty years ago. She was seventy-four, and as she drew nearer to the dividing line between life and death the bodily veil grew thin. It was not loss she faced now, but a temporary separation. And as she looked back over the long years she felt, as she had never felt in the days of her youth, that the sum total of life was good. 'There is so much to live for,' she wrote on May 12, 1895. 'I have lost much in failures and disappointments, as well as in grief but, do you know, life is more precious to me now in my old age.'

Claydon continued to be her second home, but after Sir Harry's death her visits became less frequent. The affection, the welcome was there. 'We are crazy with joy that you give us so blessed a hope of seeing you in November,' wrote Margaret Verney in October. But the renewal of physical vigour which had been so extraordinary a part of her Indian summer was beginning to fail. Gradually her life closed in; after 1896 she never left South Street, and she spent thereafter the whole of her life in her bedroom.

But it was only her body which had failed, for her mind and spirit remained as vigorous as ever. Indeed, she seemed to gain, as if in compensation, added confidence and hope. 'Yes, one does feel the passing away of so many who seemed essential to the world. I have no one now to whom I could speak of those who are gone. But all the more I am eager to see successors,' she wrote in a private note dated 'All Saints. All Souls. November 2nd, 1896.' She was still actively occupied. 'I am *soaked* in work,' she wrote to Douglas Galton in January 1897. The War Office consulted her, and she had influence there. Lord Lansdowne, her friend and Jowett's, was Secretary of State for War.

She maintained connexions with India, corresponding with the Viceroy, Lord Elgin, continuing to receive from the India Office all papers on Indian sanitary matters, and entertaining a large number of Indian gentlemen, educationalists, doctors, and administrators. In 1898 she received the Aga Khan. 'He was,' she wrote in a private note, 'a most interesting man, but you could never teach him sanitation. ... I told him as well as I could all the differences, both in town and country, during my life. "Do you think you are improving?" he asked. By improving he meant believing more in God.'

Year by year her legend steadily grew. The world had taken her figure to its heart, but in an extraordinary, an unprecedented, way. No crowd of admirers waited outside her house in South Street; indeed, the greater part of the world supposed she was dead, had supposed she was dead for the past forty years. Even the survivors of the men she had nursed did not know what had become of her. 'I should have communicated with you sooner,' wrote the organizer of an annual banquet of Inkerman survivors in 1895, 'but I did not know your address.' But whether she was dead or alive was unimportant: the image of her lived with vivid life. Not only in England but in the United States of America, in Turkey, Japan, in Brazil, her name had a magic possessed by no other.

She herself regarded her own legend with impatience. She despised the judgement of the crowd; she disapproved on moral grounds of personal influence. Shortly before his death Sir Harry Verney succeeded in persuading her to be photographed; he introduced the photographer unexpectedly during his morning visit to her room at Claydon and she gave way. On July 27, 1895, she wrote to Margaret Verney: 'Mr Payne – is that the name? – of Aylesbury, who did the photograph of me for Sir Harry, has written to ask me to give my consent to his publishing and selling my photograph. ... *I really cannot;* there is a perfect fury this year of writing to me for my photograph, autograph and a "few lines". And a very large number of these are from America – as many as 17 by one mail. The greatest number I throw away hardly reading and never answering. But if I have someone I *must* answer I can only say I have *no* photograph of myself and I don't know where to buy one.'

The year of Queen Victoria's Diamond Jubilee, 1897, added enormously to her legend. The Victorian Era Exhibition included a section representing the progress of trained nursing, and it was planned round Miss Nightingale; she was asked for Crimean relics, for pictures of Scutari, for her portrait, for the loan of her bust by Steell. She refused. 'Oh the absurdity of people and their vulgarity!' she wrote. 'The relics, the representations of the Crimean War! What are they? They are first the tremendous lessons we have had to learn from its tremendous blunders and ignorances. And next they are Trained Nurses and the progress of Hygiene. These are the "representations" of the Crimean War. And I will not give my foolish portrait (which I have not got) or anything else as "relics" of the Crimea. It is too ridiculous. . . .'

However, one of the organizers of the exhibition was Lady Wantage, and Lady Wantage was exceptionally pretty and charming. She called, and Miss Nightingale, always susceptible to charm, gave way. She wished to substitute a few hard facts about the work of the Royal Sanitary Commissions for Crimean relics, but Lady Wantage, wrote Miss Nightingale, 'would not take them] . . . she stuck to her point and she is *so* charming.' Miss Nightingale lent the bust by Steell and tracked down her Crimean carriage. 'O my dear Harry,' she wrote to Henry Bonham Carter in March 1897, 'that wretched Russian car with wretched but active boy and pony, all dismantled, *hangs* round my *neck*. . . . It was discovered all to pieces in an Embley farmhouse when Embley was sold. I never cared *what* became of it.'

The exhibition was the scene of extraordinary demonstrations. Her relics were treated by the crowds as holy. Flowers were laid daily before the bust by an unknown hand; old soldiers, it was said, had been seen to come forward and kiss the carriage. It was canonization, but of an unwilling saint. She was disgusted. In October 1897, when the exhibition was closing, she wrote to Louis Shore Nightingale, Shore's son: 'Now I must ask you about my bust. (Here I stop to utter a great many bad words not fit to put on paper. I also utter a pious wish that the bust may be smashed.) I should not have remembered it but that I am told somebody came every day to bedeck it with fresh flowers. I utter a pious wish that that person may be – saved. . . . What *is* to be done about the bust?'

Her life had turned to a golden evening, and it seemed the golden evening might last for ever. Year after year slid by, and still she faced life with relish; still the vigour of her mind was unimpaired. Then the darkest of shadows fell across the tranquil radiance as she began slowly to go blind.

Since 1867 she had had occasional pain in her eyes especially

after working at night. After 1884 her sight began to trouble he
seriously. 'Please remember I have *no* eyes,' she wrote to D
Sutherland in 1885, 'or rather I *have* eyes but they are neuralgi
You must not please tell me to look in the book but mark th
passages for me.' In 1888 she told Douglas Galton, 'my eyes a
so bad now. I can hardly see by candle light.' In February 188
she had become 'too blind to read newspapers'. Three month
later she asked Douglas Galton to take over the writing of a
article she had been invited to contribute to *Chambers's Encycl
paedia* 'because I have no longer eyes to write'.

Her spirit remained undimmed. 'No, no a thousand times no.
am not growing apathetic,' she wrote to Sir Robert Rawlinson i
1889. As late as 1898 she reread Shakespeare and made copiou
notes. In her letters her phrases were vigorous as ever. 'Do yo
know the taste of your heart in your mouth?' she asked Margar
Verney in 1891. She received a present, in 1893, 'with a poud pu
of gratitude such as the best fish elicits from the cat'. She said (
Lord Shaftesbury, 'He would have been in a lunatic asylum if h
had not devoted himself to reforming lunatic asylums.'

She was fully conscious, however, of disquieting symptoms. A
her sight grew worse, she wrote fewer private notes, but in 189
she wrote 'Want of memory', and in 1896 'How to preserve m
sight!' It was the only mention of her growing blindness she eve
made. If fear clutched at her, she concealed it. In earlier life sh
had talked a great deal about her health, in her old age she nev
mentioned it, but she must have remembered that Fanny ha
become childish and blind.

Slowly, inexorably, the curtain descended. She had alway
written with astonishing legibility and firmness (every line of h
enormous letters is as easy to read as if it were print), but now th
indelible pencil she took to using began to waver, the lines ra
across the page, the letters were formed with difficulty. Still he
vitality, her gaiety were unquenched. Margaret Verney's daug
ter, Ellin, married and in 1899 had her first child, a girl. In
spirited correspondence Miss Nightingale did her utmost to ha
the child named Balaclava, one of the most beautiful names, sh
declared, in the world. As late as 1900 she wrote to one
Margaret Verney's younger daughters: 'I am sorry to see the ti
leaving Italian for German. There are as many divine things
one page of Dante as in the whole of Goethe. Still it is no use,
Canute said, to kick against the tide. ... As for riding,

426

hockey", no games will equal it for improving the circulation
over and exercising the muscles and animal courage. A live
horse and the sympathy of "the horse and its rider" is worth all
the bats and (deaf and dumb) balls put together. So "drat"
hockey and long live the horse! Them's my sentiments.'

Year by year in a steady procession her old friends left the
mortal stage. In 1898 Sir Robert Rawlinson died – he had been
Sanitary Commissioner in the Crimean War and had remained
her close friend ever since. In 1899 she lost Sir Douglas Galton.
In 1902 William Rathbone; 'one of God's best and greatest sons',
she wrote on his funeral wreath. Still her optimism remained un-
minished; still she looked forward with an undaunted spirit.
Lady Stephen, one of Shore's two daughters, was sitting, when a
girl, with Miss Nightingale, who was lying back on her pillows,
and they were speaking of one of the friends she had lost. Lady
Stephen said that after a busy life he was at rest. Miss Nightingale
at once sat bolt upright. 'Oh no,' she said with conviction, 'I am
sure it is an *immense* activity.'

In 1901 darkness closed in on her. Her sight failed completely,
and, except with the greatest difficulty, she could no longer read
or write. At the same time her mind began to fail; she was not
always aware of her surroundings and lay for hours in a state of
coma. She fought to keep her grip on life. Her memory was failing
but she concealed it. Before she had an important visitor she
would have herself coached up to make a good impression. In
1903, before Lord Kitchener called on her, Miss Nightingale sent
her companion to look up all the facts about Lord Kitchener's
latest policy and memorized them just before the interview. Lord
Kitchener remarked on leaving that it was astonishing how Miss
Nightingale in her old age followed what was going on.

Every day she had *The Times* read to her. She also enjoyed biographies and articles from reviews which recorded action. One of
her favourite books was Theodore Roosevelt's *Strenuous Life*.
No longer able to act herself, she enjoyed hearing of action by
others. Sometimes instead of being read to, she would recite
poetry to herself, passages from Shakespeare, Milton, Shelley,
and the Italian poets; sometimes she would sing airs from the
operas she had loved in her youth in a voice still surprisingly full
and sweet.

A time came when there was no more reading, no more reciting
or singing. In 1906 it was necessary to tell the India Office that it

427

was useless to send papers on sanitary matters any longer to Miss Nightingale; the power of apprehension had almost left her; she was quite blind, and her memory had failed. She saw very few people and she no longer recognized visitors: she took them for friends of her youth and asked for Sir Harry Verney, who had been dead for twelve years. Hour after hour she lay inert, unconscious, her hands, still pretty in old age, folded peacefully outside the bedclothes. Words no longer reached her, although when her young relatives sang hymns she seemed to recognize familiar tunes and be pleased.

And now when she had passed beyond the power of the world to please or pain, a shower of honours fell on her. In November 1907 the Order of Merit was bestowed on her by King Edward VII, the first time it had ever been given to a woman. Since no ceremony was possible, the Order was left at South Street by the King's representative. It was not even certain that she understood the honour she had received. An explanation was attempted, but she hardly seemed to grasp it. 'Too kind, too kind,' she murmured. In the following year she received the Freedom of the City of London. The Roll of Honour was brought to her bedside, and her hand was guided to sign two wavering initials 'F. N.', but it was evident that she did not understand what she was signing.

The legend surrounding her silent inert figure burst into new life. Many people reading the news of these signal honours were taken aback to find that Florence Nightingale was still alive. A flood of congratulations poured in; there were poems, songs, illuminated addresses, flowers. The Mayor of Florence sent official congratulations, the Florence Nightingale Society of America, the Ladies of the Red Cross Society of Tokio, sent tributes to 'the great and incomparable name of Florence Nightingale'; thousands of women who had been christened Florence in her honour banded together to send a joint message. Crimean veterans assured her that she had never been forgotten.

In June 1907 the International Conference of Red Cross Societies had held a conference in London and sent a message to 'Miss Florence Nightingale, the pioneer of the first Red Cross movement, whose heroic efforts on behalf of suffering humanity will be recognized and admired by all ages as long as the world shall last.' Now local branches sent messages; regiments remembered her, the Commander-in-Chief, Lord Roberts, wrote a warm personal message in his own hand; Queen Alexandra wrote, and the

Kaiser sent a bouquet of flowers: '*very* beautiful and *very* large,' wrote Miss Nightingale's companion to Henry Bonham Carter in December 1907, 'lily of the valley and splendid pink carnations with *yards* of pink ribbon to match. Do you think the Emperor will wish the Press Association informed?' This was the Emperor's wish.

May 1910 was the Jubilee of the founding of the Nightingale Training School, and to mark the occasion a meeting was held in New York in the Carnegie Hall at which the Public Orator, Mr Choate, delivered an eulogium on the great record and noble life of Miss Florence Nightingale. There were now over one thousand training schools for nurses in the United States alone.

She knew nothing. Slowly, with heartbreaking slowness, death approached. Intervals of consciousness became less and less frequent. After February 1910 she no longer spoke. The iron frame which had endured the cold and fevers of the Crimea, which had been taxed and driven and misused in forty years of gigantic labours, still lived on, deprived of memory, of sensation, of sight, but still alive.

The end came on August 13, 1910. She fell asleep about noon and did not wake again.

In an immensely long will, which finds a place in collections of legal curiosities, she divided her possessions with meticulous detail, distributing prints, books, furniture, and mementoes in hundreds of personal bequests. She expressed a wish 'that no memorial whatever should mark the place where lies my Mortal Coil'; if this proved impossible she wished her body 'to be carried to the nearest convenient burial ground accompanied by not more than two persons without trappings'. A simple cross without her name, only with initials, and date of birth and death was to mark the spot. She also directed that her body should be given 'for dissection or post-mortem examination for the purposes of Medical Science'.

This was not done. But in deference to her wishes the offer of a national funeral and burial in Westminster Abbey was declined. She was buried in the family grave at East Wellow, and her coffin was carried by six sergeants of the British Army. Her only memorial is a line on the family tombstone, 'F. N. Born 1820. Died 1910.' She had lived for ninety years and three months.

Sources

(1) MSS.

The Nightingale Papers
The Verney Nightingale Papers
The Herbert Papers
The Mohl Nightingale Correspondence
The Correspondence of Miss Hilary Bonham Carter
The Correspondence of Mr Frederick Verney
The Leigh Smith Papers

(2) GOVERNMENT PUBLICATIONS

Place of publication: London, unless otherwise stated

Report upon the State of the Hospitals of the British Army in the Crimea and Scutari, together with an Appendix. Presented to both Houses of Parliament by command of Her Majesty, 1855. (The Hospitals Commission.)

Report to the Right Hon. Lord Panmure, G.C.B., Etc., Minister at War, of the Proceedings of the Sanitary Commission Dispatched to the Seat of War in the East, 1855–56. Presented to both Houses of Parliament by command of Her Majesty, March 1857. (The Sanitary Commission.)

First, Second and Third Report from the Select Committee on the Army before Sebastopol; with the Proceedings of the Committee. Ordered by the House of Commons to be printed, March 1, 1855. (The Roebuck Committee.)

Report of the Commission of Inquiry into the Supplies of the British Army in the Crimea, with the Evidence Annexed. Presented to both Houses of Parliament by Command of Her Majesty, 1856. (The McNeill and Tulloch Commission.)

Report of the Commissioners appointed to inquire into the Regulations affecting the Sanitary Condition of the Army, the Organization of Military Hospitals, and the Treatment of the Sick and Wounded; with Evidence and Appendix. Presented to both Houses of Parliament by Command of Her Majesty, 1858. (The Royal Sanitary Commission.)

Royal Commission on the Sanitary State of the Army in India. Report of the Commissioners. Précis of Evidence. Minutes of Evidence. Addenda, 1863. (The Indian Sanitary Commission.)

Suggestions in Regard to Sanitary Works required for Improving Indian Stations. Prepared by the Barrack and Hospital Improvement Commission, 1864.

Memorandum of Measures Adopted for Sanitary Improvements in India up to the End of 1867; together with Abstracts of the Sanitary Reports Hitherto Forwarded from Bengal, Madras and Bombay. Printed by order of the Secretary of State for India in Council, 1868. Ditto to the end of June 1869. Ditto to the end of June 1870. Ditto to the end of June 1872.

(3) WRITINGS BY MISS NIGHTINGALE

The Institution of Kaiserswerth on the Rhine for the Practical Training of Deaconesses under the direction of the Rev. Pastor Fliedner, embracing the support and care of a Hospital, Infant and Industrial Schools, and a Female Penitentiary. Printed by the Inmates of the London Ragged Colonial Training School, 1851.

Letters from Egypt. Privately printed, 1854.

Statements Exhibiting the Voluntary Contributions Received by Miss Nightingale for the Use of the British Hospitals in the East, with the Mode of their Distribution, in 1854, 1855, 1856. Harrison and Sons, 1857.

Notes on Matters Affecting the Health, Efficiency, and Hospital Administration of the British Army. Founded Chiefly on the Experience of the Late War. Presented by Request to the Secretary of State for War. Privately printed for Miss Nightingale. Harrison and Sons, 1858.

Subsidiary Notes as to the Introduction of Female Nursing into Military Hospitals in Peace and in War. Presented by Request to the Secretary of State for War. Privately printed for Miss Nightingale. Harrison and Sons, 1858.

A Contribution to the Sanitary History of the British Army during the Late War with Russia. Harrison and Sons, 1859.

Notes on Hospitals. John W. Parker and Sons, 1859. 3rd edition, almost completely rewritten, 1863. Longmans, Green and Co.

Suggestions for Thought to the Searchers after Truth among the Artizans of England. Privately printed for Miss Nightingale. 3 vols. Eyre and Spottiswoode, 1860.

Notes on Nursing: What it is, and what it is not. By Florence Nightingale. 2nd ed. Harrison and Sons, 1860.

Army Sanitary Administration and its Reform under the late Lord Herbert. M'Corquodale and Co., 1862.

Observations on the Evidence Contained in the Stational Reports Submitted to the Royal Commission on the Sanitary State of the Army in India. By Florence Nightingale. (Reprinted from the Report of the Royal Commission), Edward Stanford, 1863. (The *Observations.*)

Introductory Notes on Lying-in Institutions. Together with a Proposal for Organizing an Institution for Training Midwives and Midwifery Nurses. By Florence Nightingale. Longmans, Green and Co., 1871.

Life or Death in India. A paper read at the meeting of the National Association for the promotion of Social Science, Norwich, 1873. With an Appendix on life or death by irrigation, 1874.

The Zemindar, the Sun, and the Watering Pot as Affecting Life or Death in India. Unpublished, proof copies among the Nightingale Papers, 1873–76.

On Trained Nursing for the Sick Poor. By Florence Nightingale. The Metropolitan and National Nursing Association, 1876.

Miss Florence Nightingale's Addresses to Probationer-Nurses in the 'Nightingale Fund' School at St Thomas's Hospital and Nurses who were formerly trained there, 1872–1900. Printed for private circulation.

Florence Nightingale's Indian Letters. A glimpse into the agitation for tenancy reform. Bengal, 1878–82. Edited by Priyaranjan Sen. Calcutta, 1937.

(4) AUTHORITIES

The British Expedition to the Crimea, by W. H. Russell, LL.D., *The Times* special correspondent. G. Routledge and Co., 1858.

A Memoir of Baron Bunsen. Drawn chiefly from family papers by his widow, Frances, Baroness Bunsen. 2 vols. Longmans, Green and Co., 1868.

Considerations on the Military Organization of the British Army: Respectfully addressed to General His Royal Highness The Duke of Cambridge, K.G., G.C.B., G.C.M.G., Commander-in-Chief, etc. etc. etc. By General Sir Robert Gardiner, K.C.B., Royal Artillery. Byfield, Hawksworth and Co., 1858.

Constantinople During the Crimean War. By Lady Hornby. Richard Bentley, 1863.

Conversations with M. Thiers, M. Guizot, and other Distinguished Persons during the Second Empire. By the late William Nassau Senior, edited by his daughter, M. C. M. Simpson. 2 vols. Hurst and Blackett, 1878.

The Crimean Commission and the Chelsea Board: being a Review of the Proceedings and Report of the Board, by Colonel Tulloch, late Commissioner in the Crimea. Harrison and Sons, 1857.

Emma Darwin. A Century of Family Letters, 1792–1896. Edited by her daughter, Henrietta Litchfield. 2 vols. John Murray, 1915.

The Autobiography of Elizabeth Davis, a Balaclava Nurse. Edited by Jane Williams. 2 vols. Hurst and Blackett, 1857.

Delane of 'The Times'. By Sir Edward Cook. Constable and Co. Ltd., 1915.

Diary of the Crimean War. By Frederick Robinson, M.D., Assistant Surgeon, Scots Fusilier Guards. Richard Bentley, 1856.

Eastern Hospitals and English Nurses: the Narrative of Twelve Months Experience in the Hospitals of Koulali and Scutari. By a Lady Volunteer. Hurst and Blackett, 1856.

England and Her Soldiers. By Harriet Martineau. Smith, Elder and Co., 1859.

English Note-books. By Nathaniel Hawthorne. Kegan Paul, Trench and Company, 1883.

Experiences of a Civilian in Eastern Military Hospitals. By Peter Pincoffs, M.D., late Civil Physician to the Scutari Hospitals. Williams and Norgate, 1857.

Experiences of an English Sister of Mercy. By Margaret Goodman. Smith, Elder and Co., 1862.

The Life and Correspondence of Sir Bartle Frere. By John Martineau. 2 vols. John Murray, 1895.

Mrs Gaskell and Her Friends. By Elizabeth Sanderson Haldane. Hodder and Stoughton, 1930.

The Life of William Ewart Gladstone. By John Morley. 3 vols. Macmillan and Co., 1903.

Thomas Grant, First Bishop of Southwark. By Grace Ramsay (K. O'Meara). Smith, Elder and Co., 1874.

The Greville Memoirs, 1814–60. Edited by Lytton Strachey and Roger Fulford. 8 vols. Macmillan and Co., 1938.

Memoir of Sidney Herbert. Sidney Herbert, Lord Herbert of Lea. A Memoir by Lord Stanmore. John Murray, 1906.

Hospitals and Sisterhoods. John Murray, 1854.

India Called Them. By William Henry Beveridge. George Allen and Unwin, 1947.

The Invasion of the Crimea: Its origin, and an account of its progress down to the death of Lord Raglan. By A. W. Kinglake, in 9 vols. 6th edition. William Blackwood and Sons, 1887.

The Letters of Queen Victoria. A selection of Her Majesty's correspondence between the years 1837 and 1861. Published by authority of His Majesty the King. Edited by Arthur Christopher Benson, M.A., and Viscount Esher. 3 vols. John Murray, 1907. Second Series, edited by George Earle Buckle, 1861–85. 3 vols. John Murray. 1926. Third series, edited by George Earle Buckle, 1866–1901. 3 vols. John Murray, 1930.

Life and Death of Athena, an Owlet from the Parthenon. Privately printed, 1855.

The Life of His Royal Highness the Prince Consort. By Sir Theodore Martin, K.C.B. 5 vols. Smith, Elder and Co., 1875–80.

The Life and Letters of Sir John Hall, M.D., K.C.B., F.R.C.S. By S. M. Mitra. Longmans, Green and Co., 1911.

Life and Letters of Benjamin Jowett. By Evelyn Abbott and Lewis Campbell. 2 vols. John Murray, 1897.

The Life of Lord Lawrence. By Reginald Bosworth Smith. Nelson and Sons, 1908.

Lord Lawrence. By Sir Richard Temple. Macmillan and Co., 1890.

The Light Cavalry Brigade in the Crimea. Extracts from the Letters and Journal of the late General Lord George Paget, K.C.B., during the Crimean War. John Murray, 1881.

The Life and Struggles of William Lovett in his Pursuit of Bread, Knowledge, and Freedom. With some short account of the different associations he belonged to, and of the opinions he entertained. Trübner and Co., 1876.

Memoir of the Rt Hon. Sir John McNeill, G.C.B., and of his Second Wife, Elizabeth Wilson. By their Granddaughter. John Murray, 1910.

Memories of the Crimea. By Sister Mary Aloysius. Burns and Oates, Limited, 1897.

The Letters of John Stuart Mill. Edited by Hugh S. R. Elliot. 2 vols. Longmans, Green and Co., 1910.

Richard Monckton Milnes. The Life, Letters and Friendships of Richard Monckton Milnes, First Lord Houghton. By T. Wemyss Reid. Cassell and Co. Ltd., 1890.

Madame Mohl. Her Salon and her Friends. A study of social life in Paris. By K. O'Meara. R. Bentley and Son, 1885.

Letters and Recollections of Julius and Mary Mohl. By M. C. M. Simpson. Kegan, Paul and Co., 1887.

The Correspondence of John Lothrop Motley, D.C.L. Edited by George William Curtis. John Murray, 1889.

A Narrative of Personal Experiences and Impressions during Residence on the Bosphorus throughout the Crimean War. By Lady Alicia Blackwood. Hatchard, 1881.

The Life of Florence Nightingale. By Sir Edward Cook. 2 vols. Macmillan and Co., 1913.

The Life of Florence Nightingale. By Sarah A. Tooley. Cassell and Co. Ltd., 1910.

Florence Nightingale, 1820–56. I. B. O'Malley. Thornton Butterworth, 1931.

A History of Nursing. The Evolution of Nursing Systems from the Earliest Times to the Foundation of the First English and American Training Schools for Nurses. By M. Adelaide Nutting, R.N., and Lavinia L. Dock, R.N. G. P. Putnam's Sons, 1907.

Palmerston. By Philip Guedalla. Ernest Benn, 1926.

The Panmure Papers, being a selection from the correspondence of Fox Maule, second Baron Panmure, afterwards eleventh Earl of Dalhousie, K.T., G.C.B. Edited by Sir George Douglas, Bart., M.A., and Sir George Dalhousie Ramsay, C.B., late of the War Office, with a supplementary chapter by the late Rev. Principal Rainy, D.D. Hodder and Stoughton, 1908.

Pioneer Work in Opening the Medical Profession to Women, and Autobiographical Sketches. Elizabeth Blackwell, M.D. Longmans, Green and Co., 1895.

Henry Ponsonby. His Life from his letters, by Arthur Ponsonby. Macmillan and Co., 1943.

Reminiscences, Julia Ward Howe, 1819–99. Houghton Mifflin Company, Boston, 1900.

Scutari and its Hospitals. By the Hon. and Rev. Sydney Godolphin Osborne. Dickinson Brothers, 1855.

Sevastopol. Our Tent in the Crimea; and Wanderings in Sevastopol. By Two Brothers. Richard Bentley, 1856.

Seventy-one Years of a Guardsman's Life. By General Sir George Higginson, G.C.B., etc. John Murray, 1916.

The Earl of Shaftesbury, K.G. In Memoriam. October 1–9, 1885. Ragged School Union.

The First Woman Doctor. The Story of Elizabeth Blackwell, M.D. By Rachel Baker. Julian Messner Inc, New York, 1944.

The Life and Work of the Seventh Earl of Shaftesbury, K.G. By Edwin Hodder. Cassell and Co. Ltd., 1887.

The History of St Thomas' Hospital. By F. G. Parsons, D.Sc., F.R.C.S., F.S.A. 3 vols. Methuen and Co. Ltd., 1932.

Sketch of the History and Progress of District Nursing. By William Rathbone. Macmillan and Co., 1890.

Soyer's Culinary Campaign. Being Historical Reminiscences of the Late War, with The Plain Art of Cookery for Military and Civil Institutions, the Army, Navy, Public, etc. etc., by Alexis Soyer, author of *The Modern Housewife, Shilling Cookery for the People,* etc. G. Routledge and Co., 1857.

The Story of the Highland Brigade in the Crimea. Founded on letters written during the years 1854, 1855, and 1856. By Lieutenant-Colonel Anthony Sterling. Remington and Co. Ltd., 1895.

With Lord Stratford in the Crimean War. By James Henry Skene. Richard Bentley and Son, 1883.

Life of Stratford de Redcliffe (Viscount Canning): from his Memoranda and Private and Official Papers, by S. Lane Poole. 2 vols. Longmans, Green and Co., 1888.

A History of the British Army. By the Hon. J. W. Fortescue. Vol. xiii. Macmillan and Co., 1930.

Index

438

443

Fontana Modern Novels

Doctor Zhivago *Boris Pasternak*

The world-famous novel of life in Russia during and after the Revolution. 'Doctor Zhivago will, I believe, come to stand as one of the great events in man's literary and moral history' *Edmund Wilson, New Yorker.* 'One of the most profound descriptions of love in the whole range of modern literature' *Stuart Hampshire, Encounter*

The First Circle *Alexander Solzhenitsyn*

The unforgettable novel of Stalin's post-war Terror. 'The greatest novel of the 20th Century' *Spectator.* 'An unqualified masterpiece —this immense epic of the dark side of Soviet life' *Observer*

The Leopard *Giuseppe di Lampedusa*

'Perhaps the greatest novel of the century' *L. P. Hartley.* 'Incontestably a masterpiece' *Listener.* '*The Leopard* has certainly enlarged my life . . . Reading and rereading it has made me realise how many ways there are of being alive' *E. M. Forster*

The Once and Future King *T. H. White*

T. H. White's classic re-creation of the Arthurian Legend. 'A glorious dream of the Middle Ages as they never were but as they should have been' *New York Times.* 'A magnificent and tragic tapestry . . . Irresistible' *J. W. Lambert, Sunday Times*

The Towers of Trebizond *Rose Macaulay*

'An achievement at once dazzling and intensely distinguished . . . exquisitely comic, deeply serious, sad and challenging . . .' *Evening News.* 'I would put it among the twenty best novels of the century' *Anthony Burgess*

ontana Modern Novels

he Mandarins *Simone de Beauvoir*

magnificent satire by the author of *The Second Sex*. 'The *Mandarins* gives us a brilliant survey of the post-war French intellectual . . . a dazzling panorama' *Paul Johnson, New Statesman*. A superb document . . . a remarkable novel' *Iris Murdoch, Sunday Times*

es Belles Images *Simone de Beauvoir*

Her totally absorbing study of upper-class Parisian life. 'A brilliant sortie into Jet Set France' *Daily Mirror*. 'As compulsively readable as it is profound, serious and disturbing' *Queen*

he African Child *Camara Laye*

The story of the author's childhood among the Malinke tribe. 'A remarkable book. Camara Laye is an artist and has written a book which is a work of art' *Times Literary Supplement*. 'Memorable for the affection, honesty and intimacy that touch an African people into living warmth' *The Times*

A Dream of Africa *Camara Laye*

The sequel to *The African Child*, by 'the first writer of genius to come out of Africa' *B.B.C*. 'Gently lays bare the soul of a young Guinean student in Paris returning home to the enveloping tribal pattern . . . A proud, poetic, visionary story' *Irish Times*

The Gab Boys *Cameron Duodu*

The perceptive novel of a group of young Ghanaians growing up under a new dictatorship. 'One of the most readable and instructive African books I've come across' *Observer*. 'Mr. Duodu lets off shafts at Civil Service corruption, the inadequacies of education and the absurdities of British life as seen by Africans . . . Distinctly entertaining' *Sunday Times*

Fontana Modern Novels

At Lady Molly's *Anthony Powell*

The fourth novel in his famous "Music of Time" series, describ[
by John Davenport in the *Observer* as 'the most exciting experim[
in post-war English fiction.' 'I enjoyed *At Lady Molly's* even m[
than its predecessors, which is saying a lot' *Spectator*

Casanova's Chinese Restaurant *Anthony Powell*

The fifth "Music of Time" novel. 'Brilliant literary comedy as v[
as a brilliant sketch of the times' *Time Magazine*. 'Anthony Pov[
is perhaps the most distinctive, and also one of the most brilli[
novelists now writing' *Evening Standard*

What's Become of Waring *Anthony Powell*

A hilarious parody of the English literary scene by one of [
greatest living satirists. 'Exceedingly funny and brilliantly easy[
read. Mr. Powell manages to convey a very agreeable friendlin[
and sympathy without any taint of sentimentality' *Maurice Rich[
son, Observer*. 'Brilliant and accomplished' *L. P. Hartley*

The Tin Men *Michael Frayn*

'*The Tin Men* are computers and the executives who nanny the[
Mr. Frayn is brilliant at getting exactly right how technocrats [
things absolutely wrong. Goes straight into the Evelyn Wau[
class' *Sunday Times*. 'Dazzlingly funny' *Observer*. 'As brilliant [
all Michael Frayn's work' *P. G. Wodehouse*